T0337885

INORGANIC MATERIALS SYNTHESIS AND FABRICATION

INORGANIC MATERIALS SYNTHESIS AND FABRICATION

John N. Lalena
University of Maryland University College–Europe

David A. Cleary
Gonzaga University

Everett E. Carpenter
Virginia Commonwealth University

Nancy F. Dean
Formerly, Honeywell Electronic Materials

WILEY-INTERSCIENCE

A JOHN WILEY & SONS, INC., PUBLICATION

Published by John Wiley & Sons, Inc., Hoboken, New Jersey.
Published simultaneously in Canada.

For general information on our other products and services or for technical support, please contact our Customer Care Department within the United States at (800) 762-2974, outside the United States at (317) 572-3993 or fax (317) 572-4002.

Wiley also publishes its books in a variety of electronic formats. Some content that appears in print may not be available in electronic formats. For more information about Wiley products, visit our web site at www.wiley.com.

Library of Congress Cataloging-in-Publication Data:

Inorganic materials synthesis and fabrication / John N. Lalena ... [et al.].
    p. cm.
  Includes index.
  ISBN 978-0-471-74004-9 (cloth)
 1. Materials. 2. Inorganic compounds. I. Lalena, John N.
  TA403.6.I55 1996
  620.1′1—dc22

                                       2007033398

Printed in the United States of America

10 9 8 7 6 5 4 3 2 1

CONTENTS

PREFACE

With our first textbook, *Principles of Inorganic Materials Design,* we set out to fulfill our stated goal of preparing a single-source presentation of inorganic materials design. Accordingly, the primary emphasis was on structure–property correlation. A comprehensive treatment of the broader general topic of inorganic materials science necessitates that we discuss chemical, synthetic, and fabrication processes, topics we now take up in detail.

Customarily, chemists have been interested in the submicroscopic length scale, studying the compositions and structures of solids, their relationships to properties, and processes that bring about changes in those entities. The physicists who deal with condensed matter have had similar goals working at the electronic length scale, where they have been concerned primarily with describing various physical properties quantitatively. The focus of materials scientists and engineers, on the other hand, has evolved from studying microstructural features and processes to a state in which they now also draw on the body of knowledge acquired by chemists and physicists in order to design improved materials for utilization in specific engineering applications. Today, the artificial demarcation between the various disciplines is beginning to vanish. Becoming consistently common objectives are synthesis and fabrication, which inherently require consideration of details spanning multiple length scales.

To serve the need to educate science and engineering students in this area, in this book we take an interdisciplinary approach, as we did in our first book, but with the focus shifted to describing how chemical reactions proceed between *single-phase* inorganic solids (molecular and nonmolecular) and other substances that result in the transformation of the solid, to a new single phase with composition differing from that of the original material. Such reactions are conveniently categorized here as being of the solid–vapor, solid–liquid, or solid–solid type. In general, synthetic schemes may be thought of as bottom-up processes where the chemical transformation occurs at the interface between the reacting phases.

This book is about the preparation of single-phase inorganic materials. The design concerns and associated process flows for multiphase/heterostructure devices (e.g., composite materials, semiconductor integrated circuits) are better covered in specialized engineering texts and so are omitted here. However, we cover top-down materials fabrication processes, such as plastic deformation and consolidation processing in some detail, topics that have traditionally been restricted to materials science and engineering courses. As with our first textbook, this book takes on a distinct historical tone and includes short biographical sketches of some of the people who have made seminal contributions to the field over the past century. We believe that students appreciate learning about their heritage, and the history of science is, after all, a worthy scholarly endeavor in and of itself.

To maximize the level of expertise applied to this endeavor, invitations were extended to two additional authors, Professor E. E. Carpenter and Dr. N. F. Dean, both of whom enthusiastically agreed to participate, and we are very grateful for their contributions. The four of us believe that this book would make an excellent companion to *Principles of Inorganic Materials Design,* together serving not only as strong introductory texts to materials science and engineering, chemistry, and physics students, but also as welcomed reference sources for working professionals. Some of the topics covered in our first book actually fit equally well here, being, in fact, essential components to discussions on the interplay between chemical structure, reaction energetics, and kinetics. We are very grateful to the publisher for allowing us to reproduce those portions in this work.

J. N. LALENA
D. A. CLEARY
E. E. CARPENTER
N. F. DEAN

1 Crystallographic and Microstructural Considerations

The bulk properties of a solid, such as conductivity, magnetism, and second harmonic generation, hinge on the solid's structure, which, in turn, is normally that arrangement of a material's fundamental particles (molecules, atoms, or ions), with the lowest potential energy as a function of all the atomic positions. We have learned through computational approaches that interatomic potential energy is a function of short-range, long-range, and many-body interactions. For molecular-based substances, one must also consider van der Waals interactions, hydrogen bonds, and capillary and hydrophobic forces. At first glance, these noncovalent forces may seem of secondary importance, but their influence on governing the particular structure adopted by a substance, and hence its resulting properties, can be striking. Organic chemists have long been aware of this. For example, the double-helical structure of DNA is, to a large degree, a consequence of hydrogen bonding between the base pairs. In recent years, supramolecular chemists have focused attention on noncovalent structure-directing intramolecular and intermolecular interactions in the spontaneous formation of ordered aggregates, appropriately termed *chemical self-assembly processes*. It has been applied increasingly to the synthesis of hybrid organic–inorganic materials. However, due to certain limitations, self-assembly of purely inorganic substances, particularly materials in the micro- to millimeter size range (Section 1.5.2), is still in its infancy.

Externally, crystals typically possess morphologies showing the highest symmetry consistent with the chemical and spatial growth constraints imposed. The crystal faces tend to follow the *holohedral*, or *holosymmetric* symmetry, of the crystal class (i.e., they belong to the point group with the highest symmetry of its crystal system). However, as the relatively recently coined terms *crystal engineering* and *grain boundary engineering* (for polycrystalline materials) imply, to attain the optimal properties needed in any given engineering or functional application, it is often desirable—indeed, sometimes necessary—to control morphology (crystal shape, orientation) and/or microstructure (grain size, shape, and texture).

Perhaps the most common scenario is the need to produce crystals with specific crystallographic orientations. For example, large single-crystalline silicon wafers

Inorganic Materials Synthesis and Fabrication, By John N. Lalena, David. A. Cleary, Everett E. Carpenter, and Nancy F. Dean
Copyright © 2008 John Wiley & Sons, Inc.

with particular crystallographic orientations are required for the fabrication of microelectronic devices. The atomic-scale surface topography of each orientation gives rise to a unique interfacial structure between the silicon substrate and films, either deposited or grown thermally. The interface is a significant portion of the total film volume in today's devices, where film thicknesses are in the nanometer range. With metal–oxide semiconductor field-effect transistor (MOSFET) devices, (100)-oriented silicon slices are currently used because they give more reliable gate oxides on thermal oxidation. However, for bipolar devices, the (111) orientation is used.

There are numerous other situations in which particular crystal orientations are required as starting material for the fabrication of devices. For example, high-quality quartz crystals are grown from seeds with specific orientations to maximize the obtainable number of parallelepiped units, called *cuts*, of that particular type for the production of crystal oscillators. The mode of vibration, and hence the corresponding frequency band for which it may be used, is dependent on the unit's crystallographic orientation and thickness, while the angle of the cut relative to the crystal face controls the frequency deviation over a specified temperature range. In a similar fashion, potassium titanyl phosphate, which is widely used for sum frequency generation (e.g., frequency doubling) requires cuts of specific crystallographic orientations that are determined solely by the nonlinear optical process for which it will be used.

Specific crystallographic orientations and symmetry-dependent properties may also be exploited in synthetic schemes in which solids are used as starting materials. Properties and phenomena are facilitated along, or may even be restricted to, specific crystallographic directions or orientations. Examples of anisotropy so utilized include:

- *Transport properties.* Ionic conduction and atomic diffusion are usually easier along certain crystallographic directions or planes.
- *Epitaxial and topotactic controlled reactions.* There are preferred crystallographic orientations for substrates on which single-crystalline films are grown via *homoepitaxy* (the film is the same substance and has the same orientation as the substrate) or heteroepitaxy (the film is a different substance and may have a different orientation than the substrate). Because of lattice matching between reactant and product phases, epitaxial reactions (surface structure controlled), as well as topotactic reactions (bulk crystal structure controlled), proceed under milder synthetic conditions. The kinetic control afforded by the surface step structure in these processes makes possible the obtainment of phases that are thermodynamically metastable but kinetically stable. For example, $YMnO_3$ crystallizes in the hexagonal system at atmospheric pressure, but cubic perovskite films have been grown on $NdGaO_3$ substrates.
- *Chemical activity (functionality).* Some crystal facets may be significantly more chemically reactive than others. For example, the (100) face of vanadyl

pyrophosphate is active for the oxidation of *n*-butane to maleic anhydride, whereas the other faces are not.

- *Magnetic anisotropy.* In conjunction with slip casting, magnetic alignment of anisotropic particles may be used to produce polycrystalline materials with a preferred orientation, or texture.

- *Plasticity.* Slip, the gliding motion of full planes of atoms or partial planes, called *dislocation*, allows for the deformation processing of polycrystalline metals (forging, extrusion, rolling, swaging, and drawing). Slip occurs much more readily across close-packed planes in close-packed directions.

Because the central topic of this book is synthesis and fabrication, the correlation of crystal *structure* with physical properties is not of primary importance. Nonetheless, as the preceding paragraphs imply, a discussion of the correlation between crystal *symmetry* and physical properties would be wise. The question is: At what point should we make that presentation? Crystal symmetry/physical property correlations were elucidated by physicists in the early to mid-nineteenth century, concurrent with mineralogists' work on the classification of crystals based on geometric form, decades before space group theory and its experimental confirmation by x-ray diffraction made possible crystal structure determination. It therefore seems fitting to place that treatment here, followed by morphological crystallography, the theory of space lattices, surface and interfacial structure, and finally, the crystallographic and morphological considerations pertinent to various preparative routes, which is in roughly the same chronological order as their development.

It must first be recognized that morphological (external) symmetry is not always identical with a specimen's true crystallographic symmetry. As a result, discrepancies between the observed and expected symmetry of a physical property may arise when the symmetry of the crystal is deduced solely from its external appearance. *Consequently, it is more appropriate to deduce the structure and symmetry of a crystal from its physical properties.* An auspiciously unambiguous relationship exists between crystallographic symmetry and physical properties. This principle, which is a basic postulate of crystal physics, was first recognized by Franz Ernst Neumann (1798–1895) in the 1830s (Neumann, 1833) and was later formalized by Neumann's students Woldemar Voigt (1850–1918) and Bernhard Minnigerode (1837–1896) in the 1880s. It may be stated as follows: *The symmetry of a physical phenomenon is [at least] as high as the crystallographic symmetry.* Similarly, the orientation of the principal axes of the matter tensor representing the physical property must also be consistent with the crystal symmetry (Section 1.1). In addition to physical properties such as electronic and thermal conductivity, mechanical properties such as elasticity, hardness, and yield strength also comply with the crystallographic symmetry.

There are two important points to remember regarding the applicability of Neumann's principle. First, forces *imposed* on a crystal, including mechanical stresses and electric fields, can have any arbitrary direction or orientation. These types of forces are represented mathematically with field tensors, not matter tensors, and so are not subject to Neumann's principle. Second, Neumann's principle applies strictly to physical properties only. Nevertheless, because of crystalline anisotropy, chemical activity (e.g., oxidation rate, corrosion rate, etch rate) is also found to be either direction- or orientation-dependent. The degree of surface chemical activity generally correlates with the density of bonds (particularly dangling bonds), which, in turn, is influenced by energy-lowering surface reconstructions, a topic we cover in Section 2.3.1.3. For example, the silicon (111) face etches more slowly than, but oxidizes nearly twice as rapidly as, the Si(001) face.

1.1. RELATIONSHIP BETWEEN PHYSICAL PROPERTIES AND CRYSTALLOGRAPHIC SYMMETRY

Neumann's principle asserts that the symmetry elements of any physical property of a crystal must include, at least, all the symmetry elements of the point group of the crystal. Single crystals are generally not isotropic. Hence, the physical properties of single crystals generally will be anisotropic, that is, dependent on the direction in which they are measured. For example, only with cubic crystals or polycrystals possessing a random crystallite orientation are the directions of heat flow (the flux) and the temperature gradient (the driving force) parallel. Accordingly, it is necessary to use mathematical expressions known as *tensors* to explain anisotropic properties in the most precise manner.

A tensor is an object with many components that look and act like components of ordinary vectors. The number of components necessary to describe a tensor is given by p^n, where p is the number of dimensions in space and n is called the *rank*. For example, a zero-rank tensor is a scalar, which has $3^0 = 1$ component regardless of the value of p. A first-rank tensor is a vector; it has three components in three-dimensional space (3^1), the projections of the vector along the axes of some reference frame (e.g., the mutually perpendicular axes of a Cartesian coordinate system). Although the magnitude and direction of a physical quantity, intuitively, do not depend on our arbitrary choice of a reference frame, a vector is defined by specifying its components from projections onto the individual axes of the reference system. Thus, a vector can be defined by the way these components change, or transform, as the reference system is changed by a rotation or reflection. This is called a *transformation law*. For example, a vector becomes the negative of itself if the reference frame is rotated $180°$, whereas a scalar is invariant to coordinate system changes (Lalena and Cleary, 2005).

Second-rank tensors such as transport properties relate two first-rank tensors, or vectors. Thus, a second-rank tensor representing a physical property has nine components (3^2), usually written in 3×3 matrixlike notation. Each component is associated with two axes: one from the set of some reference frame and one from the material frame. Three equations, each containing three terms on the right-hand side, are needed to describe a second-rank tensor exactly. For a general second-rank tensor τ that relates two vectors, $\mathbf{p}$ and $\mathbf{q}$, in a coordinate system with axes x_1, x_2, x_3, we have

$$
\begin{aligned}
p_1 &= \tau_{11}q_1 \quad \tau_{12}q_2 \quad \tau_{13}q_3 \\
p_2 &= \tau_{21}q_1 \quad \tau_{22}q_2 \quad \tau_{23}q_3 \\
p_3 &= \tau_{31}q_1 \quad \tau_{32}q_2 \quad \tau_{33}q_3
\end{aligned}
\tag{1.1}
$$

The tensor, with components τ_{ij}, is written in matrixlike notation as

$$
\begin{bmatrix}
\tau_{11} & \tau_{12} & \tau_{13} \\
\tau_{21} & \tau_{22} & \tau_{23} \\
\tau_{31} & \tau_{32} & \tau_{33}
\end{bmatrix}
\tag{1.2}
$$

Note that each component of $\mathbf{p}$ is related to all three components of $\mathbf{q}$. Thus, each component of the tensor is associated with a pair of axes. For example, τ_{32} gives the component of $\mathbf{p}$ parallel to x_3 when $\mathbf{q}$ is parallel to x_2. In general, the number of indices assigned to a tensor component is equal to the rank of the tensor. Tensors of all ranks, like vectors, are defined by their transformation laws. For our purposes, we need not consider these.

Fortunately, several simplifications can be made (Nye, 1957). Transport phenomena, for example, are processes whereby systems transition from a state of nonequilibrium to a state of equilibrium. Thus, they fall within the realm of irreversible or nonequilibrium thermodynamics. *Onsager's theorem*, which is central to nonequilibrium thermodynamics, dictates that as a consequence of time-reversible symmetry, the off-diagonal elements of a transport property tensor are symmetrical (i.e., $\tau_{ij} = \tau_{ji}$). This is known as a *reciprocal relation*. The Norwegian physical chemist Lars Onsager (1903–1976) was awarded the 1968 Nobel Prize in Chemistry for reciprocal relations. Thus, the tensor above can be rewritten as

$$
\begin{bmatrix}
\tau_{11} & \tau_{12} & \tau_{13} \\
\tau_{12} & \tau_{22} & \tau_{23} \\
\tau_{13} & \tau_{23} & \tau_{33}
\end{bmatrix}
\tag{1.3}
$$

Note the perhaps subtle, but very important, change in subscripts from Eq. 1.2, leaving us with merely six independent components.

Finally, symmetrical tensors can also be diagonalized. For second-rank tensors, three mutually perpendicular unit vectors can be found that define three *principal axes*, such that if these axes are used as coordinate axes, the matrices are diagonal. This leaves

$$
\begin{bmatrix}
\tau_{11} & 0 & 0 \\
0 & \tau_{22} & 0 \\
0 & 0 & \tau_{33}
\end{bmatrix}
\tag{1.4}
$$

Because of this further simplification, only three independent quantities in a symmetrical second-rank tensor are needed to define the magnitudes of the principal components. The other three components (from the initial six), however, are still needed to specify the directions of the axes with respect to the original coordinate system.

In the case of physical properties, crystal symmetry imposes even more restrictions on the number of independent components (Nye, 1957). A tensor representing a physical property must be invariant with regard to every symmetry operation of the given crystal class. Tensors that must conform to the crystal symmetry in this way are called *matter tensors*. The orientation of the principal axes of a matter tensor must also be consistent with the crystal symmetry. The principal axes of crystals with orthogonal crystallographic axes will be parallel to the crystallographic axes. In the monoclinic system, the x and z crystallographic axes are orthogonal to each other but nonorthogonal to y. For triclinic crystals, there are no fixed relations between either the principal axes or crystallographic axes, and no restrictions on the directions of the principal axes. The effects of crystal symmetry on symmetrical second-rank matter tensors are given below.

For cubic crystals and nontextured polycrystals, we have

$$\begin{bmatrix} \tau_{11} & 0 & 0 \\ 0 & \tau_{11} & 0 \\ 0 & 0 & \tau_{11} \end{bmatrix} \tag{1.5}$$

For tetragonal, trigonal, and hexagonal crystals,

$$\begin{bmatrix} \tau_{11} & 0 & 0 \\ 0 & \tau_{11} & 0 \\ 0 & 0 & \tau_{33} \end{bmatrix} \tag{1.6}$$

For orthorhombic crystals,

$$\begin{bmatrix} \tau_{11} & 0 & 0 \\ 0 & \tau_{22} & 0 \\ 0 & 0 & \tau_{33} \end{bmatrix} \tag{1.7}$$

For monoclinic crystals,

$$\begin{bmatrix} \tau_{11} & 0 & \tau_{13} \\ 0 & \tau_{22} & 0 \\ \tau_{13} & 0 & \tau_{33} \end{bmatrix} \tag{1.8}$$

For triclinic crystals,

$$\begin{bmatrix} \tau_{11} & \tau_{12} & \tau_{13} \\ \tau_{12} & \tau_{22} & \tau_{23} \\ \tau_{13} & \tau_{23} & \tau_{33} \end{bmatrix} \tag{1.9}$$

The diagonal elements in the tensors above follow from the indistinguishability of the axes in their respective crystal classes. For example, if we denote the normalized unit length of the three crystallographic axes in each crystal class with the letters a, b, c and denote the angles between these three axes with the Greek letters α, β, γ, we can see that there are three indistinguishable orthonormal dimensions (orthogonal axes normalized to the same unit length) in the cubic class ($a = b = c$; $\alpha = \beta = \gamma = 90°$); two orthonormal dimensions in the tetragonal class ($a = b \neq c$; $\alpha = \beta = \gamma = 90°$); two orthonormal dimensions in the trigonal class ($a = b = c$; $\alpha = \beta = \gamma \neq 90°$); two orthonormal dimensions in the hexagonal class ($a = b \neq c$; $\alpha = \beta = 90°$, $\gamma = 120°$); no orthonormal dimensions in the orthorhombic class ($a \neq b \neq c$; $\alpha = \beta = \gamma = 90°$); no orthonormal dimensions in the monoclinic class ($a \neq b \neq c$; $\alpha = \gamma = 90°$, $\beta \neq 90°$); and no orthonormal dimensions in the triclinic class ($a \neq b \neq c$; $\alpha \neq \beta \neq \gamma \neq 90°$). The off-diagonal elements in the monoclinic and triclinic crystals give the additional components necessary to specify the tensor. Notice that a cubic single crystal is isometric and so has isotropic properties. The same is also true for polycrystals with a random crystallite orientation (e.g., powders), regardless of the crystal class to which the substance belongs. Thus, a single scalar quantity is sufficient for describing the conductivity in crystals of the cubic class and nontextured polycrystalline materials (Lalena and Cleary, 2005).

It is sometimes possible to use the anisotropy in certain physical properties advantageously during fabrication processes. For example, the magnetic susceptibility, which describes the magnetic response of a substance to an applied magnetic field, is a second-rank matter tensor. It is the proportionality constant between the magnetization of the substance and the applied field strength. When placed in a magnetic field, a crystal with an anisotropic magnetic susceptibility will rotate to an angle in order to minimize the magnetic free-energy density. This magnetic alignment behavior can aid in texture control of ceramics and clays if the particles are sufficiently dispersed to minimize particle–particle interactions, which can be accomplished with slip casting or another powder suspension process (Section 7.3.2). The route has been used to prepare many bulk substances and thin films, including some with only a small anisotropic paramagnetic or diamagnetic susceptibility, such as gadolinium barium copper oxide, zinc oxide, and titanium dioxide (anatase), with textured (grain-aligned) microstructures and correspondingly improved physical properties (Lalena and Cleary, 2005).

1.1.1. Geometrical Representation of Tensors

The components of a symmetrical second-rank tensor, referred to its principal axes, transform like the three coefficients of the general equation of a second-degree surface (a quadric) referred to its principal axes (Nye, 1957). Hence, if all three of the quadric's coefficients are positive, an ellipsoid becomes the geometrical representation of a symmetrical second-rank tensor property (e.g., electrical and thermal conductivity, permittivity, permeability, dielectric and magnetic susceptibility). The ellipsoid has inherent symmetry *mmm*. The relevant features are that (1) it is centrosymmetric, (2) it has three mirror planes perpendicular to the

TABLE 1.1   Relationships Between the Quadric and Crystal Axes for Symmetrical Second-Rank Tensors

Crystal Class	Tensor Components to Be Measured	Restrictions on Principal Directions and Principal Values
Cubic	One value	Principal directions must be parallel to crystal axes; all three principal values equal in magnitude (isotropic)
Tetragonal	Two values	Principal directions must be parallel to crystal axes; two principal values equal in magnitude
Orthorhombic	Three values	Principal directions aligned with crystal axes; no restrictions on values
Hexagonal	Two values	One principal direction aligned with three-fold crystal axis; two principal values equal in magnitude
Trigonal	Two values	One principal direction aligned with three fold crystal axis; two principal values equal
Monoclinic	Three values plus one angle	One principal direction aligned with two-fold crystal axis; no restrictions on values
Triclinic	Three values plus three directions	No restrictions on directions or values; need to measure three directions and three values

principal directions, and (3) its twofold rotation axes are parallel to the principal directions. We have already stated that the orientation of the principal axes of the matter tensor representing a physical property must be consistent with the crystal symmetry. What this means is that the exact shape of the tensor property and its quadric, as well as the orientation of the quadric with respect to the crystal, are all restricted by the crystallographic symmetry. Table 1.1 lists the relationships between the quadric and the crystal axes for the various crystal classes.

1.1.2. Measurement of Physical Properties of Crystals

The most straightforward method for the measurement of a physical property on a single crystal along its principal axes often involves the initial removal of a uniform regularly shaped section (e.g., a parallelepiped or circular disk) from the crystal. This, in turn, usually requires that the single crystal first be oriented, for example, by using a Laue back-reflection technique. Consider the coefficient of thermal expansion (CTE) along the two independent crystallographic axes of a crystal belonging to the tetragonal crystal class, which may be measured using a thermomechanical analyzer (TMA). Measurement of this property is accomplished by first orienting the crystal, followed by cutting out a rectangular parallelepiped specimen (e.g., using a diamond saw blade), and finally, taking the length change along the crystal axes (a, c) directly from the TMA measurements along both the shorter and longer dimensions of the parallelepiped. Of course, the CTE, as with other intensive properties (which are independent of the mass of the sample) such as conductivity, cannot really be measured directly but rather, must be calculated from measuring the corresponding extensive property.

Often, it is not possible to obtain single crystals that are large enough to be worked with in a convenient manner. In those cases, physical properties must be measured on polycrystalline samples. There is always discrepancy, or disagreement, between measured physical properties of single crystals and polycrystals due to microstructural effects. Hence, physical properties measured from polycrystalline samples are sometimes considered less reliable from a reproducibility standpoint.

1.2. MORPHOLOGICAL CRYSTALLOGRAPHY

Before the advent of diffractometry, crystallographers could only visually examine crystals. However, an entire classification system was developed that is still in use today for describing a crystal's morphology, or external appearance, and for assigning a crystal's external symmetry to a point group. Mineralogists in the early nineteenth century focused on the theory of crystal forms from purely geometrical points of view. Although these systems do not make reference to primitive forms or any other theory of internal structure, they are perfectly valid approaches. The theory of crystal forms was developed independently by Christian Samuel Weiss (1780–1856), professor of geology and mineralogy at the University of Berlin, and by the mineralogist Friedrich Karl Mohs (1773–1839), in 1816 and 1822, respectively. Weiss was able to derive four definite crystal form classes: tessular (isometric, or cubic), pyramidal (tetragonal), prismatic (orthorhombic), and rhombohedral (hexagonal), by choosing as coordinate axes lines drawn joining opposite corners of certain forms. The monoclinic and triclinic systems were considered by Weiss to be hemihedral and tetragonal modifications of the prismatic system, but were considered distinct in Mohs' more advanced treatment (Lalena, 2006).

Weiss and Mohs also developed notation systems relating each face to the coordinate axes, but these were surpassed in use by a system first introduced by the British polymath William Whewell (1794–1866) during a crystallography fellowship period in 1825, and later incorporated in an 1839 book by his student William Hallowes Miller (1801–1880). The notation system, now named after Miller, is discussed further in Section 1.3.2.

Just as there is a correspondence between the symmetry of crystals and that of their physical properties, there is also a connection between the symmetry exhibited by a crystal at the macroscopic and microscopic length scales, in other words, between the "external" crystal morphology and true "internal" crystal structure. Under favorable circumstances, the point group (but not the space group) to which a crystal belongs can be determined solely by examination of the crystal morphology, without the need of confirmation by x-ray diffraction. It is not always possible because although many crystal forms (a collection of equivalent faces related by symmetry) are normally apparent in a typical crystal specimen, some forms may be absent or show unequal development.

If a crystal is grown in a symmetrical growth environment (e.g., freely suspended in a liquid), its morphological symmetry is exactly that of the isogonal point group (the point group with the same angular relations as that) of the space lattice. Morphological symmetry may depart from the true point group symmetry of the space lattice because of differing intrinsic growth rates of the various faces or because of nonsymmetrical growth conditions (Buerger, 1978). Even the presence of dislocations is believed to influence the growth rates and thus the development of various forms (Dowty, 1976). A comprehensive treatment of geometric crystallography, including morphological and internal symmetry, their interrelationship, and a systematic derivation of all the crystallographic point groups and space groups can be found in Buerger's classic textbook *Elementary Crystallography* (Buerger, 1978).

1.2.1. Single-Crystal Morphology

There are four commonly used terms for describing morphology which should be understood: zone, form, habit, and twin. A *zone* is a volume enclosed by a set of faces that intersect one another along parallel edges. The zone axis is the common edge direction. For example, the crystallographic axes and the edges of a crystal are all zone axes. A crystal *form* is a collection of equivalent faces related by symmetry (e.g., a polyhedron). One can choose the directions of three edges of a crystal as coordinate axes (x, y, z) and define unit lengths (a, b, c) along these axes by choosing a plane parallel to a crystal face that cuts all three axes. For any other crystal face, integers (h, k, l) can be found such that the intercepts the face makes on the three axes are in the ratios $a:h$, $b:k$, $c:l$. Together, these three integers describe the orientation of a crystal face. The integers are prime and simple (small) and they may be positive or negative in sign.

In a cube (a hexahedron), all the faces are equivalent. The six faces have indices (100), ($\bar{1}$00), (010), (0$\bar{1}$0), (001), and (00$\bar{1}$), but the set is denoted as {100}, signifying the entire cube, whereas (100) signifies just one face. In a similar fashion, an octahedron has the form symbol {111} and consists of the following eight faces: (111), ($\bar{1}$11), (1$\bar{1}$1), (11$\bar{1}$), ($\bar{1}\bar{1}$1), ($\bar{1}$1$\bar{1}$), (1$\bar{1}\bar{1}$), and ($\bar{1}\bar{1}\bar{1}$). One or more crystal forms are usually apparent in the crystal morphology, and these may be consistent with the point group symmetry of the lattice. A crystal of α-quartz (low quartz), for instance, may display five external forms showing trigonal point group symmetry. Symmetry considerations limit the number of possible types of crystal forms to 47. However, when we look at crystals from the lattice-based viewpoint, there are only seven crystal systems. This is because there are 15 different forms, for example, in the cubic (isometric) crystal system alone. The 47 forms are listed in Table 1.2, grouped by the crystal systems to which they belong. Included in the table are representative examples of minerals exhibiting these form developments.

TABLE 1.2 The 47 Possible Forms Distributed Among the Various Crystal Systems

Crystal Class	Symmetry Elements[a]	Forms Occurring in the Respective Crystal Class	Representative Inorganic/Mineral Substances
Triclinic			
1	E	Pedion	$Ca_8B_{18}Cl_4 \cdot 4H_2O$
$\overline{1}$	E, i	Pinacoid	$MnSiO_3$
Monoclinic			
2	E, C_2	Sphenoid, pedion, pinacoid	$FeAl_2(SO_4)_4 \cdot 22H_2O$
m	E, σ_h	Sphenoid, pedion, pinacoid	$CaMg(AsO_4)F$
$2/m$	E, C_2, i, σ_h	Prism, pinacoid	As_2S_3
Orthorhombic			
222	E, C_2, C_2', C_2'	Disphenoid, prism, pinacoid	$ZnSO_4 \cdot 7H_2O$
$mm2$	$E, C_2, \sigma_v, \sigma_v$	Pyramid, prism, dome, pinacoid, pedion	$BaAl_2Si_3O_{10} \cdot 2H_2O$
mmm	$E, C_2, C_2', C_2', \sigma_v, \sigma_v, \sigma_h$	Dipyramid, prism, pinacoid	Sulfur
Tetragonal			
4	$E, 2C_4, C_2$	Tetragonal pyramid, tetragonal prism, pedion	None known
$\overline{4}$	$E, 2S_4, C_2$	Tetragonal disphenoid, tetragonal prism, pinacoid	$Ca_2B(OH)_4AsO_4$
$4/m$	$E, 2C_4, C_2, i, 2S_4, \sigma_h$	Tetragonal dipyramid, tetragonal prism, pinacoid	$PbMoO_4$
422	$E, 2C_4, C_2, 2C_2', 2C_2''$	Tetragonal trapezohedron, tetragonal dipyramid, ditetragonal prism, tetragonal prism, pinacoid	$Pb_2CO_3Cl_2$
$4mm$	$E, 2C_4, C_2, 2\sigma_v, 2\sigma_c$	Ditetragonal pyramid, tetragonal pyramid, ditetragonal prism, tetragonal prism, pedion	$Pb_2Cu(OH)_4Cl_2$
$\overline{4}\,2m$	$E, C_2, 2C_2', 2\sigma_d, 2S_4$	Tetragonal scalenohedron, tetragonal disphenoid, tetragonal bipyramid, ditetragonal prism, tetragonal prism, pinacoid	Cu_2FeSnS_4

(continued overleaf)

TABLE 1.2 *(continued)*

Crystal Class	Symmetry Elements[a]	Forms Occurring in the Respective Crystal Class	Representative Inorganic/Mineral Substances
4/mmm	E, $2C_4$, C_2, $2C_2'$, $2C_2''$, i, $2S_4$, σ_h, $2\sigma_v$, $2\sigma_d$	Ditetragonal dipyramid, tetragonal dipyramid, ditetragonal prism, tetragonal prism, pinacoid	Rutile
Trigonal (rhombohedral)			
3	E, $2C_3$	Trigonal pyramid	$NaIO_4 \cdot 3H_2O$
$\bar{3}$	E, $2C_3$, i, $2S_6$	Rhombohedron, hexagonal prism, pinacoid	$FeTiO_3$
32	E, $2C_3$, $3C_2'$	Trigonal trapezohedron, rhombohedron, trigonal dipyramid, ditrigonal prism, hexagonal prism, trigonal prism, pinacoid	Low quartz
3m	E, $2C_3$, $3\sigma_v$	Ditrigonal pyramid, trigonal pyramid, hexagonal pyramid, ditrigonal prism, trigonal prism, hexagonal prism, pedion	$KBrO_3$
$\bar{3}$m	E, $2C_3$, $3C_2'$	Hexagonal scalenohedron, rhombohedron, hexagonal dipyramid, dihexagonal prism, hexagonal prism, pinacoid	Corundum
Hexagonal			
6	E, $2C_6$, $2C_3$, C_2	Hexagonal prism, pedion	Nepheline
$\bar{6}$	E, $2C_6$, σ_h, $2S_3$	Trigonal dipyramid, trigonal prism, pinacoid	None
6/m	E, $2C_6$, $2C_3$, C_2, i, $2S_3$, $2S_6$, σ_h	Hexagonal dipyramid, hexagonal prism, pinacoid	Apatite
622	E, $2C_6$, $2C_3$, C_2, $3C_2'$, $3C_2''$	Hexagonal trapezohedron, hexagonal dipyramid, dihexagonal prism, hexagonal prism, pedion	High quartz

6mmm	E, $2C_6$, $2C_3$, C_2, $3\sigma_v$, $3\sigma_d$	Dihexagonal pyramid, hexagonal pyramid, dihexagonal prism, hexagonal prism, pedion	Wurtzite
$\bar{6}$m2	E, $2C_3$, $3C_2'$, σ_h, $2S_3$, $3\sigma_v$	Ditrigonal dipyramid, trigonal dipyramid, hexagonal dipyramid, ditrigonal prism, hexagonal prism, trigonal prism, pinacoid	BaTiSi$_3$O$_9$
6/mmm	E, $2C_6$, $2C_3$, C_2, $3C_2'$, $3C_2''$, i, $2S_3$, $2S_6$, σ_h, $3\sigma_v$, $3\sigma_d$	Dihexagonal dipyramid, hexagonal dipyramid, dihexagonal prism, hexagonal prism, pinacoid	Beryl
Cubic			
23	E, $8C_3$, $3C_2$	Tetartoid, deltohedron, tristetrahedron, pyritohedron, tetrahedron, dodecahedron, cube	NaBrO$_3$
m3	E, $8C_3$, $3C_2$, i, $8S_6$, $3\sigma_h$	Diploid, trisoctahedron, trapezohedron, pyritohedron, octahedron, dodecahedron, cube	Pyrite
432	E, $8C_3$, $3C_2$, $6C_2$, $6C_4$	Gyroid, trisoctahedron, trapezohedron, tetrahexahedron, octahedron, dodecahedron, cube	None
$\bar{4}$3m	E, $8C_3$, $3C_2$, $6\sigma_d$, $6S_4$	Hextetrahedron, deltohedron, tristetrahedron, tetrahexahedron, tetrahedron, dodecahedron, cube	Sphalerite
m3m	E, $8C_3$, $3C_2$, $6C_2$, $6C_4$, $6\sigma_d$, i, $8S_6$, $3\sigma_h$, $6S_4$	Hexoctahedron, trisoctahedron, trapezohedron, tetrahexahedron, octahedron, dodecahedron, cube	Diamond

Source: Buerger (1978), pp. 112–168.

[a] E, identity element; i, inversion center; C_n, n-fold proper rotation axis; S_n, n-fold improper rotation (rotoreflection) axis; σ_v, vertical mirror plane (reflection plane contains principal axis); σ_h, horizontal mirror plane (reflection plane perpendicular to principal axis); σ_d, diagonal mirror plane (reflection plane contains principal axis and bisects the angle between the twofold axes normal to the principal axis).

The cube and octahedron are both referred to as *closed forms* because they are comprised of a set of equivalent faces that enclose space completely. All 15 forms in the isometric system, which include the cube and octahedron, are closed. One of the isometric forms, the hexoctahedron, has 48 faces. Six forms have 24 faces (tetrahexahedron, trisoctahedron, trapezohedron, hextetrahedron, gyroid, and diploid). Five forms have 12 faces (dodecahedron, tristetrahedron, pyritohedron, deltahedron, and tetartoid). The final three isometric and closed forms are, perhaps, more familiar to the chemist. These are the tetrahedron (4 faces), cube (6 faces), and octahedron (8 faces). Nonisometric closed forms include the dipyramids (6, 8, 12, 16, or 24 faces), scalenohedrons (8 or 12 faces), the rhombic and tetragonal disphenoids (4 faces), the rhombohedron (6 faces), the ditrigonal prism (6 faces), the tetragonal trapezohedron (8 faces), and the hexagonal trapezohedron (12 faces). *Open forms* do not enclose space. These include the prisms with 3, 4, 6, 8, or 12 faces, parallel to the rotation axis. These parallel faces are equivalent but do not enclose space. Other open (and nonisometric) forms include the pyramid (3, 4, 6, 8, or 12 faces), domes (2 faces), sphenoids (2 faces), pinacoids (2 faces), and pedions (1 face). Open forms may only exist in combination with a closed form or with another open form.

As mentioned earlier, a crystal's external morphology may not be consistent with the true point group symmetry of the space lattice. This may be due to (1) one or more forms being absent or showing anisotropic development of equivalent faces, or (2) the true symmetry of the unit cell simply not being manifested macroscopically upon infinite translation in three dimensions. Let's first consider anisotropic development of faces. The growth rates of faces—even crystallographically equivalent faces and faces of crystals belonging to the isotropic (cubic) crystal class—need not be identical. This may be due to kinetic or thermodynamic factors.

One possible reason is a nonsymmetrical growth environment. For example, the nutrient supply may be blocked from reaching certain crystal faces by foreign objects or by the presence of habit-modifying impurities. Since visible crystal faces correspond to the slow-growing faces, the unblocked faces may grow so much faster that only the blocked faces are left visible where as the fast-growing faces transform into vertices. Consider pyrite, which belongs to the cubic system. Crystal growth relies on a layer-by-layer deposition on a nucleus via an *external* flux of adatom species, which may very well be anisotropic. Hence, unequal development of crystallographically equivalent {100} faces can lead to pyrite crystals exhibiting acicular and platelike morphologies instead of the anticipated cube shape. In fact, not all crystals exhibit distinct polyhedral shapes. Those that do not are termed *nonfaceted crystals*. The word *habit* is used to describe the overall external shape of a crystal specimen. Habits, which can be polyhedral or nonpolyhedral, may be described as cubic, octahedral, fibrous, acicular, prismatic, dendritic (treelike), platy, blocky, or bladelike, among many other terms.

As a second example of a mineral with several possible form developments, let's consider, in a little more detail, quartz (Figure 1.1). Quartz belongs to the symmetry class 32, which has two threefold rotation axes and three twofold axes.

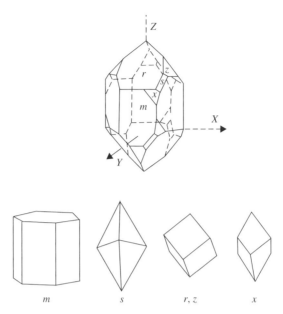

Figure 1.1   Top: Quartz crystal exhibiting the true symmetry of the crystal class to which quartz belongs. Bottom: The forms comprising such a quartz crystal. From left to right, the hexagonal prism, trigonal dipyramid, rhombohedron, and trigonal trapezohedron.

Five forms must necessarily be present to reveal this symmetry: the $\{10\bar{1}0\}$, the $\{10\bar{1}1\}$, the $\{01\bar{1}1\}$, the $\{11\bar{2}1\}$, and the $\{51\bar{6}1\}$. These correspond, respectively, to a hexagonal prism; a dominant, or "positive," rhombohedron; a subordinate, or "negative," rhombohedron; a trigonal (triangular) dipyramid; and a trigonal trapezohedron. In mineralogy, these are labeled, in the order given, with the lowercase letters m, r, z, s, and x. Three orthogonal crystallographic axes are defined as: X, bisecting the angle between adjacent hexagonal prism faces; Y, which runs through the prism face at right angles to X; and Z, an axis of threefold symmetry.

As illustrated in Figure 1.1, both rhombohedra (r and z) cap, or terminate, the quartz crystal on each end. Each rhombohedron has a set of three faces. By convention, the larger set of three faces is considered the positive rhombohedron. When present, the trigonal trapezohedron (x) is seen at the junction of two prism faces (m) and the positive rhombohedron and it displays a trapezohedral planar shape. The trigonal pyramid (s) is at the junction of the positive rhombohedron and the prism, which is in line vertically with the negative rhombohedron. It typically forms an elongated rhombus-shaped face. However, in some specimens, one or more of the aforementioned forms are missing or show development inconsistent with the true point group symmetry of quartz. In fact, most quartz crystals do not display the trigonal dipyramid or trapezohedron faces, the former being especially rare. With these two forms absent, the rhombohedra may exhibit either equal or unequal development. The latter case implies the highest apparent (but still false) crystal symmetry, as the hexagonal prism appears to be terminated at both ends with hexagonal pyramids. It is also possible for the hexagonal prism to be absent, in which case the combination of the two rhombohedra results in a hexagonal dipyramid (or bipyramid), termed a *quartzoid*.

We mentioned earlier that the true symmetry of the unit cell may not simply be manifested macroscopically upon infinite translation in three dimensions. Buerger has illustrated this with the mineral nepheline, $(Na,K)AlSiO_4$ (Buerger, 1978). The true symmetry of the nepheline crystal lattice, the symmetry of the unit cell, consists merely of a sixfold rotation axis (class 6) as would be exhibited by a hexagonal prism with nonequivalent halves. That is, there is no mirror plane perpendicular to the rotation axis. However, the absence of this mirror plane is obviously not macroscopically visible in the hexagonal prism form development of nepheline, implying a higher apparent symmetry ($6/mmm$).

The situation with crystallized iodoform, CHI_3, is similar. The molecules are strictly pyramidal, but the crystal contains complementary "positive" and "negative" pyramids capping a hexagonal prism, as do the minerals zinkenite ($Pb_9Sb_{22}S_{42}$) and finnemanite [$Pb_5(AsO_3)_3Cl$]. In fact, no crystals showing form development consistent with class 6 symmetry have been observed. It is observed, rather, that form developments tend to follow the holohedral or holosymmetric symmetry of the crystal class (i.e., the point group with the highest symmetry of its crystal system). This is most commonly manifested by equal development of complementary forms in the merosymmetric classes (i.e., those with less symmetry than the lattice).

Often, the habit assumed by a particular specimen is under kinetic control, being dependent primarily on the growth environment. If crystals grow into one another, as they do in solidification products and polycrystals, it is possible that no forms will be well developed. However, when multiple crystal forms are present, it is possible that some forms (sets of equivalent faces) might have intrinsically higher growth rates than others. Form developments tend to lower the surface energy of the crystal. Surface energy is a function of several parameters:

- The distance of the face from the center of the crystal
- The face orientation, or Miller (hkl) indices
- Surface roughness
- The radius of curvature

Under equilibrium growth conditions, the fastest crystal growth will occur in the direction perpendicular to the face with the highest surface energy. As a consequence of this tendency, the total area of high-energy faces is reduced while that of low-energy faces is increased, which effectively results in an overall lowering of the surface energy of the crystal. For nonfaceted crystals, a smaller radius of curvature possesses a higher energy than one with a larger radius of curvature. The effect of surface energy on crystal morphology is discussed in detail in Section 2.3.

Two methods have traditionally been used to rationalize morphology resulting from the differing growth rates of various faces in terms of the lattice symmetry of the crystal. The first is the *Donnay–Harker method* (Donnay and Harker, 1937), which extends the work of Auguste Bravais (1811–1863) and Georges Friedel (1865–1933). The Bravais–Friedel empirical rule (Bravais, 1866; Friedel, 1907) was that the relative morphological importance of crystal forms is in the same order as the concentration of lattice points on their crystal faces, the *reticular density*. The dominant faces represent those planes cutting through the greatest densities of lattice points. Ranking the crystal forms in order of decreasing importance is also equivalent to ranking them in order of increasing reticular area or to decreasing interplanar spacings, which is proportional to the Miller indices of the faces. Hence, those faces with low Miller indices tend to be the morphologically most important. Because the interplanar spacing is the same for all crystal classes in a system, the morphological importance of the various forms tends to follow the holosymmetric symmetry of the system (Buerger, 1978). Joseph D. H. Donnay (1902–1994) and David Harker (1906–1991) later refined the idea to account for all equivalence points, including not only lattice points but also nonlattice sites of identical internal coordinates (Donnay and Harker, 1937). Although it is often successful, the Donnay–Harker method takes no account of atomic arrangement or bonding and is not applicable to the prediction of surface structure (Dowty, 1976).

Bravais rule is an empirical observation, not a scientific law. Recently, it has been proposed that it may, arguably, be viewed as a manifestation of the Curie–Rosen principle of maximum symmetry from information theory, which states that *the degree of symmetry in an isolated system either remains constant or increases as the system evolves*. To understand the relevance of this principle here, it is necessary to define precisely what symmetry is, correlate it with *information entropy*, and establish the link between entropy in information theory and entropy in thermodynamics, which deals with physical nature. Shu-Kun Lin has attempted this (Lin, 1996). Symmetry is regarded as *invariance* of an object when it is subjected, as a whole, to certain transformations in the space of the variables describing it (e.g., rotation, reflection, translation). The degree of symmetry correlates with the degree of *information entropy*, or information loss (Lin, 1996, 2001; Rosen, 1996). For example, a high degree of information entropy is associated with a high degree of similarity. The maximum possible information entropy, indistinguishability (total loss of information), corresponds to perfect symmetry, or highest symmetry, and low orderliness. By contrast, low symmetry corresponds to distinguishability and high orderliness. The principle of maximum symmetry would then seem to imply that crystals should exhibit the highest possible symmetry consistent with the growth constraints imposed. Unfortunately, the link between informational entropy and thermodynamic entropy may only be philosophical; it is still a hotly debated topic. Nevertheless, the principle of maximum symmetry serves as a useful heuristic.

The second, more recent approach at explaining the different rates of growth along different crystallographic directions, addresses all of these aspects, and is known both as the *Hartman–Perdok theory* and the *periodic bond chain* (PBC) *method* (Hartman and Perdok, 1955a,b,c). It was developed originally by Piet Hartman and Wiepko Gerhardus Perdok (1914–2005), professor of applied crystallography at the University of Groningen, and was built on earlier work by Paul Niggli (1883–1953) and Robert Luling Parker (1893–1973). Hartmann and Perdok showed that crystal structure determines which directions will exhibit rapid growth rates due to a large energy gain which they called the *attachment energy*. As a result of this rapid growth, those faces grow themselves out of existence, becoming "consumed" by the slow-growing faces. In the PBC method, one first identifies infinite stoichiometric chains of atoms connected by strong bonds (the periodic bond chains). These chains do not have dipole moments perpendicular to their crystallographic direction. The strong bonds are the short bonds in the first coordination sphere formed during crystallization. Crystal faces either have two or more PBCs (flat, or F, faces), one PBC (stepped, or S, faces), or none (kinked, or K, faces). The F faces grow the slowest and are thus the morphologically important ones, while the K and S faces grow so quickly they should not be present in the crystal forms. We discuss the PBC method in more detail in Section 4.2.4. For crystal growth to proceed, it is necessary for the surface to capture and incorporate adatoms. Hence, another factor is the availability of surface sites that readily accommodate new growth units. The theory most successful at dealing with this issue is the terrace–ledge–kink (TLK) model, which is discussed further in Sections 1.4.1 and 2.3.1.2.

1.2.2. Twinned Crystals

A *twin* is a symmetrical intergrowth of two or more crystals, or "individuals," of the same substance (Figure 1.2). A *simple twin* contains two individuals; a *multiple twin* contains more than two individuals. The *twin element* is the geometric element about which a twin operation is performed, relating the different individuals in the twin. The twin element may be a reflection plane (*contact twins*) or a rotation axis (*penetration twins*). The twin operation is a symmetry operation for the twinned edifice only, not for the individuals. In *twinning by merohedry*, the twin and the individual lattice point group, as well as their translational symmetry, coincide. If both the point group and translational symmetries of the twin and individual differ, it is referred to as *twinning by reticular merohedry*. Most commonly, twinning is by *syngonic merohedry*, in which the twin operation is a symmetry element of the holohedral point group (one of the seven point groups exhibiting the complete symmetry of the seven crystal systems) while the point group of the individual crystals is a subgroup, exhibiting less then complete (holohedral) symmetry. With *metric merohedry*, the individual lattice has an accidentally specialized metric corresponding to a higher holohedry and a twin operation exists belonging only to the higher holohedry. For example, a twin operation belonging to an orthorhombic lattice may exist for a twinned edifice comprised of two monoclinic crystals. The empirical rule of merohedral twinning was developed originally by Auguste Bravais, François-Ernest Mallard (1833–1894), and, later, Georges Friedel.

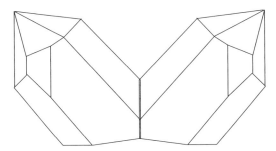

Figure 1.2   "Japan law" contact twin quartz crystal. This type of twinning was discovered in 1829 by C. S. Weiss in quartz crystal from the La Gardette mine in France. However, because of the abundance of these specimens in Japan, they are now known as *Japanese twins*.

Like a grain boundary, the twin boundary is a higher-energy state relative to the crystal. However, because a twin boundary is highly ordered, it is of much lower energy than a typical nontwin grain boundary. Recognizing this, Buerger later proposed that *if the crystal structure is of such a nature that, in detail, it permits a continuation of itself in alternative twin junction configuration, without involving violation of the immediate coordination requirements (the first coordination sphere) of its atoms, the junction has low energy and the twin is energetically possible* (Buerger, 1945).

Twins may also be divided into the following categories based on their origin: *transformation twins* and *growth twins, gliding twins. Growth twins* originate at the nucleation stage under conditions of supersaturation, where there is greater likelihood for the arrival of clusters of atoms, already coordinated, at the twin position. Such twins persist and grow if subsequent clusters of adatoms continue to arrive in that fashion. *Glide twinning* is caused by a specific type of structural shear in plastic deformation. The lower-energy nontwined crystals absorb part of the energy supplied in the plastic deformation process, and if the crystal structure permits it, a layer of atoms glide into a twin position. With continued stress, gliding takes place in the next layer. Because gliding in all the parallel layers does not take place simultaneously, twin lamella form. Calcite is an example of a crystal that forms glide twins readily at low differential stresses (~ 10 MPa). Twinning is possible along three glide planes. *Transformation twinning* occurs during the transformation from a high-temperature phase to a lower-symmetry low-temperature phase: for example, when sanidine (monoclinic $KAlSi_3O_8$) is cooled to form microcline (triclinic $KAlSi_3O_8$). In such a process, there is spontaneous formation of nuclei in different orientations, which subsequently grow into one another. Each member of the aggregate is either in parallel or twinned orientation with respect to other members. This follows from the fact that they could be brought into coincidence by one of the possible symmetry operations of the high-temperature phase that vanished in the formation of the low-temperature phase, which has a symmetry that is a subgroup of the high-temperature phase (Buerger, 1945).

Photo courtesy of the Emilio Segrè Visual Archives. Copyright © The Massachusetts Institute of Technology. Reproduced with permission.

Martin Julian Buerger (1903–1986) received his Ph.D. in mineralogy in 1929 from the Massachusetts Institute of Technology MIT, where he remained as an assistant professor. As a graduate student, he attended lectures on x-ray diffraction delivered by W. L. Bragg in 1927, at which point he began his lifelong investigations into crystal structure–property correlation. In the 1930s, Buerger established an x-ray laboratory at MIT, out of which he revised or invented many different types of x-ray cameras and diffractometers as well as carrying out many crystal structure determinations. He also wrote 12 books on crystallography. In addition to being a pioneer in crystal structure analysis, Buerger contributed to a range of other topics within crystallography, crystal growth, polymorphism, twinning, and early (predislocation model) plastic deformation theories. He served as president of both the American Society for X-ray and Electron Diffraction and the Crystallographic Society of America, which later were combined into the American Crystallographic Association. Buerger helped to organize the International Union of Crystallography. He also served as editor-in-chief of *Zeitschrift für. Kristallographie*, and he contributed diagrams, notes, and corrections to the *International Tables of Crystallography*. In 1956, Buerger was the second person to be appointed institute professor at MIT, after John Clarke Slater. He received the Day Medal of the Geological Society of America (1951), the Roebling Medal of the Mineralogical Society of America (1958), and was the first recipient of the Fankuchen Award in 1971. Buerger became university professor emeritus at MIT in 1968 and then accepted an appointment at the University of Connecticut, becoming emeritus there in 1973. The ACA established the triannual *M. J. Buerger Award* in his memory in 1983. Buerger was elected to the U.S. National Academy of Sciences in 1953 and was a member of several foreign academies. He also has a mineral named after him: buergerite, $NaFe_3Al_6(BO_3)_3[Si_6O_{18}]O_3F$.

Source: Obituary by Leonid V. Azároff, *J. Appl. Crystallogr*, **1986**, *19*, 205.

1.2.3. Polycrystalline Texture

Polycrystalline substances, in bulk form and thin films, are composed of crystalline grains, which are sometimes times called *crystallites*. The *microstructure* of a polycrystalline material refers to the crystallite morphology (particle size and shape) and orientation distribution of the grains. The orientation distribution is called *texture*. For example, the grains may all be oriented at random, they may exhibit some preferred orientation (e.g., alignment), or subregions may be present called *domains*, each of which has its own independent orientation. The most common method of illustrating texture is through the use of stereographic projections called *pole figures*, which show the inclination to the normal of a particular crystal plane relative to a reference plane. This is developed much more fully in Section 7.1.1. A succinct discussion can also be found in our companion book, *Principles of Inorganic Materials Design* (Lalena and Cleary, 2005).

As physical properties are anisotropic for all single crystals except for those of the cubic class, texture has important consequences on the symmetry of the physical properties exhibited by polycrystalline materials. It turns out that for a bulk sample containing a sufficiently large number of grains, a completely random grain orientation distribution displays macroscopically isotropic properties as well. Hence, for all cases other than cubic single crystals and nontextured polycrystals, the magnitude of the physical property (e.g., conductivity) will depend on the crystallographic direction. Materials processing is aimed at achieving textures that impart optimal properties for specific design applications.

1.3. SPACE LATTICES

Although the exact morphology (the collection of particular crystal forms, their shapes, and texture) adopted by a specimen is under kinetic control, the number of possible crystal forms is restricted by symmetry, as stated earlier. The number of faces that belong to a form is determined by the symmetry of the point group of the lattice. Put another way, the symmetry of the internal structure *causes* the symmetry of the external *forms*. In this section we consider that internal symmetry in more detail.

Matter is composed of spherical-like atoms. No two atomic cores—the nuclei plus inner shell electrons—can occupy the same volume of space, and it is impossible for spheres to fill all space completely. Consequently, spherical atoms coalesce into a solid with void spaces called *interstices*. A mathematical construct known as a *space lattice* may be envisioned, which is comprised of equidistant lattice points representing the geometric centers of structural motifs. The lattice points are equidistant since a lattice possesses translational invariance. A motif may be a single atom, a collection of atoms, an entire molecule, some fraction of a molecule, or an assembly of molecules. The motif is also referred to as the *basis* or, sometimes, the *asymmetric unit*, since it has no symmetry of its own. For example, in rock salt a sodium and chloride ion pair constitutes the asymmetric unit. This ion pair is repeated systematically, using point symmetry and translational symmetry operations, to form the space lattice of the crystal.

A space lattice may be characterized by its space group symmetry, which consists of point group symmetry (rotations and reflection) and translational symmetry. The morphological symmetry must conform to the angular components of the space group symmetry operations minus the translational components. That is, the symmetry of the external morphology is that of the point group *isogonal* with (possessing the same angular relations as) its space group. It is thus possible under favorable circumstances to determine the point group by an examination of the crystal morphology.

Inorganic solids may be grouped into three structure categories, which are, in the order of decreasing thermodynamic stability: single crystalline, polycrystalline, and amorphous (glassy). Amorphous materials possess no long-range structural order or periodicity. By contrast, a single crystal is composed of a periodic three-dimensional array of atoms or molecules whose positions may be referenced to a lattice, which, as already stated, possesses translational invariance. Some authors also classify recursive patterns known as *fractals* as a distinct structural class. With fractals, the structure is self-similar, looking identical at all length scales. This is a violation of translational invariance and is, in fact, a new type of symmetry called *scale invariance*. Some porous solids (e.g., silica aerogels) have fractal geometry. Many objects in nature (e.g., cauliflower, tree branches, snowflakes, rocks, dendrites) approximate a fractal since the morphologies of their whole are superficially similar to those of the parts comprising them. A polycrystalline material is an aggregate of *crystallites*, or grains, whose lattices exhibit the same translational invariance as their single-crystal counterparts, but the grains are not usually morphologically as well formed. Furthermore, a polycrystalline specimen, as a whole, possesses a *microstructure* that is characterized by the size, shape, and orientation distribution (texture) of the individual crystallites. The atomic-scale structure of the interface between any two crystallites is a function of the relative orientations of the crystallites.

The atoms in a crystalline substance occupy positions in space that can be referenced to *lattice points*, which crystallographers refer to as the *asymmetric unit* (physicists call it the *basis*). Lattice points represent the smallest repeating unit, or chemical point group. For example, in NaCl, each Na and Cl pair may be represented by a lattice point. In structures that are more complex, a lattice point may represent several atoms (e.g., polyhedra) or entire molecules. The repetition of lattice points by translations in space forms a *space lattice*, representing the extended crystal structure.

1.3.1. Two-Dimensional Lattice Symmetry

It is important to study two-dimensional (2D) symmetry because of its applicability to lattice planes and the surfaces of three-dimensional (3D) solids. In two dimensions, a lattice point must belong to one of the 10 point groups listed in Table 1.3 (by the international symbols) along with their symmetry elements. This group, called the *two-dimensional crystallographic plane group*, consists of combinations of a single rotation axis perpendicular to the lattice plane with or

TABLE 1.3   The 10 Two-Dimensional Crystallographic Plane Point Groups and Their Symmetry Elements (International Symbols)[a]

Point Group	Symmetry Elements
Oblique System (Lattice = parallelogram)	
1	1
2	2
Rectangular System (Lattice = Rectangle or Centered Rectangle)	
m	*m*
2*mm*	2, *m* (two perpendicular planes)
Square System (Lattice = Square)	
4	4
4*mm*	4, *m* (two doubly degenerate sets)
Hexagonal System (Lattice = 120° Rhombus)	
3	3
3*m*	3, *m* (triply degenerate set)
6	6
6*mm*	6, *m* (two triply degenerate sets)

[a]*n* (where n = 1, 2, 3, 4, 6), *n*-fold rotation axis perpendicular to the plane; *m* mirror plane (perpendicular to the plane and containing the rotation axis).

without parallel reflection planes (perpendicular to the lattice plane and containing the rotation axis). Note that only those point groups with one-, two-, three-, four-, or sixfold rotations are found in 2D (and 3D) lattices. This can be understood by considering an analogous task of completely tiling a floor with regular polygon tiles. Rhombuses, rectangles, squares, triangles, and hexagons may be used, but not pentagons, heptagons, or higher polygons.

An initial point may be translated in a regular way in two dimensions, thereby generating the *plane lattice*, or *net*, in which each point of the array has exactly the same environment. The points of this plane lattice may be imagined connected together with translation vectors, such that the following equation (in which u and v are integers) holds:

$$\mathbf{R} = u\,\mathbf{a} + v\,\mathbf{b} \qquad (1.10)$$

The translation vectors can be chosen so that a *primitive unit cell* is defined that generates the plane lattice when repeated in two-dimensional space. A primitive unit cell of any dimension contains only one lattice point, which is obtained by sharing lattice points at each of its corners with the neighboring unit cells. By convention, the lattice vectors are chosen to be the shortest and most nearly equal as is possible. There are only five unique ways of choosing such translation vectors for a 2D lattice. These are the five *two-dimensional Bravais lattices*, named after the French physicist and mineralogist Auguste Bravais, who derived them in 1850 (Bravais, 1850). The unit cells for each lattice may be described by three parameters: two translation vectors (a, b) and one interaxial angle, usually symbolized as γ. These five lattices (oblique, rectangular, centered-rectangular,

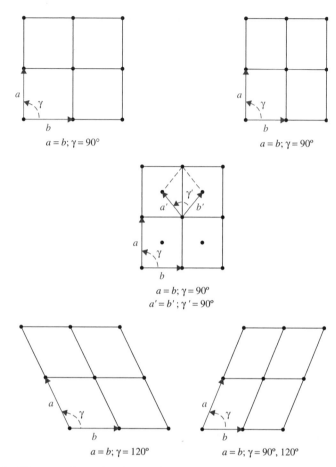

Figure 1.3 Five two-dimensional Bravais lattices. Clockwise from upper left: square, rectangular, oblique, hexagonal, and (center) centered rectangular.

square, and hexagonal) are shown in Figure 1.3. It should be noted that a centered lattice need not be used. An equivalent primitive unit cell can be chosen by defining a new set of nonorthogonal lattice vectors of equal length starting with the same origin and ending on the centered atom, as illustrated in the middle figure. When multiple unit cells are possible, the unit cell with maximum symmetry and smallest area is customarily chosen.

If the 10 point groups allowed are arranged in nonredundant patterns allowed by the five 2D Bravais lattices, 17 unique two-dimensional space groups, called *plane groups*, are obtained (Fedorov, 1891a). Surface structures are usually referred to the underlying bulk crystal structure. For example, translation between lattice points on the crystal lattice plane beneath and parallel to the surface (termed the *substrate*) can be described by an equation identical to Eq 1.10:

$$\mathbf{R} = u'\mathbf{a} + v'\mathbf{b} \qquad (1.11)$$

In many cases, the following relation, in which the surface vectors are denoted $\mathbf{a}_s$ and $\mathbf{b}_s$, is found to hold:

$$\mathbf{a}_s = p\mathbf{a} \quad \text{and} \quad \mathbf{b}_s = q\mathbf{b} \tag{1.12}$$

When both p and q are unity, the surface unit cell and the *projection* of the substrate unit cell are the same. This type of surface is designated as 1×1. If both dimensions of the surface's unit cell are twice those of the substrate, the surface is designated 2×2, if only one dimension is n times that of the substrate cell, the surface is $n \times 1$. There are also various "root-three" $\left(\sqrt{3} \times \sqrt{3} \right)$ structures resulting from adsorbed gases on metal surfaces or metals on semiconductor surfaces. The notation for surface structures with adsorbates gets quite complex when rotations relative to the substrate structure are involved. However, we are not concerned with those here.

1.3.2. Three-Dimensional Lattice Symmetry

In an analogous fashion, motifs in three-dimensional lattices must belong to one of the 32 three-dimensional crystallographic point groups, which may be derived by inclusion of the additional symmetry elements found in three-dimensional lattices: namely, centers of inversion, rotoinversion axes, perpendicular reflection planes, and additional axes of rotation. An object that can be replicated to fill all space (a lattice point) in such a way that all the objects so generated are in identical environments (a lattice) can possess only one of these 32 unique combinations of 3D point symmetry, which are the same as the 32 types of external symmetry exhibited by crystals derived by Johann Friedrich Christian Hessel (1796–1872), a physician and professor of mineralogy at the University of Marburg (Hessel, 1830, 1897). These 32 crystallographic point groups and their symmetry elements are listed in the first and second columns of Table 1.2, subdivided into the seven crystal systems. As with 2D nets, only those point groups with one-, two-, three-, four-, or sixfold rotations are found in 3D lattices. The finite number of possible crystallographic point groups also limits the number of external crystal forms to 47.

Equation 1.10, for the translation vectors in a 2D lattice, can be modified in a simple fashion for three dimensions:

$$\mathbf{R} = u\mathbf{a} + v\mathbf{b} + w\mathbf{c} \tag{1.13}$$

In direct analogy with two dimensions, we can define a *primitive unit cell* that when repeated by translations in space, generates a 3D space lattice. There are only 14 unique ways of connecting lattice points in three dimensions, which define unit cells (Bravais, 1850). These are the 14 *three-dimensional Bravais lattices*. The unit cells of the Bravais lattices may be described by six parameters: three translation vectors (a, b, c) and three interaxial angle (α, β, γ). These six parameters differentiate the seven crystal systems: triclinic, monoclinic, orthorhombic, tetragonal, trigonal, hexagonal, and cubic.

There are 230 3D *space groups*, obtained by arranging the 32 crystallographic point groups in the patterns allowed by the 14 3D Bravais lattices. This was shown independently, using different methods, by the British crystallographer William Barlow (1845–1934), the Russian mathematician Evgraf Stepanovich Fedorov (1853–1919), and the German mathematician Arthur Moritz Shönflies (1853–1928) (Fedorov, 1891b; Schönflies, 1891; Barlow, 1894). It was once believed that geometric crystallography was a "closed" or "dead" science, since all possible crystallographic point groups and space groups had been derived. However, the discovery in the 1980s of *quasicrystalline* solids that contain five-, eight-, 10-, and 12-fold rotational symmetry seems to have changed that view.

The notation system developed by Schönflies for designating point group symmetry (Table 1.3) is still widely used by spectroscopists. However, crystallographers use the *Hermann–Mauguin*, or *International*, *notation* for space group symmetry. This system was developed by Carl Hermann (1898–1961) and Charles Mauguin (1878–1958) (Hermann, 1928, 1931; Mauguin, 1931). Each space group is isogonal with one of the 32 crystallographic point groups. However, space group symbols reveal the presence of two additional symmetry elements, formed by the combination of point group symmetry (proper rotations, improper rotations, and reflection) with the translational symmetries of the Bravais lattices. The two types of combinational symmetry are the *glide plane* and *screw axis*. The first character of an international space group symbol is a capital letter designating the Bravais lattice centering type (primitive $= P$; all-face-centered $= F$; body-centered $= I$; face-centered $= C, A$; rhombohedral $= R$). This is followed by a modified point group symbol giving the symmetry elements (axes and planes) that occur for each of the lattice symmetry directions for the space group (Lalena, 2006).

Within the unit cell, atoms may be located at *general positions* or, if they lie on a symmetry element (inversion center, rotation axis, mirror plane), at *special positions*. A general position is left invariant only by the identity operation. Each space group has only one general position but the position may have multiple equivalent coordinates. For example, for a phase crystallizing in the space group *Pmmm*, an atom located in the general position x,y,z will, by symmetry, also be found at seven other coordinates: $-x, -y,z$; $-x,y$; $-z$; $x, -y$; $-z$; $-x, -y, -z$; $x,y, -z$; $x, -y,z$; and $-x,y,z$. The general position is said to have a *multiplicity* of eight. For primitive cells, the multiplicity of the general position is equal to the order of the point group of the space group; for centered cells, the multiplicity is equal to the product of the order of the point group and the number of lattice points per cell (Lalena, 2006).

A set of symmetry-equivalent coordinates is said to be a special position if each point is mapped onto itself by one other symmetry operation of the space group. In the space group *Pmmm*, there are six unique special positions, each with a multiplicity of four, and 12 unique special positions, each with a multiplicity of two. If the center of a molecule happens to reside at a special position, the molecule must have at least as high a symmetry as the site symmetry of the special position. Both general and special positions are also called *Wyckoff*

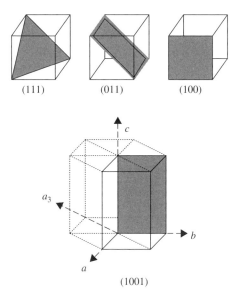

Figure 1.4   Examples of lattice planes and their Miller indices. (After Lalena and Cleary, 2005. Copyright © John Wiley & Sons, Inc. Reproduced with permission.)

positions, in honor of the American crystallographer Ralph Walter Graystone Wyckoff (1897–1994). Wyckoff's 1922 book, *The Analytical Expression of the Results of the Theory of Space Groups*, contained tables with the positional coordinates, both general and special, permitted by the symmetry elements. This book was the forerunner of *International Tables for X-ray Crystallography*, which first appeared in 1935 (Lalena, 2006).

Often, it is necessary to refer to a specific family of lattice planes or a direction in a crystal. A family of lattice planes is a set of imaginary parallel planes that intersect the unit cell edges. Each family of planes is identified by its Miller indices. The Miller indices are the reciprocals of the fractional coordinates of the three points where the first plane away from the origin intercepts each of the three axes. The letter h refers to the intersection of the plane on a; k the intersection on b; and l the intersection on c. Some examples are illustrated in Figure 1.4. These indices are the same as those introduced earlier for denoting the external faces of crystals. When referring to a specific plane in a family, the numbers are grouped together in parentheses, (hkl). Any family of planes always has one member that passes through the origin of the unit cell. The plane used in determining the Miller indices is always the first one away from the origin, which may be obtained by moving in either direction.

Note that a Miller index of zero implies that the plane is parallel to that axis, since it is assumed that the plane will intersect the axis at $1/\infty$. A complete set of equivalent planes is denoted by enclosing the Miller indices in braces: $\{hkl\}$. For example, cubic systems (100), ($\overline{1}$00), (010), ($0\overline{1}0$), (001), and ($00\overline{1}$) are equivalent and the set is denoted as $\{100\}$. The maximum possible number of (hkl) combinations that are equivalent occurs for cubic symmetry and is 48. In hexagonal cells, four indices are sometimes used, ($hkil$), where the relation $h + k + i = 0$ always holds. The value of the i index is the reciprocal of the fractional intercept of the plane on the a_3 axis, as illustrated in Figure 1.4. It is derived in exactly the same way as the others. Sometimes, hexagonal indices are written with the i index as a dot, and in other cases it is omitted entirely.

To specify a crystal direction, a vector is drawn from the origin to a point P. This vector will have projections u' on the **a** axis, v' on the **b** axis, and w' on the **c** axis. The three numbers are divided by the highest common denominator to give the set of smallest integers, u, v, and w. The direction is then denoted [uvw]. Sets of equivalent directions are labeled uvw. For cubic systems, the [hkl] direction is always orthogonal to the (hkl) plane of the same indices.

1.4. SURFACE AND INTERFACE STRUCTURES

We assume in the following discussion that the solid surface under consideration is of the same chemical identity as the bulk, that is, free of any oxide film or passivation layer. Crystallization proceeds at the interfaces between a growing crystal and the surrounding phase(s), which may be solid, liquid, or vapor. Even what we normally refer to as a crystal "surface" is really an interface between the crystal and its surroundings (e.g., vapor, vacuum, solution). An *ideal surface* is one that is the perfect termination of the bulk crystal. Ideal crystal surfaces are, of course, highly ordered since the surface and bulk atoms are in coincident positions. In a similar fashion, a *coincidence site lattice* (CSL), defined as the number of coincident lattice sites, is used to describe the goodness of fit for the crystal–crystal interface between grains in a polycrystal. We'll return to that topic later in this section.

1.4.1. Surface Structure

Again, we assume that the solid surface in question is untarnished. Even so, most surfaces are not ideal. They undergo energy-lowering processes known as *relaxation* or *reconstruction*. The former process does not alter the symmetry, or structural periodicity, of the surface. By contrast, surface reconstruction is a surface symmetry-lowering process. With reconstruction, the surface unit cell dimensions differ from those of the *projected* crystal unit cell. It will be recalled that a crystal surface must possess one of 17 two-dimensional space group symmetries. The bulk crystal, on the other hand, must possess one of 230 space group symmetries.

The *terrace–ledge–kink* (TLK) *model* (Kossel, 1927; Stranski, 1928) is commonly used to describe equilibrium solid surfaces. This model was proposed by the German physicist Walther Kossel (1888–1956), who had contributed to the theory of ionic bonding earlier in the century, and by the Bulgarian physical chemist Iwan Nicholá Stranski (1897–1979). It categorizes ideal surfaces or

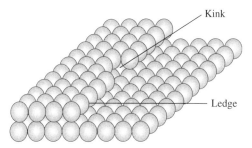

Figure 1.5 Atomic-scale illustration of the terrace–ledge–kink (TLK) model. The top partial layer of spheres represent a terrace, and the bottom layer, the underlying surface atoms.

interfaces as singular, vicinal, and general. An interface is regarded as *singular* with respect to a degree of freedom if it is at a local minimum of energy with respect to changes in that degree of freedom. A *vicinal* interface, by contrast, possesses an interfacial free energy *near* a local minimum with respect to a degree of freedom. A *general* interface, on the other hand, is not at a minimum of energy with respect to any of its degrees of freedom.

Singular surfaces tend to have dense, relatively close packed atomic planes in the crystalline phase parallel to the surface plane and are thus nominally flat. Singular faces have low energy and low entropy. With vicinal surfaces (Figure 1.5), the surface is inclined at a small angle to a low-index plane. Consequently, vicinal surfaces consist of flat two-dimensional one-atom-thick steps of constant width called *terraces*, the edges of which are called *ledges*. Additionally, *kinks*, caused by the removal of atoms in the ledges, are usually present.

A screw dislocation that intersects a crystal surface, termed a *Frank source* after British physicist Sir Charles Frank (1911–1998), is one way in which a surface step may be introduced on real surfaces. In Section 2.5.1 we will see that vicinal surfaces grow layerwise by step propagation, the terraces advancing by the motion of the ledges, which provide preferred binding sites for adatoms. In the case of the Frank source, adatoms attach around the existing screw dislocation, creating an initial double step, which grows in an Archimedean spiral pattern (Burton et al., 1951), as illustrated in Figure 1.6. The Burton–Cabrera–Frank (BCF) model of spiral growth has been applied to a variety of substances, including epitaxially grown semiconductor GaAs films, pyroelectric triglycine sulfate crystals, and amino acids crystallized from solution. Recently, strain fields of dislocations at the surface of Si(001) films have been engineered to nucleate Ge quantum dots preferentially at specific surface locations (Hannon et al., 2006).

1.4.2. Solid–Vapor and Solid–Liquid Interfacial Structures

We can specify the thickness of the crystal interface as that depth over which the structural order of the crystal transitions to that of the surrounding phase(s). On the solid side of the interface, the depth can vary from one lattice spacing for ordered interfaces to several lattice spacings for disordered surfaces

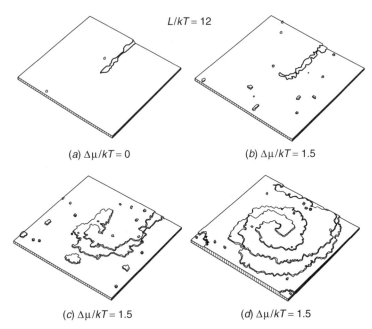

$L/kT = 12$

(a) $\Delta\mu/kT = 0$ (b) $\Delta\mu/kT = 1.5$

(c) $\Delta\mu/kT = 1.5$ (d) $\Delta\mu/kT = 1.5$

Figure 1.6   The presence of a screw dislocation provides an initial step for the spiral growth pattern illustrated in this early Monte Carlo simulation. (After J. D. Weeks and G. H. Gilmer, *Advances in Chemical Physics*. Copyright © 1979 John Wiley & Sons, Inc. Reproduced with permission.)

and interfaces. Ordered interfaces are typically observed in crystal growth from solutions. Disordered interfaces are much more diffuse, being several lattice spacings thick, and thus exhibiting a more gradual transition from the long-range order of the crystal lattice to the incoherency of the liquid or gas. Such "rough" interfaces are common in solidification (molten phase) processes used for the production of polycrystalline metals and alloys. The liquid phase itself at a solid–liquid interface exhibits crystal-induced structural ordering. The liquid's molecules tend to arrange in layers parallel to the solid surface, which, owing to its atomic-scale corrugation, produces density modulations in the liquid along the direction perpendicular to the solid surface. In general, our understanding of interfacial structure between particles in condensed phases (solid–liquid and solid–solid) is much less developed than that of solid–vapor interfaces. This is partly due to the difficulty associated with experimentally probing "buried" interfaces.

1.4.3. Solid–Solid Interfacial Structure

The modern method for quantifying the goodness of fit between two adjacent grains in a pure polycrystalline substance (homophase interfaces) or in a multiphase solid (heterophase interfaces) counts the number of lattice points (*not* atomic positions) in each grain that coincide. In special cases, for example when the grain boundary plane is a twin plane, the lattice sites for each of the adjacent crystals coincide *in* the boundary. These are called *coherent boundaries*. Consider a pair of adjacent crystals. We mentally expand the two neighboring crystal lattices until they interpenetrate and fill all space. Without loss of generality, it is assumed that the two lattices possess a common origin. If we now hold one crystal fixed and rotate the other, it is found that a number of lattice sites for each crystal, in addition to the origin, coincide with certain relative orientations. The set of coinciding points form a *coincidence site lattice* (CSL), which is a sublattice for both individual crystals (Lalena and Cleary, 2005).

To quantify the lattice coincidence between the two grains, A and B, the symbol Σ customarily designates the reciprocal of the fraction of A (or B) lattice sites that are common to both A and B:

$$\Sigma = \frac{\text{number of crystal lattice sites}}{\text{number of coincidence lattice sites}} \quad (1.14)$$

For example, if one-third of the A (or B) crystal lattice sites are coincidence points belonging to both the A and B lattices, then $\Sigma = 1/\frac{1}{3} = 3$. The value of Σ also gives the ratio between the areas enclosed by the CSL unit cell and crystal unit cell. The value of Σ is a function of the lattice types and grain misorientation. The two grains need not have the same crystal structure or unit cell parameters. Hence, they need not be related by a rigid body rotation. The boundary plane intersects the CSL and will have the same periodicity as that portion of the CSL along which the intersection occurs (Lalena and Cleary, 2005).

The simple CSL model is directly applicable to the cubic crystal class. The lower symmetry of the other crystal classes necessitates the more sophisticated formalism known as the *constrained coincidence site lattice*, or CCSL (Chen and King, 1988). In this book we treat only cubic systems. Interestingly, whenever an *even* value is obtained for Σ in a cubic system, it will always be found that an additional lattice point lies in the center of the CSL unit cell. The true area ratio is then half the apparent value. This operation can always be applied in succession until an odd value is obtained: thus, Σ is always *odd* in the cubic system. A rigorous mathematical proof of this would require that we invoke what is known as *O-lattice theory* (Bollman, 1967). The O-lattice takes into account all equivalence points between two neighboring crystal lattices. It includes as a subset not only coinciding lattice points (the CSL) but also all nonlattice sites of identical internal coordinates. However, expanding on that topic would take us well beyond the scope of this book. The interested reader is referred to Bhadeshia (1987) or Bollman (1970).

Single crystals and bicrystals with no misorientation (i.e., $\theta = 0$) are, by convention, denoted $\Sigma 1$. In practice, small- or low-angle grain boundaries with a misorientation angle less than 10 to $15°$ are also included under the $\Sigma 1$ term. Since Σ is always odd, the coincidence orientation for high-angle boundaries with the largest fraction of coinciding lattice points is $\Sigma 3$ (signifying that one-third of the lattice sites coincide). Next in line would be $\Sigma 5$, then $\Sigma 7$, and so on.

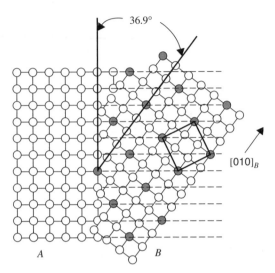

Figure 1.7   View down the [001] direction of a tilt boundary between two crystals (*A*, *B*) with a misorientation angle of 36.9° about [001]. The grain boundary is perpendicular to the plane of the page. Every fifth atom in the [010] direction in B is a coincidence point (shaded). The area enclosed by the CSL unit cell (bold lines) is five times that of the crystal unit cell, so $\sum = 5$. (After Lalena and Cleary, 2005. Copyright © John Wiley & Sons, Inc. Reproduced with permission.)

Figure 1.7 shows a tilt boundary between two cubic crystals. The grain boundary plane is perpendicular to the plane of the page. In the figure we are looking down one of the ⟨100⟩ directions, and the [100] axis about which grain *B* is rotated is also perpendicular to the page and passes through the origin. At the precise misorientation angle of 36.9°, one-fifth of the *B* crystal lattice sites are coincidence points, which also belong to the expanded lattice of crystal *A*; this is a Σ5 CSL misorientation. The set of coincidence points forms the coincidence site lattice, the unit cell of which is outlined. Note that the area enclosed by the CSL unit cell is five times that enclosed by the crystal unit cell.

For tilt boundaries, the value of Σ can also be calculated if the plane of the boundary is specified in the coordinate systems for both adjoining grains. This method is called the *interface-plane scheme* (Wolfe and Lutsko, 1989). In a crystal, lattice planes are imaginary sets of planes that intersect the unit cell edges. The tilt and twist boundaries can be defined in terms of the Miller indices for each of the adjoining lattices and the twist angle, Φ, of both plane stacks normal to the boundary plane, as follows:

$(h_1k_1l_1) = (h_2k_2l_2);$ $\Phi = 0$ symmetric tilt boundary
$(h_1k_1l_1) \neq (h_2k_2l_2);$ $\Phi = 0$ asymmetric tilt boundary
$(h_1k_1l_1) = (h_2k_2l_2);$ $\Phi > 0$ low-angle twist boundary
$(h_1k_1l_1) \neq (h_2k_2l_2);$ $\Phi > 0$ high-angle twist boundary

Thus, the value of the CSL-Σ value is obtained for symmetric tilt boundaries between cubic crystals as follows:

$$\sum = \begin{cases} h^2 + k^2 + l^2 & \text{for } h^2 + k^2 + l^2 = \text{odd} \\ \frac{h^2 + k^2 + l^2}{2} & \text{for } h^2 + k^2 + l^2 = \text{even} \end{cases} \tag{1.15}$$

For asymmetric tilt boundaries between cubic crystals, Σ is calculated from (Randle, 1993)

$$\sum = \sqrt{\frac{h_1^2 + k_1^2 + l_1^2}{h_2^2 + k_2^2 + l_2^2}} \tag{1.16}$$

For example, it can be shown that in Figure 1.7 the grain boundary plane cuts the B unit cell at (340) in the B coordinate system and the A unit cell at (010) in the A coordinate system. Thus, Eq. 1.16 yields $\Sigma = (25/1)^{1/2} = 5$.

In polycrystals, misorientation angles rarely correspond to *exact* CSL configurations. There are ways of dealing with this deviation, which set criteria for the proximity to an exact CSL orientation that an interface must have to be classified as belonging to the class $\Sigma = n$. The Brandon criterion (Brandon et al., 1964) asserts that the maximum deviation permitted is $v_0 \Sigma^{-1/2}$. For example, the maximum deviation that a $\Sigma 3$ CSL configuration with a misorientation angle of $15°$ is allowed to have and still be classified as $\Sigma 3$ is $15°(3)^{-1/2} = 8.7°$. The coarsest lattice characterizing the deviation from an exact CSL orientation, which contains the lattice points for each of the adjacent crystals, is referred to as the *displacement shift complete* (DSL) *lattice*.

Despite the difficulties associated with characterizing inexact CSL orientations, the CSL concept is useful because grain boundary structure, which depends on the orientation relationship between the grains and hence the CSL, directly influences intragranular properties such as chemical reactivity (e.g., corrosion resistance), segregation, and fracture resistance. *Grain boundary engineering* is a relatively new field that concentrates on controlling the intragranular structure, or CSL geometry, to improve these properties: in turn, improving bulk materials performance (Watanabe, 1984, 1993). For the most part, this means introducing a large fraction of low-Σ boundaries, particularly twin boundaries. It is believed, however, that optimal grain boundary properties may be restricted to narrow regions (small deviations) about exact CSL orientations (Lalena and Cleary, 2005).

1.5. CONTROLLED CRYSTAL GROWTH AND MICROSTRUCTURAL EVOLUTION

Consistently reliable approaches for the de novo prediction of a material's crystal structure (unit cell shape, size, and space group), morphology (external symmetry), microstructure, as well as its physical properties, remain elusive for molecular and nonmolecular materials. Nevertheless, empirical synthetic guide-lines do exist for preparing general single-crystal morphologies and polycrys-talline microstructures, even though we are far from capable of reproducing the vast number of morphologies and degrees of perfection nature has produced. Non-molecular and molecular materials are often prepared by conventional crystalliza-tion processes from, respectively, melts or solutions. Crystallization from a liquid melt phase with the same chemical identity as the solid produced is normally termed *solidification*. Crystallization processes from a liquid solvent include those from homogeneous supersaturated solutions and the low-temperature processes known as the *sol–gel, hydrothermal*, and *solvothermal techniques*. A relative newcomer to materials science and engineering is self-assembly. Now we discuss the crystallographic and morphological control attainable with all the crystal-lization processes and self-assembly processes, which makes these strategies attractive approaches to materials synthesis.

1.5.1. Conventional Crystallization from Melts and Solvents

There are many analogies between the crystallization process from a molten state and from a solvent. These are discussed fully in Chapter 4. It is often desirable for us to assist the growth of a particular face of a single-crystalline material. For example, as mentioned earlier, because of the desirable properties displayed by certain crystallographic planes in silicon and other semiconductors (e.g., GaAs, InSb, CdTe) exceedingly high-purity single-crystalline substrates with precise orientations are used in microelectronic devices. Such substrates are obtained from oriented single-crystalline ingots, or boules, which may be produced by a variety of solidification processes to be introduced shortly. These and other conventional crystal growth methods are discussed in more detail in subsequent chapters.

In general, the task of controlling single-crystal growth along certain direc-tions may be relatively simple or, conversely, more difficult, depending on the crystal growth method used and the preferential growth directions of the crys-tal. Defect-free morphologically flawless crystals are rarely found in nature. The quartz crystals used as crystal oscillators, for example, are therefore produced by a *hydrothermal recrystallization* process. In fact, C. E. Schafhäult prepared tiny synthetic, or "cultured," quartz crystals hydrothermally as early as 1845. Hydrothermal chemical synthesis of phases from precursors is also possible, which we discuss in Section 4.4.

To obtain single crystals by hydrothermal recrystallization, a *charge*, which consists of fragments of natural quartz crystals, is dissolved in a solvent, such as 0.5 to 1.0 molar alkaline aqueous solution of sodium carbonate or sodium hydroxide, in the dissolving chamber of an autoclave under high temperature and pressure ($\sim$340 to 380°C, 70 to 150 MPa). Under these conditions, of course, the solution vaporizes. An oriented seed crystal (generally Y-bar or Z-plate with a long [1010] direction) is hung in the upper region above the solution, called the *growing chamber*, which is separated from the lower chamber by a metal baffle.

The growing chamber is maintained at a lower temperature than the dissolving chamber ($\Delta T = \sim 10$ *to* $40°$C, depending on the pressure). The temperature difference inside this closed cylinder of fluid results in a vertical, closed-loop heat convection pattern that forces the less dense warm fluid to rise, thus constituting what is called a *phase-change thermosyphon*. Because the growing chamber is at a lower temperature, that region becomes supersaturated and the quartz deposits, or recrystallizes, on the seed. Quartz crystals of over 1500 kg mass, taking over a month to obtain, are produced routinely in this manner in industry. After growth, a diamond saw is used to cut sections from the quartz crystal for use as crystal oscillators, a very laborious task!

The primary experimental parameter affecting growth rate is ΔT, but growth rate is also influenced by the pressure as well as solvent type and concentration. As explained in Section 1.2.1, the growth rate is asymmetric for the various crystal faces. For example, in the early stages of the growth process, the faces of the trigonal trapezohedra (Figure 1.1) become inhibited by those of the hexagonal prism. Even crystallographically equivalent faces (e.g., x) may have different growth rates, due to differing levels of impurity segregation or surface adsorption of vapor species, which act as growth-rate inhibitors.

Other crystal growth methods, where a seed crystal is in physical contact with a high-temperature melt under ambient pressure, include the *submerged-seed solution growth* (SSSG) process, sometimes called *volumetric growth*; and the *top-seeded solution growth* (TSSG) process, in which the seed just touches the melt surface. These methods have been used to chemically synthesize single crystals from metal oxide precursors: for example, calcium-doped yttrium barium copper oxide (YBCO) from Y_2O_3, $BaCO_3$, $CaCO_3$, and CuO, with MgO as the seed (Lin et al., 2002), and $<001>$ seed-oriented potassium titanyl phosphate (KTP) from KH_2PO_4, TiO_2, and K_2HPO_4 (Kannan et al., 2002).

In TSSG, the product crystallizes on the seed as it is pulled away from the constant-temperature melt surface or, alternatively, as the melt is slowly cooled (0.5 to 2 K/day), while maintaining a temperature gradient between the bottom of the crucible and the melt surface. In the crystal pulling version, the pulling shaft may be cooled internally, which maintains the melt surface at a slightly lower temperature than the melt at the bottom of the crucible and allows for heat convection transport. The crucible and the seed may also be rotated in opposite directions to prevent secondary nucleation from heat convection. The growth habit of the synthetic crystals produced from TSSG, which typically produces fewer morphologically developed crystal faces than are produced by SSSG, can be varied by changing the process conditions and/or the size and orientation of the seed crystal. Finally, to obtain large single crystals from incongruent melts, as in Ca-doped YBCO and KTP growth, it is important to keep the melt supersaturated and to apply an extremely slow growth rate, typically around 1 mm/day (Lin et al., 2002).

A very important industrial crystallization process, which is the most common method used for the production of high-purity oriented single-crystalline semiconductor ingots, is the *Czochralski method* (Czochralski, 1918), named

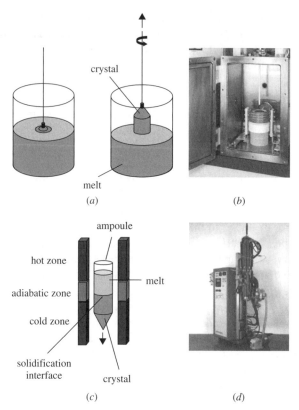

Figure 1.8 (*a*) Czochralski crystal pulling technique; (*b*) commercial Czochralski crystal growth system; (*c*) Bridgman crystal growth technique; (*d*) Bridgman crystal growth laboratory furnace. (Courtesy of Thermal Technology Incorporated, Santa Clara, CA.)

after metallurgist Jan Czochralski (1885–1953) and illustrated in Figure 1.8*a*. A single-crystal dislocation-free seed, or nucleus, with the desired orientation is placed over a congruent melt held at a temperature close to the liquid–solid phase transition. After dipping the seed crystal in the melt, it is pulled upward under continuous rotation. The cylindrical axis of the ingot coincides with the desired crystal growth axis (e.g., silicon $\langle 001 \rangle$ or $\langle 111 \rangle$). Slices, called *wafers*, on which the microelectronic devices are fabricated are then sawed from the ingot. Semiconductor crystal growth is indeed a very important industrial process, but it should be realized that the Czochralski method is not limited to the growth of oriented semiconductor ingots. It has been used to grow ruby, sapphire, spinel, yttrium–aluminum–garnet (YAG), gadolinium–gallium–garnet (GGG), alexandrite, and $La_3(Ga,Ta)_6O_{14}$, among others. However, it is not suitable for growing crystals from incongruently melting substances.

It is also possible to produce single-crystal ingots as well as highly oriented textured polycrystalline materials (i.e., those in which the grains exhibit a preferred orientation) by directional solidification. With the *Bridgman technique*

(Bridgman, 1926), named after its inventor, Nobel laureate Percy Williams Bridgman (1882–1961), the solidification interface is advanced from a seed crystal into the melt as shown in Figure 1.8c. This is accomplished by first placing the substance inside the hot zone of a furnace with a prescribed temperature gradient to create a melt and then gradually moving the melt into the cooler end to induce crystallization. In the modified *Stockbarger method* (Stockbarger, 1936), named after Donald C. Stockbarger (1895–1952), the isothermal hot and cold zones are separated by an adiabatic zone created by an insulating baffle designed to maintain a steep axial temperature gradient. The charge is gradually moved through the adiabatic zone so that the solidification interface remains essentially parallel with the plane of the baffle. It should be noted that the Bridgman method may be utilized in a vertical or horizontal fashion. However, it is necessary to use ampoules that do not bond to the crystal when it freezes. In the case of silicon, none are available. For gallium arsenide, fused silica is satisfactory (Runyan and Bean, 1990).

It is also possible to obtain polycrystalline materials, with grains on the order of millimeters to centimeters in width, at the cooler end in directional solidification techniques. The grains are approximately columnar along the direction of the temperature gradient. The longitudinal axis of the columnar grains corresponds to the direction of highest growth rate in the material. Columnar grains emerge from the sites of preferred nucleation at the cold mold walls and grow in the direction of the highest temperature gradient. Even in conventional casting, where a melt is poured into a mold of some desired shape and size to solidify, there is normally a columnar zone in the center of the cast where elongated crystals with a preferred growth direction have eliminated the randomly oriented grains near the cooler mold walls (the equiaxed zone). The texture can thus be interpreted in terms of a growth selection process: Random nucleation takes place at the mold walls. During the subsequent solidification, selective growth of grains with a specific crystal direction parallel to the highest temperature gradient or heat flow occurs. Consequently, when one examines the microstructures of polycrystalline materials obtained by most industrial solidification processes, it is often found that the individual grains (crystallites) are not morphologically similar to naturally occurring single crystals of the same substance.

The discussion of solidification microstructure above applies to pure substances and single-phase solid solutions (alloys). Although this book is concerned primarily with single-phase materials, it would now be beneficial to describe briefly the microstructures of a special type of multiphase alloy known as a *eutectic*. Eutectics are generally fine grained and uniformly dispersed. Like a pure single phase, a eutectic melts sharply at a constant temperature to form a liquid of the same composition. The simplest type of eutectic is the binary eutectic system containing no solid-phase miscibility and complete liquid phase (melt) miscibility between the two components (e.g., the Ag–Si system). On cooling, both components form nuclei and solidify simultaneously at the eutectic composition as two separate pure phases. This is termed *coupled growth* and it leads to a periodic concentration profile in the liquid close to the interface that decays in the direction perpendicular to the interface much faster than in single-phase solidification.

There are several methods for solidifying eutectics at conventional cooling rates, including the laser floating zone method, the edge-defined film-fed growth technique, the Bridgman method, and the micro-pulling down method. Generally, high volume fractions of both phases will tend to promote lamellar structures. If one phase is present in a small volume fraction, that phase tends to solidify as fibers. However, some eutectic growths show no regularity in the distribution of the phases. Eutectic microstructures normally exhibit small interphase spacing and the phases tend to grow with distinctly shaped particles of one phase in a matrix of the other phase. The microstructure will be affected by the cooling rate; it is possible for a eutectic alloy to contain some dendritic morphology, especially if it is cooled relatively rapidly. The microstructures of *hypo-* or *hypereutectic* compositions normally consist of large particles of the primary phase (the component that begins to freeze first) surrounded by fine eutectic structure. Often, the primary particles will show a dendritic origin, but they can transform into *idiomorphic grains* (having their own characteristic shape), reflecting the phase's crystal structure (Baker, 1992).

Photo courtesy of Dr. Pawel Tomaszewski, Institute of Low Temperature and Structure Research, Polish Academy of Sciences. Reproduced with permission.

Jan Czochralski (1885–1953) moved from his homeland of Poland to Germany in 1904 and received the chemist-engineer degree in 1910 from the Charlottenburg Polytechnic in Berlin, having worked in various positions at pharmacies and laboratories while attending school. His interests were in physical metallurgy, especially solidification, crystallization, corrosion, and plastic deformation. During his professional career, Czochralski published nearly 100 papers. The first, published in 1913 with his supervisor, Wichard von Moellendorff at the Allgemeinen Elecrizitaets-Gesellschaft (The General Electricity Company), was concerned with the movements of atoms during plastic deformation of single crystals, noteworthy in preceding the edge dislocation theory of Orowan, Taylor, and Polanyi by 20 years. In 1916, Czochralski serendipitously discovered a monocrystalline growth method when he accidentally dipped his pen into a crucible of molten tin rather than his inkwell. He immediately pulled his pen out to discover that a thin thread of solidified metal was hanging from the nib. He replaced the nib with a capillary and verified that the crystallized metal was a single crystal. This method was to eventually pave the way for the processing of modern semiconductors after Bell Labs used the technique in 1950 to produce single crystals of germanium. Czochralski turned down an offer by Henry Ford to work in the United States, instead accepting a proposal by the Polish president in 1929 to return to Poland from Germany, where he had resided for 24 years. After receiving an honorary doctorate that same year from Warsaw University of Technology, Czochralski became a professor in the department of chemistry in 1930, eventually forming the laboratory, and later institute, of metallurgy and metals science. However, he was stripped of his professorship in 1945, due to accusations that he collaborated with the German occupation forces during World War II. Czochralski was later cleared of any wrongdoing, but he spent the remainder of his days running a small private firm that produced cosmetics and household chemicals.

Source: Professor Czochralski — Distinguished Scientist and Inventor, by K. J. Kurzydlowski and S. Mankowski; Jan Czochralski — Achievements, by Anna Pajaczkowska; and Professor Jan Czochralski (1885–1953) and His Contribution to the Art and Science of Crystal Growth, by Pawel Tomaszewski; *J. Am. Assoc. Crystal Growth*, **1998**, *2*(2), 12–18.

Metal–metal eutectics have been studied for many years due to their excellent mechanical properties. Recently, oxide–oxide eutectics were identified as materials with potential use in photonic crystals. For example, rodlike micrometer-scaled microstructures of terbium–scandium–aluminum garnet : terbium–scandium perovskite eutectics have been solidified by the micro-pulling-down method (Pawlak et al., 2006). If the phases are etched away, a pseudohexagonally packed dielectric periodic array of pillars or periodic array of pseudohexagonally packed holes in the dielectric material is left.

1.5.2. Molecular Self-Assembly in Materials Synthesis

The term *self-assembly* refers to the spontaneous formation (without human intervention) of ordered aggregates in systems at equilibrium or far from equilibrium. The driving force for a self-assembly process need not be chemical. For example, galaxies self-assemble via gravitational attractions. The physicochemical rules governing self-assembly are, of course, similar over various system length scales, although not necessarily identical. Here we consider only chemical forces. Chemical (molecular) self-assembly, henceforth referred to merely as self-assembly, is observed in nature over a wide range of length scales itself. Nevertheless, nanostructures are the most explored; self-assembly of micrometer-to-millimeter-sized and larger components from molecular-sized building blocks is a relatively new area of materials research, as is the self-assembly of nonmolecular inorganic solids. The application of self-assembly in materials science and engineering, recently coined *mesoscale self-assembly* by Harvard University chemistry professor George Whitesides (Boncheva and Whitesides, 2005), also encompasses the fabrication of *mesocrystals* from nanoparticles that are, themselves, usually prepared by template-directed self-assembly. At the present time, the self-assembly of nonmolecular inorganic solids is mostly limited in application to monatonic substances and binary compounds (e.g., oxides, nitrides, sulfides), although some progress is being made in higher-order systems.

Self-assembly is essentially chemical fabrication. Like macroscale fabrication techniques, self-assembly allows a great deal of design flexibility in that it affords the opportunity to prepare materials with custom shapes or morphologies. The advantages of self-assembly include an increased level of architecture control and access to types of functionality unobtainable by most other types of liquid-phase techniques. For example, it has been demonstrated that materials with nonlinear optical properties (e.g., second harmonic generation), which require noncentrosymmetric structures, can be self-assembled from achiral molecules.

Control of lattice architecture and/or morphology may be attained in self-assembly by geometrical constraints of the building blocks, or by restriction of the space in which the assembly occurs. The latter approach is a type of *templating*. For the most part, advantage has been taken of this approach toward the synthesis of nanostructures. Various types of agents possessing the desired shape and size may act as templates, with a function similar in principle to that of molds used in industrial casting techniques. The pores of membranes produced by the sol–gel technique (Section 4.3), for example, have been used as templates for the synthesis of semiconductor oxide micro- and nanostructures. Ionic liquids (ILs), a class of reaction media with self-organized solvent structures (Section 4.4), have also been used in the synthesis of self-assembled, highly organized nanostructures with exceptional quality. The capability of self-assembly for facilitating obtainment of unique architectures is perhaps best exemplified by the low-surface-area hollow porous spheres comprised of interconnected hexagonal noble metal (e.g., Ag, Au) rings, which have been prepared on drops of liquid. The hexagonal rings were first fabricated by a multistep process involving photolithography and electrodeposition. Self-assembly of the rings into a sphere was then driven by capillary forces of attraction, while subsequent electroplating of silver on the rings welded them together (Huck et al., 1998).

It is instructive to compare these porous spheres (i.e., where all the edges of each hexagonal ring are adjoined to one other hexagonal ring to construct the sphere) with spherical particles prepared by alternative routes. Consider the non-hollow spherical conglomerations of nanoscale cobalt hexagonal flakes formed by hydrothermal reduction, in which the flakes are radially directed toward the center of the sphere. As with the noble metal spheres, the building blocks (flakes) must first be prepared. For the cobalt spheres, the nanoplatelets were chemically synthesized from surfactant-containing aqueous solutions of cobalt chloride. Assembly of the flakes into spheres via hydrothermal reduction is thought to involve interfacial tension and the hydrophilic surfaces of the cobalt (Hou et al., 2005). Small hollow spheres of titanium dioxide have also been prepared by deposition on carbon spheres followed by calcination to remove the sacrificial carbon core (Zhang et al., 2006). Oxides of iron, nickel, cobalt, cerium, magnesium, and copper have been prepared as hollow spheres by a hydrothermal approach, but the spherical shells obtained are comprised of nanoparticles with high surface areas. The various spherical particles just discussed, all of which have potential technological uses (e.g., catalysis, drug-carrying and controlled release, magnetic recording, microcapsule reactors) are illustrated in Figure 1.9.

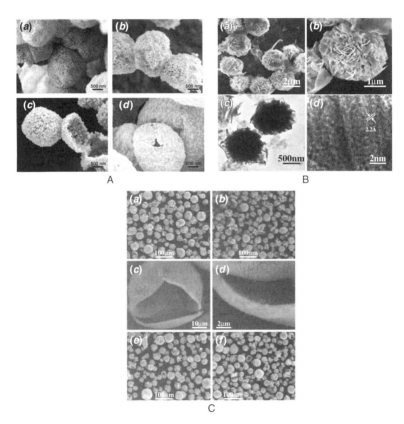

Figure 1.9 Various mesoscale spherical particles. Panel A: scanning electrom micrographs(GEMs) of (*a*) NiO, (*b*) Co$_3$O$_4$, (*c*) CeO$_2$, and (*d*) MgO hollow spheres. (Reprinted with permission from Titirici et al., 2006. Copyright © American Chemical Society.) Panel B: SEM images at (*a*) a survey, (*b*) higher resolution, transmission electrom micrographis and at (*c*) low resolution and (*d*) high resolution of Co spheres. (Reprinted with permission from Hou et al., 2005. Copyright © American Chemical Society.) Panel C: SEM images of hollow titania microparticles at (*a*) 400°C, (b–d) 400°C, (*e*) 600°C, and (*f*) 700°C. (Reprinted with permission from Zhang et al., 2006. Copyright © American Chemical Society.)

Control of lattice architecture may also be attained in self-assembly by a variety of other means. A frequently used method involves the use of structure-directing groups, also called *templating agents* (or again, just *templates* for short). Examples of *mesoscale* nonmolecular inorganic materials that have been synthesized by templated self-assembly include micrometers-long semiconducting CuS microtubes (assisted by acetic acid), $BaSO_4$ bundles (assisted by the sodium salt of polyacrylic acid), mesoporous silica with controlled pore sizes and shapes (aerosol-assisted), hollow titanium dioxide colloidal crystals (surface tension–assisted), and various silicon arrays.

In the simplest cases (e.g., constrained dimensionality), lattice architecture can be anticipated by considering both the molecular symmetries and the propagation of noncovalent interactions between structure-directing groups, which may be portions of the interacting molecules themselves (Ward, 2005). Consider the guanidinium carbocation $[C(NH_2)_3^+]$ and the sulfonate anion (RSO_3^-), each possessing threefold rotational symmetry (trigonal stereochemistry) about the central atom: ideally, D_{3h} and C_{3v} point group symmetries, respectively. Guanidinium organomonosulfonate salts (one sulfonate anion per organic group R) would be expected to self-assemble into a network of $N-H\cdots O-S$ hydrogen bonds that can be viewed as a sheet comprised of ribbons fused along their edges, suggesting the two-dimensional quasihexagonal motif illustrated in Figure 1.10a. Three-dimensional frameworks can be assembled from guanidinium organodisulfonate salts (two sulfonate groups attached to each organic group R), where the organic R groups then function as pillars supporting inclusion cavities between adjacent sheets. The constrained dimensionality thus enforces host–guest systems with lamellar architectures (Figure 1.10b), which crystallize in orthorhombic or monoclinic space groups. These systems are of interest for their potential use in optoelectronics, chemical separations, and confined chemical reactions.

The use of organic templating agents highlights the pivotal role of weak and noncovalent intermolecular forces in driving self-assembly. As indicated in the preceding paragraph, molecules, which are not necessarily components of the intended structure, may be used as templates for directing self-assembly. Such "guest" species are integral to promoting assembly of frameworks containing large open spaces, as with the guanidinium organodisulfonate salts, supplying the cohesive energy required for crystallization. Successful production of desired lattice architecture often hinges on the de novo design of these structure-directing agents themselves. The goal is to choose molecules with particular structural motifs conducive to the formation of the desired framework in the product. Unfortunately, the frameworks may collapse upon attempted removal of the template (e.g., by heating). This means that one may possibly be faced with the difficult task of phase-separating the templating agent from the desired product if removal of the template is necessary.

Despite these difficulties, mesoscale self-assembly of inorganic materials holds promise. In fact, it is believed that the importance of self-assembly in the manufacturing of electronics, photonics, optics, and robotics mesoscale components could conceivably supersede its importance in the molecular and nanoscale sciences. Some hybrid organic–inorganic systems have attracted attention because

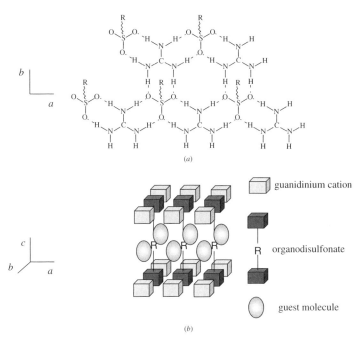

Figure 1.10 (*a*) Guanidinium organomonosulfonate salts (one sulfonate anion per organic group R) self-assembled into a quasihexagonal two-dimensional network of N–H···O-S hydrogen bonds. (From Russell et al., 1994.) (*b*) Three-dimensional lamellar frameworks assembled from guanidinium organodisulfonate salts (two sulfonate groups attached to each organic group R), where the organic R groups function as pillars supporting inclusion cavities capable of holding guest molecules between adjacent sheets.

of their novel properties. For example, organic–inorganic perovskites have unique magnetic, optical, thermochromic, and electrical properties and structural flexibility (Li et al., 2006). It has been shown that inorganic superlattices (composite heterostructures that may be described as periodic multilayers where the unit cell consists of chemically different successive layers) can also be templated by molecular assemblies. Moreover, it has been demonstrated that the structures of these superlattices can kinetically control some intralayer solid-state reactions. This is described a bit more in Section 2.5.3. There is also evidence suggesting that self-assembly is important at some point during the formation of the open-framework architectures of metal phosphates, templated by organic amines. Self-assembly is probably involved in the transformation of one-dimensional chains (containing vertex-shared four-membered rings) and ladder structures (containing edge-shared four-membered rings) into two-dimensional layer structures and three-dimensional channel-containing structures (Rao, 2001).

In contrast to mesoscale assemblies, mesocrystals are considered intermediates on the formation pathway to single crystals (Cölfen and Yu, 2005). The full growth mechanism has been reported for very few mesocrystals, two of which are copper oxalate and calcite. Nanocrystals of copper oxalate have been found to arrange into mesocrystals that can be modified morphologically by the addition of face-selective hydroxymethylpropylcellulose. In other studies, crystallization of calcite ($CaCO_3$) from concentrated calcium chloride solutions by the CO_2 gas diffusion technique in the presence of polystyrene sulfonate (Wang et al., 2005), and by copolymer latex particles functionalized by surface carboxylate groups (Lu et al., 2005), is found to be capable of yielding crystals with irregular morphology. Morphology can be varied systematically from the typical calcite rhombohedra via rounded edges and truncated triangles to finally, concavely bended lenslike superstructures. The crystals are apparently well-faceted in light microscopy. However, electron microscopy analysis has confirmed that they are highly porous and composed of almost perfectly 3D-aligned calcite nanocrystals scaffolded to the final curved superstructures (i.e., that they are mesocrystals). At high supersaturation, superstructures with changed symmetry are found.

1.6. STRUCTURES OF GLASSY AND QUASICRYSTALLINE PHASES

In Chapter 4 we discuss the thermodynamic and kinetic aspects of rapidly solidifying certain types of liquid phases, whereupon glassy and quasicrystalline solids are obtained. Here, we describe those structures briefly. Glasses are monolithic amorphous materials, absent long-range structural order, as well as grain boundaries and other crystalline defects. The term *long-range order* (LRO) refers to regularity in the arrangement of the material's atomic or molecular constituents on a length scale a few times larger than the size of these groups. A glass also possesses *short-range order* (SRO) and *medium-range order* (MRO). Natural glasses form when certain types of rock melt as a result of volcanic activity, lightning strikes, or meteorite impacts, followed by very rapid cooling and solidification. Stone-age peoples are believed to have used some of these naturally formed amorphous materials as tools for cutting. An example is *fused quartz*, or *quartz glass*, formed by the melting and rapid cooling of pure silica. Upon rapid solidification, the short-range order (geometry) of the SiO_4 tetrahedra is preserved but not the long-range crystalline order of quartz. The structure of quartz involves corkscrewing (helical) chains of SiO_4. The corkscrew takes four tetrahedra, or three turns, to repeat, each tetrahedron essentially being rotated $120°$. The chains are aligned along one axis of the crystal and interconnected to two other chains at each tetrahedron. The Si–O–Si bond angle between interconnected tetrahedra is nominally about $145°$ degrees. As in quartz, every oxygen atom in fused quartz is also bonded to two silicon atoms (each SiO_4 tetrahedron is connected to four other tetrahedra), but the tetrahedra are polymerized into a network of rings of different sizes, occurring in a wide range of geometries. Hence, in quartz glass there is a distribution of Si–O–Si bond and Si–O–Si–O torsion angles.

The very high temperatures required to melt quartz were not attainable by early craftspersons. Hence, they prepared sodium silicate glass by mixing together and melting sodium carbonate with sand. The structure of this *water glass* is similar to that of quartz glass except that with the random insertion of sodium ions within the network, nonbridging oxygen atoms (i.e., oxygen atoms bonded to a single silicon atom) are produced. The first glassy phase prepared in this manner was either in Egypt or Mesopotamia in the form of ceramic glazes. Around the middle of the second millennium B.C., glassware was being made, principally in the form of pots and vases. The first glassmaking manual, dating back to 650 B.C., appears to have come from Assyria. The next major breakthrough in the evolution of the art was glassblowing, which originated sometime around A.D.1 in the Sidon–Babylonian region. Glassmaking continued to evolve throughout Europe, with many key advances necessary for mass production made in the latter stages of the Industrial Revolution. We have much more to say about the glass formation process in Chapter 4.

A fairly recent event that attracted much attention within the crystallographic community was the discovery of metastable quasicrystals of an Al–Mn alloy with fivefold rotational symmetry in the 1980s at the U.S. National Bureau of Standards (now the National Institute for Standards and Technology) by Shechtman et al. (1984). Crystals are three-dimensional and therefore are limited to having only have one-, two-, three-, four-, or sixfold rotation axes; all other rotational symmetries are forbidden. Unlike crystals, which may or may not display rotational order, quasicrystals possess *only* rotational order and no three-dimensional translational periodicity. However, the long-range structure is coherent enough to scatter incoming waves so as to produce an interference pattern. Hence, the quasicrystalline state is manifested by the occurrence of sharp diffraction spots, indicative of long-range order, but with the presence of a noncrystallographic rotational symmetry.

Aperiodic plane-filling (two-dimensional) patterns, based on various basis sets of tiles, have actually been studied for some time. In fact, it is believed that Islamic architects and mathematicians created the first such patterns over 500 years ago (Lu and Steinhardt, 2007). In his 1964 mathematics doctoral thesis at Harvard, Robert Berger developed a set of 20,426 essentially square tiles called *Wang dominos* (squares with different colored edges or geometrically altered to prevent periodic arrangements) that tiled nonperiodically. (He later reduced this to a set of 104.) Stanford University professor Donald Knuth (b. 1938) found a set of 92 in 1968. In 1971, Raphael Robinson (1911–1995), professor of mathematics at the University of California–Berkeley, devised a set of six non-Wang aperiodic tiles. In 1973, the Oxford mathematician Roger Penrose (b. 1931) discovered a nonperiodic tiling in two dimensions with pentagonal symmetry by combining a set of six tiles (Figure 1.11). He subsequently reduced this to a set of four "kite" and "dart" tiles (altered so that they form an aperiodic set) and finally, a tiling based on a set of two rhombuses (Penrose, 1974). Each rhombus is a symmetrical tile, made by the combination of the two triangles found in the geometry of the pentagon. Kites and darts may be obtained from rhombuses, and vice versa. Penrose tilings may be considered two-dimensional quasicrystals.

Since the initial discovery of metastable quasicrystals, many ternary intermetallic compounds have been produced in the quasicrystalline state, which are thermodynamically stable at room temperature. These have been obtained

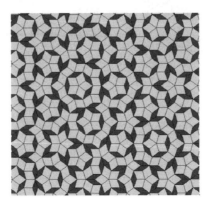

Figure 1.11   Portion of a Penrose tiling based on two rhombuses. Penrose tilings are nonperiodic tilings of the plane and are two-dimensional analogs of quasicrystals. (Diagram created by the free Windows application "Bob's Rhombus Walker," v. 3.0.19, JKS Software, Stamford, CT.)

primarily by solidifying phases with equilibrium crystal structures containing icosahedrally packed groups of atoms (i.e., phases containing icosahedral point group symmetry) at conventional cooling rates. The quasicrystalline phases form at compositions close to the related crystalline phases. We discuss quasicrystalline formation in Chapter 4. For now, we restrict the discussion to the structural aspects of these phases.

The icosahedron is one of the five Platonic solids, or regular polyhedra. A regular polygon is one with equivalent vertices, equivalent edges, and equivalent faces. The icosahedron has 20 faces, 12 vertices, 30 edges, and six fivefold proper rotation axes (collinear with six tenfold improper rotation axes). Icosahedral coordination ($Z = 12$) and other coordination polytetrahedra with coordination numbers $Z = 14$, 15, and 16 are a major component of some liquid structures, more stable than a close-packed structure, as demonstrated by F. C. Frank and J. Kasper (Frank and Kasper, 1958a, b). When these liquid structures are solidified, the resulting structure has icosahedra threaded by a network of wedge disclinations, having resisted reconstruction into crystalline units with 3D translational periodicity (Mackay and kramer, 1985; Turnbull and Aziz, 2000). Stable ternary intermetallic icosahedral quasicrystals are known from the systems Al–Li–Cu, Al–Pd–Mn, and Zn–Mg–Ln. Several other ternary systems yield metastable icosahedral quasicrystals (Lalena and Cleary, 2005).

Some stable ternary intermetallic phases have been found that are quasiperiodic in two dimensions and periodic in the third. These are from the systems Al–Ni–Co, Al–Cu–Co, and Al–Mn–Pd. They contain decagonally packed groups of atoms (local tenfold rotational symmetry). It should be noted that there are also known metastable quasicrystals with local eightfold rotational symmetry (octagonal) and 12-fold rotational symmetry (dodecagonal) as well. The dodecahedron is also one of the five Platonic solids (Lalena and Cleary, 2005).

Photo courtesy of the Royal Society. Copyright © The Royal Society. Reproduced with permission.

Sir Frederick Charles Frank (1911–1998) received his Ph.D. in 1937 from Oxford University, followed by a postdoctoral position at the Kaiser Wilhelm Institut für Physik in Berlin. During World War II, Frank was involved with the British Chemical Defense Research Establishment, and because of his keen powers of observation and interpretation, he was later transferred to Scientific Intelligence at the British Air Ministry. In 1946, Frank joined the H. H. Wills Physics Laboratory at the University of Bristol under its director, Nevill Mott, who encouraged him to look into problems concerned with crystal growth and the plastic deformation of metallic crystals. A stream of successes followed, establishing his scientific fame, as evidenced by many eponyms: the Frank–Read source, the Frank dislocation, Frank's rule, Frank–Kasper phases. His theoretical work has been the foundation of research by innumerable scientists from around the world. Frank was awarded the Order of the British Empire (OBE) Medal in 1946, elected a Fellow of the Royal Society (FRS) in 1954, and was knighted in 1977.

Source: Obituary of Sir Charles Frank OBE FRS, 1911–1998, prepared by A. R. Lang for the British Crystallographic Asociation. Copyright © 1998.

For a periodic space-filling lattice, one can translate the pattern by a certain distance in a certain direction, and every point in the translated portion will coincide with a point in the original lattice. However, with a space-filling quasilattice, this operation is not possible. One can take a bounded region and translate it to coincide exactly with some other part of the original pattern. Such a representative portion constitutes a kind of "unit cell" of the quasilattice. There are two different complementary packable rhombohedra, an acute-angled and an obtuse-angled rhombohedra that can be combined in a space-filling manner to form rhombic dodecahedra, icosahedra, triacontahedra, and ultimately, quasicrystals (Lalena, 2006).

A relationship actually exists between periodic and quasiperiodic patterns such that any quasilattice may be formed from a periodic lattice in some higher dimension (Cahn, 2001). The points that are projected to the "physical" three-dimensional space are usually selected by cutting out a "slice" from the higher-dimensional lattice. Therefore, this method of constructing a quasiperiodic lattice is known as the *cut-and-project method*. In fact, the pattern for *any* three-dimensional quasilattice (e.g., icosahedral symmetry) can be obtained by a suitable projection of points from some six-dimensional periodic space lattice into a three-dimensional subspace. The idea is to project part of the lattice points of the higher-dimensional lattice to three-dimensional space, choosing the projection such that one preserves the rotational symmetry. The set of points so obtained are called a *Meyer set* after French mathematician Yves Meyer (b. 1939), who first studied cut-and-project sets systematically in harmonic analysis (Lalena, 2006).

Without relying on the cut-and-project technique, Rabson et al. (1991) have computed all the three-dimensional quasicrystallographic space groups with n-fold axial point groups and standard lattices by a method that treats crystals and quasicrystals in equivalent fashion, taking advantage of the more readily apparent symmetry in reciprocal space than in direct space. The familiar three-dimensional crystallographic space groups with axial point groups emerge simply and directly as special cases of the general n-fold three-dimensional quasicrystallographic treatment with $n = 3$, 4, and 6. Additionally, they have given a general discussion of extinctions in quasicrystals as well as a simple three-dimensional geometrical specification of the extinctions for each axial space group.

The central issue in quasicrystal structure analysis, however, has been the locations of the atoms. Unlike with the structure determination of crystals, for quasicrystals it has proven rather difficult to acquire a conclusive solution based on the diffraction intensities. This is due to the fact that several quasiperiodic structural models, differing in local atomic arrangements, can generate very similar intensity distributions (with crystals, these are known as *homometric structures*). Consequently, direct observation, such as from high-resolution electron microscopy, is essentially required for solving quasicrystalline atomic structure. As recently discussed (Lalena, 2006), the International Union of Crystallography (IUCr) was inspired by the growing numbers of quasicrystals and the success of Fourier space crystallography in explaining quasiperiodic and periodic crystals alike. In 1991, the IUCr shifted the essential attribute of crystallinity from direct space to reciprocal space by redefining the term *crystal* to mean "any solid having an essentially discrete diffraction diagram" (Lifshitz and Mermin, 1992; Lifshitz, 1995).

REFERENCES

Baker, H., Ed. *ASM Handbook*, Vol. *3*, Alloy Phase Diagrams, ASM International, Materials Park, OH, **1992**.

Barlow, W. Z. *Krystallogr. Mineral.* **1894**, *23*, 1–63.

Bhadeshia, H. K. D. H. *Worked Examples in the Geometry of Crystals*, Institute of Metals, London, **1987**.

Bollman, W. *Philos. Mag. A*. **1967**, *6*, 363.

Bollman, W. *Crystal Defects and Crystallian Interfaces*, Springer Verlag, Berlin, **1970**.

Boncheva, M.; Whitesides, G. M. *Mater. Res. Soc. Bull*. **2005**, *30*(10), 736–742.

Brandon, D. G.; Ralph, B.; Ranganathan, S.; Wald, M. S. *Acta Metall*. **1964**, *12*, 813.

Bravais, A. J. *Ecole Polytech*. **1850**, *19*, 1–128.

Bravais, A. *Études Cristallographiques*, Gauthier-Villars, Paris, **1866**.

Bridgman, P. W. *Proc. Am. Acad. Arts Sci*. **1926**, *60*, 306.

Buerger, J. M. *Am. Mineral*. **1945**, *30*, 469–482.

Buerger, J. M. *Elementary Crystallography*, MIT Press, Cambridge, MA, **1978**.

Burton, W. K.; Cabrera, N.; Frank, F. C. *Philos. Trans. R. Soc. London A*. **1951**, *243*, 299.

Cahn, J. W. *J. Res. Nat. Inst. Stand. Technol*. **2001**, *106*, 975–982.

Chen, F.-R.; King, A. H. *Acta Crystallogr. B* **1988**, *43*, 416.

Cölfen, H.; Yu, S.-H. *Mater. Res. Soc. Bull*. **2005**, *30*(10), 727–735.

Czochralski, J. *Z. Phys. Chem*. **1918**, *92*, 219–221.

Donnay, J. D. H.; Harker, D. *Am. Mineral*. **1937**, *22*, 446.

Dowty, E. *Am. Mineral*. **1976**, *61*, 448.

Fedorov, E. S. *Proc. Imp. Saint Petersburg Soc. Ser*. **1891**, *2*(28), 345–389 (in Russian).

Fedorov, E. S. *Proc. Imp. Saint Petersburg Soc. Ser*. **1891a**, *2*(28) 1–146 (in Russian).

Frank, F. C.; Kasper, J. S. *Acta Crystallogr*. **1958a**, *11*, 184.

Frank, F. C.; Kasper, J. S. *Acta Crystallogr*. **1958b**, *12*, 483.

Friedel, G. *Bull. Soc. Fr. Mineral*. **1907**, *30*, 326.

Hannon, J. B.; Shenoy, V. B.; Schwarz, K. W. *Science*. **2006**, *313*, 1266.

Hartman, P.; Perdok, W. G. *Acta Crystallogr*. **1955a**, *8*, 49–52.

Hartman, P.; Perdok, W. G. *Acta Crystallogr*. **1955b**, *8*, 521–524.

Hartman, P.; Perdok, W. G. *Acta Crystallogr*. **1955c**, *8*, 525–529.

Hermann, C. *Z. Kristallogr*. **1928**, *68*, 257.

Hermann, C. *Z. Kristallogr*. **1931**, *76*, 559.

Hou, Y.; Kondoh, H.; Ohta, T. *Chem. Mater*. **2005**, *17*, 3994.

Hessel, J. F. C. Jr. *Gehler's Physikalische Wöterbuch*, Schwikert, Leipzig, Germany, **1830**, pp. 1023–1360. Reprinted in *Ostwald's Klassiker der Exakten Wissenschaften*, Engelmann, Leipzig, Germany, 1897.

Huck, W. T. S.; Trin, J.; Whitesides, G. M. *J. Am. Chem. Soc*. **1998**, *120*, 8267.

Kannan, C. V.; Ganesamoorthy, S.; Kumaragurubaran, S.; Subramanian, C.; Sundar, R.; Ramasamy, P. *Cryst. Res. Technol*. **2002**, *37*(10), 1049–1057.

Kossel, W. *Nachr. Ges. Wiss. Goettingen*. **1927**, 135.

Lalena, J. N. *Crystallogr. Rev*. **2006**, *12*, 125.

Lalena, J. N.; Cleary, D. A. *Principles of Inorganic Materials Design*, Wiley, Hoboken, NJ, **2005**.

Li, Y. Y.; Lin, C. K.; Zheng, G. L.; Cheng, Z. Y.; You, H.; Wang, W. D.; Lin, J. *Chem. Mater*. **2006**, *18*, 3463.

Lifshitz, R. In *Lecture Notes for the International School on Quasicrystals*, Balatonfüred, Hungary, May 13–20, **1995**.

Lifshitz, R.; Mermin, N. D. *Acta Crystallogr. A*. **1992**, *48*, 928.

Lin, C. T.; Liang, B.; Chen, H. C. *J. Cryst. Growth*. **2002**, 237–239, 778–782.

Lin, S.-K. *J. Chem. Inf. Comput. Sci*. **1996**, *36*, 367–376.

Lin, S.-K. *Int. J. Mol. Sci*. **2001**, *2*, 10–39.

Lu, C.; Qi, L.; Cong, H.; Wang, X.; Yang, J.; Yang, L.; Zhang, D.; Ma, J.; Cao, W. *Chem. Mater*. **2005**, *17*, 5218–5224.

Lu, P. J.; Steinhardt, P. J. *Science*. **2007**, *315*(5815), 1106.

Mackay, A. L.; Kramer, P. *Nature*. **1985**, *316*, 17.

Mauguin, C. Z. *Kristallogr*. **1931**, *76*, 542.

Neumann, F. E. *Pogendorff Ann. Phys*. **1833**, *27*, 240.

Nye, J. F. *Physical Properties of Crystals: Their Representation by Tensors and Matrices*, Oxford University Press, London, **1957**.

Pawlak, D. A.; Kolodziejak, K.; Turczynski, S.; Kisielewski., J.; Rozniatowski, K.; Diduszko, R.; Kaczkan, M.; Malinowski, M. *Chem. Mater*. **2006**, *18*, 2450–2457.

Penrose, R. *Bull. Inst. Math. Appl*. **1974**, *10*, 266–271.

Rabson, D. A.; Mermin, N. D.; Rokhsar, D. S.; Wright, D. C. *Rev. Mod. Phys*. **1991**, *63*, 699–733.

Rao, C. N. R. *Proc. Indian Acad. Sci*. (*Chem. Soc.*). **2001**, *113*, 363–374.

Rosen, J. *Symmetry in Science: An Introduction to the General Theory*, Springer-Verlag, New York, **1996**.

Runyan, W. R.; Bean, K. E. *Semiconductor Integrated Circuit Processing Technology*, Addison-Wesley, Reading, MA, **1990**.

Russell, V. A.; Etter, M. C.; Ward, M. D. *J. Am. Chem. Soc*. **1994**, *116*, 1941.

Schönflies, A. M. *Kristallsysteme und Kristallstruktur*, Teubner, Leipzig, Germany, **1891**.

Shechtman, D.; Blech, L.; Gratias, D.; Cahn, J. W. *Phys. Rev. Lett*. **1984**, *53*, 1951.

Shewmon, P. *Diffusion in Solids*, Minerals, Metals, and Materials Society, Warrendale, PA, **1989**.

Stockbarger, D. C. *Rev. Sci. Instrum*. **1936**, *7*, 133.

Stranski, I. N. Z. *Phys. Chem*. **1928** *136*, 259.

Titirici, M.-M.; Antonietti, M.; Thomas, A. *Chem Mater*. **2006**, *18*, 3808.

Wang, T.; Cölfen, H.; Antonietti, M. *J. Am. Chem. Soc*. **2005**, *127*(10), 3246–3247.

Ward, M. D. *Mater. Res. Soc. Bull*. **2005**, *30*(10), 705–712.

Watanabe, T. *Res. Mech*. **1984**, *11*, 47.

Watanabe, T. In *Grain Boundary Engineering*, U., Erb, and G., Palumbo, Eds., CIM, Montreal, Canada, **1993**.

Wolfe, D.; Lutsko, J. F. Z. *Kristall*. **1989**, *189*, 239.

Zhang, D.; Yang, D.; Zhang, H.; Lu, C.; Qi, L. *Chem. Mater*. **2006**, *18*, 3477.

2 Chemical Energetics and Atomistics of Reactions and Transformations in Solids

In the physical sciences we are taught early that what distinguishes the three states of matter (i.e., gases, liquids, and solids) from one another are the differing packing densities and mobilities of their constituent atoms or molecules. The greatest packing densities are *usually* found in the solid state. Similarly, the fundamental units, which may be either atoms or molecules, are in relatively fixed and, hence, less mobile positions within a solid. The physical manifestations of this are that unlike the other states of matter, solids are of definite shape (i.e., solids are not fluid) and they have small compressibilities as well as small thermal expansion coefficients. The most important implication for the chemical properties of solids is that very small diffusion coefficients will be observed for a species migrating through a *typical* nonmolecular solid: say, under a chemical potential gradient as compared to when the same species diffuses through a gas or liquid. The diffusion coefficient is directly proportional to the amount of mass (flux) that moves across a cross-sectional area of a sample during a specified time period for a given concentration gradient. A high diffusion coefficient indicates a large flux, in contrast to a system having a low diffusion coefficient.

It may seem that this might preclude the occurrence of chemical reactions and phase transformations in most solids. However, there are thermodynamic driving forces present for chemical reactions and physical phase transformations in solids just as there are for liquids and gases. These forces are, in fact, fundamentally the same for all the states of matter: namely, a lowering of the system's free energy. Mass transport takes place in solids, even densely packed solids, thanks to mechanisms involving mediation by point defects, albeit at a much slower rate than in the other states of matter. Nevertheless, the *activation barrier* (sometimes called the *kinetic barrier*, or *free energy of activation*) for a phase change may be so high that at ordinary temperatures or pressures, the change need not be over an observable time scale. A material formed under conditions of extreme temperature and pressure may be *kinetically* stable when returned to ambient conditions, even if it is *thermodynamically* metastable. The archetypical example is diamond, which is metastable with respect to graphite at atmospheric pressure.

Inorganic Materials Synthesis and Fabrication, By John N. Lalena, David. A. Cleary, Everett E. Carpenter, and Nancy F. Dean
Copyright © 2008 John Wiley & Sons, Inc.

At that pressure, diamond takes billions of years to spontaneously convert to graphite. With this consideration in mind, we concede that the structure adopted by a solid at a particular temperature and pressure is the result of competition between thermodynamics and kinetics.

When we have a single-phase polycrystalline substance as opposed to a single crystal, we naturally find that homophase interfaces, or *grain boundaries*, will also contribute to the energy and kinetics of the system. Contrary to popular belief, not all grain boundaries exhibit a high degree of incoherence, in which there is little correlation between atomic positions across the boundary. That situation arises only when the mismatch between adjacent crystals is very high. Nonetheless, impurity atoms that don't "fit" in the lattices tend to segregate in grain boundaries, and the incoherency that does exist across the interface contributes to an increase in entropy compared to the grains themselves. Consequently, grain boundaries are slightly higher energy metastable regions. When annealed, polycrystals tend to evolve toward single crystals through a process of grain growth. Grain growth and hence the resulting grain morphology (size, shape, orientation) is usually under kinetic control. The morphology that appears is the one with the maximum growth rate (Lalena and Cleary, 2005).

The actual energy of a grain boundary depends on its orientation and coherency, which is usually quantified by the reciprocal of the fraction of lattice sites that are common to the crystals on both sides of the boundary. This number is given the symbol Σ. For example, if one-fifth of the A (or B) crystal lattice sites are coincidence points belonging to both the A and B lattices, then $\Sigma = 1/\frac{1}{5} = 5$. The value of Σ is a function of the lattice types and the misorientation between adjoining grains. As illustrated in Figure 1.7, the value of Σ also gives the ratio between the area enclosed by the unit cell of the coincidence site lattice (CSL) and the area enclosed by the crystal's unit cell. The grain boundary plane intersects the CSL and will have the same periodicity as that portion of the CSL, along which the intersection occurs. Generally, low-Σ boundaries tend to have lower grain boundary energy, on average. In a polycrystalline aggregate, we can specify the fraction of a particular CSL boundary type. This approach is useful because the grain boundary structure often correlates with certain materials properties, particularly conductivity, creep (time-dependent deformation at constant load), and corrosion. The incorporation of a high percentage of a specific grain boundary type in a polycrystal is referred to as *grain boundary engineering* (Lalena and Cleary, 2005).

At low temperatures, say, less than 1000 K, the most stable phase is normally a crystalline one (rather than an amorphous state), which minimizes the entropy (small ΔS). Even polycrystalline substances, with their high-energy grain boundaries, are more thermodynamically stable than amorphous materials; glasses (metallic and nonmetallic) are metastable states. For example, nanocrystalline grains (grain size <100 nm) can sometimes be obtained from a metallic glass when it is annealed at temperatures at which primary crystallization can occur. Metallic glasses appear to be even more susceptible than nonmetallic glasses to this crystallization above the glass transition temperature, which is called *devitrification*. Metallic glasses transform to more stable crystalline phases typically around 300 to 450°C.

The preceding introduction might lead one to believe that this chapter could simply be divided into two basic parts—*thermochemical considerations* and *kinetic considerations*—which would cover all the relevant subject matter. However, in the last decade, first-principles (ab initio) computations have become commonplace and their results have often confirmed predictions based on thermochemical approaches, sometimes even surpassing them in accuracy. Hence, there is a need to encompass both thermochemical and ab initio treatments. We group the latter under the heading *structural energetics* and explore this topic further in Section 2.2. We also talk about the relevant thermodynamic and kinetic factors for specific systems in their respective chapters. For now, we discuss thermodynamics and kinetics in the most general terms.

2.1. EQUILIBRIUM THERMODYNAMICS

In thermodynamics, a *reversible process* is one that can be reversed by infinitesimal changes in some property of the system without loss of energy. In chemistry, this normally means a transition (e.g., chemical reaction or phase transformation) from some initial state to some final state. If, after transitioning to the final state, the process is reversed such that the system is returned to its initial state, there would be no net change in either the system or its surroundings. An *irreversible* physical or chemical change is one that will not reverse itself spontaneously without some corresponding change in the surrounding conditions (e.g., temperature, pressure). A good example of an irreversible process is heat conduction. Heat can never flow spontaneously from a cold reservoir to a warmer one. In theory, all chemical reactions are reversible *to an extent*. However, *completely* reversible processes, which would be in a continuous state of equilibrium, are impossible since it would take an infinite time for such a process to finish. Hence, all real natural processes have some irreversible character. Nevertheless, the concept of reversibility is necessary as a mathematical tool to define the theoretical limit and to develop fundamental thermodynamic relationships.

The *second law of thermodynamics* tells us that the *criterion for a spontaneous (irreversible) physical or chemical process in a closed macroscopic system is an increase in the total entropy of the system plus its surroundings* ($\Delta S_{tot} > 0$), where the surroundings are considered the rest of the universe. By *closed* we mean the system and surroundings can only exchange energy, not matter. Thus, irreversible processes produce entropy. Similarly, the criterion for a reversible process, which returns to its initial state through a series of small equilibrium steps, is $\Delta S_{tot} = 0$. It is important to note that ΔS is defined only for reversible processes. Specifically, it is the ratio of the heat transferred between the system and surroundings during a reversible process, q_{rev}, divided by the absolute temperature at which the heat was transferred: $\Delta S = q_{rev}/T$. However, this quantity is a state function, having a unique value dependent only on the *net* process and is independent of the path or mechanism taken from initial to final state. Therefore, the same value will apply when we replace the reversible process with an irreversible process. An irreversible process increases the entropy of the system and its surroundings. By implication, the entropy of the universe (i.e., the system and its surroundings), assumed as an isolated system, tends to increase.

But what exactly is this quantity called entropy? In classical (macroscopic) thermodynamics, *entropy* is regarded as that energy unavailable for external thermodynamic work (i.e., work mediated by thermal energy). In statistical thermodynamics, entropy is defined as a measure of the number of microscopic configurations that are capable of yielding the macroscopic description of the system. For a given set of macroscopic quantities (e.g., temperature and volume), entropy measures the degree to which the probability of the system is spread out over different possible quantum states. If more states are available to the system, it will have greater entropy. Historically, entropy has also been associated with a measure of the molecular disorder, or randomness, of an isolated system. Because there are so many, somewhat nonintuitive definitions of entropy and hence statements of the second law, entropy has more recently been regarded as the *energy dispersal in the system*. Consequently, the second law can be rephrased as stating that pressure differences, density differences, and temperature differences all tend to equalize over time.

In chemical thermodynamics, we prefer to focus our attention on the system rather than the surroundings. Thus, it is convenient to consider the free-energy function, the quantity of energy available in a system for producing work. Using this state function, *the criterion for spontaneity is a decrease in the system's free energy on moving from the initial state to the final state*. For processes occurring at constant temperature and constant volume, the *Helmholtz free-energy change* is defined as the change in the internal energy of the system, ΔU_{sys} (ΔE_{sys} in some texts), *minus* the random thermal motion, or heat content, in the system ($T\Delta S_{\text{sys}}$), which cannot be used to extract work. When the pressure rather than the volume is held constant, any pressure–volume work exchanged between the system and surroundings must be added to ΔU. The sum of ΔU and $P\Delta V$ is known as the *enthalpy change*, ΔH. Thus, the *Gibbs free-energy change* for a *closed system*, which can exchange energy but not matter with its surroundings, is defined in differential form as

$$dG = dH_{\text{sys}} - T\,dS_{\text{sys}} \tag{2.1}$$

where the enthalpy and entropy changes are those of the system. Integrating Eq. 2.1 yields $\Delta G = \Delta H_{\text{sys}} - T\Delta S_{\text{sys}}$. The enthalpy change of the system is numerically equal, but opposite in algebraic sign, to that of the surroundings.

This is but one possible expression for the Gibbs free energy. We could write an expression in terms of changes in other state variables, such as temperature and pressure. Furthermore, we must account for the possibility that a component may be distributed among or transported between several phases within the system (e.g., alloys). Alternatively, many reactions of interest to the materials chemist take place in *open systems*, in which there is transport of energy and matter between the system and surroundings. For example, a gas or liquid may come into contact with a solid surface, where it creates a new substance that is transported away. Incorporating these other possible scenarios allows us to write the following very general, but very useful expression for the Gibbs free energy:

$$dG = V\,dP - S\,dT + \sum_{i=1}^{k} \mu_i\,dN_i + \sum_{i=1}^{n} X_i\,da_i + \cdots \tag{2.2}$$

Here, μ is the chemical potential for the ith component, and X can be any of a number of external forces that causes an external parameter of the system, a, to change by an amount da. Under conditions of constant temperature and pressure, the first two terms on the right side of Eq. 2.2 drop out. If there are no external forces (X) acting on the system, the equilibrium condition for the transport of matter requires that the value of the chemical potential for each component be the same in every phase (at constant T and P).

The Gibbs energy change is related to some other important physical quantities, such as the equilibrium constant for a chemical reaction and the electromotive force of an electrochemical cell, by the *Nernst* and *van't Hoff equations:*

$$\Delta G = -RT\,\ln K = \Delta H - T\Delta S \tag{2.3}$$

$$\Delta G = -n\mathcal{F}\mathcal{E} \tag{2.4}$$

where K is the equilibrium constant, R the gas constant (1.986 cal/mol), T the absolute temperature, n the number of moles of electrons, $\mathcal{F}$ the Faraday constant (the charge on a mole of any single charged entity, 96,494 C), and $\mathcal{E}$ the electromotive force for an electrochemical cell involving a chemical reaction of interest. Equation 2.3 is a combination of the Nernst equation ($\Delta G = -RT\,\ln K$), named after the 1920 Nobel laureate in chemistry, Walther Hermann Nernst (1864–1941), and the integrated form of the van't Hoff equation, $-RT\,\ln K = \Delta H - T\Delta S$, named after Jacobus H. van't Hoff (1852–1911), who received the first Nobel Prize in Chemistry in 1901. Equation 2.4, the Nernst equation for a cell, follows directly from elementary electrical theory, where the work done in transporting an electrical charge between points of different electric potential is equal to the product of the charge and the potential difference.

In a chemical reaction for which the activation energy is supplied, the total free energy of the system decreases spontaneously at constant temperature and pressure until it reaches a minimum, whether it is instantaneous or exceedingly slow. This is the *second law of thermodynamics*. A *stable state* is the state of lowest free energy. A *metastable state* is a state in which additional energy must be supplied to the system for it to reach true stability. Metastable states are separated from lower-energy states by higher-energy barriers. An *unstable state* is a state in which no additional energy need be supplied to the system for it to reach either metastability or stability; it does so spontaneously. These states

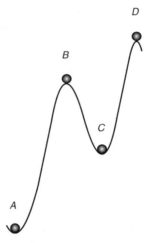

Figure 2.1 Example of the second law of thermodynamics in a mechanical system, which moves toward that state with the lowest total potential energy. The balls seek the position of lowest gravitational potential energy (height). Points B and D are unstable; the balls roll downward to points A or C. Point C is metastable, separated from the lowest energy state, point A, by a high-energy barrier.

are illustrated in Figure 2.1 for an analogous mechanical system in which the quantity of interest is gravitational potential energy. The system is a ball on a hilly landscape. Since the gravitational potential energy of the ball is proportional to the product of its mass and its height, the potential energy diagram has the same shape as the landscape shown in the figure. At the unstable points B and D, the ball free-falls spontaneously to points A/C and C, respectively, with the lost potential energy being dissipated to the surroundings as heat. At the metastable point C, the ball must be supplied energy to overcome the hurdle and reach the point of lowest gravitational energy, point A. This tendency to minimize total potential energy is due to the second law of thermodynamics. The system moves away from the state with low heat content and high potential energy toward the state with a high heat content and low potential energy since the latter state maximizes the entropy of the universe. For chemical systems, the corresponding similarly behaved quantity is the free energy rather than the gravitational potential energy.

The *phase rule* of Josiah Willard Gibbs (1839–1903) gives the general conditions for chemical equilibrium between phases in a system. At equilibrium, $\Delta G = 0$, there is no further change with time in any of the system's macroscopic properties. It is assumed that surface, magnetic, and electrical forces may be neglected. In this case, the phase rule can be written as

$$f = c - p + 2 \qquad (2.5)$$

where f, which must equal zero or a positive integer, gives the degrees of freedom (number of independent variables); c is the number of components (the number of independent constituents needed to fix the chemical composition of every phase in the system, usually elements in metallic systems); p is the number of phases (substances that are microscopically homogeneous on average and possessing uniform thermodynamic properties); and the factor 2 is for the two variables, temperature and pressure. If the effect of pressure is ignored in condensed systems with negligible vapor pressures, the factor 2 in Eq. 2.5 is replaced by 1, giving the *condensed phase rule* (Lalena and Cleary, 2005).

The phase rule(s) can be used to distinguish different types of equilibria based on the number of degrees of freedom. For example, in a unary system, an *invariant equilibrium* $(f = 0)$ exists between the liquid, solid, and vapor phases at the triple point, where there can be no changes to temperature or pressure without reducing the number of phases in equilibrium. Because f must equal zero or a positive integer, the condensed phase rule $(f = c - p + 1)$ limits the possible number of phases that can coexist in equilibrium within one-component condensed systems to one or two, which means that other than melting, only allotropic phase transformations are possible. Similarly, in two-component condensed systems, the condensed phase rule restricts the maximum number of phases that can coexist to three, which also corresponds to an invariant equilibrium. However, several *invariant reactions* are possible, each of which maintains the number of equilibrium phases at three and keeps f equal to zero (L represents a liquid and S, a solid):

Monotectic:	$L_1 + S \rightarrow L_2$		Euctectoid:	$S_1 \rightarrow S_2 + S_3$
Eutectic:	$L \rightarrow S_1 + S_2$		Syntectic:	$L_1 + L_2 \rightarrow S$
Metatectic:	$S_1 \rightarrow L + S_2$		Peritectic:	$L + S_1 \rightarrow S_2$
Monotectoid:	$S_1 + S_2 \rightarrow S_2 + S_3$		Peritectoid:	$S_1 + S_2 \rightarrow S_3$

There are other types of equilibria, in addition to the invariant type, which can be deduced from Eq. 2.5. For example, when three phases of a two-component system are in equilibrium, such as with a closed vessel containing hydrogen gas in equilibrium with a metal and the metal hydride, immersed in a water bath, it is possible to change the value of just one variable (temperature or pressure or composition) without changing the number of phases in equilibrium. This is called *univariant equilibrium* $(f = 1)$. If the composition is held constant, temperature and pressure will have a fixed relationship in a univariant system. Hence, if the pressure of hydrogen gas in the vessel is increased slightly, the temperature of its contents remains the same as heat escapes through the vessel walls to the water bath.

If the composition of hydrogen in this system were to be fixed at a lower concentration, such that there is but a single condensed phase comprised of the two components in equilibrium with hydrogen gas (i.e., a solid solution of metal and absorbed hydrogen gas, but no metal hydride), there will be two degrees of freedom $(f = 2)$. There is no fixed relationship between pressure and temperature at constant composition in such a system. Both temperature and hydrogen pressure may be varied, changing the absorption or desorption of hydrogen gas from the solid solution, but leaving fixed at two the number of phases. This is termed *bivariant equilibrium*. We shall discuss only briefly univariant equilibria in binary systems before proceeding to thermochemical calculations. But first, a word about the graphical representation of phase equilibria is in order.

The relationships between the various phases that appear within a system under equilibrium conditions are shown in *phase diagrams*. The synthetic chemist finds phase diagrams useful because they tell us if a given phase in the system under investigation is accessible under the equilibrium conditions. For this reason alone, it would be advantageous for the reader to acquire proficiency in interpreting phase diagrams. The following brief introduction is meant merely to encourage the reader to refer to any one of several available texts that offer a range from introductory to comprehensive coverage of all aspects of phase equilibria. A succinct intermediate-level treatment can be found in our companion book, *Principles of Inorganic Materials Design* (Lalena and Cleary, 2005).

Phase diagrams for single-component systems are two-dimensional plots showing the phase changes in the substance with changes in temperature (abscissa) and/or pressure (ordinate). For binary (two-component) systems, a three-dimensional plot would be required, since composition, temperature, and pressure are variable. For condensed binary systems, however, the pressure is fixed (normally to 1 atm) and the 3D diagram may be reduced to a 2D graph of composition on the abscissa (in weight percent or atomic percent) and temperature on the ordinate. The presence or absence of the various types of invariant equilibria determines the positions and shapes of the areas, called *phase fields*, in a phase diagram. Binary systems, for example, may contain both single and two-phase fields, and where a two-phase field does exist, it must be located between two single-phase fields. Figure 2.2 shows a hypothetical phase diagram for a eutectic binary system. The terminal phase fields, α and β, which are bounded by solvus lines, are single-phase solid solutions. A two-phase solid mixture $(\alpha + \beta)$,

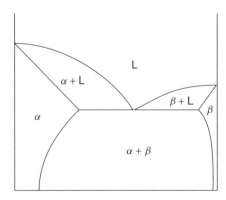

Figure 2.2 Phase diagram for a hypothetical eutectic binary system. There are three single-phase fields (α, β, and L) and three two-phase fields ($\alpha + \beta$, $\alpha + L$, and $\beta + L$).

bounded by the lower solvus lines, is in the center region beneath the solidus. Similarly, two distinct two-phase fields are located above the solidus and beneath the liquidus.

The addition of a third component to a condensed system makes two-dimensional plots of phase equilibria difficult. There are several options for graphically displaying equilibria in ternary systems. The most information is contained in an isometric projection called a *solid diagram*, with a base that is an equilateral triangle whose edges represent the compositions for each of the binary subsystems (e.g., AB, AC, BC), as illustrated in Figure 2.3*a*. The two-dimensional vertical sides, or boundaries, represent the binary-phase diagrams (Figure 2.3*c*). The entire diagram is capped by the liquidus surface. Reading values from this type of figure is not easy, so the equilibria in ternary systems are normally represented by a series of two-dimensional sections. Horizontal sections (*isotherms*) are capable of displaying subsolidus equilibria for any possible composition at a given temperature (Figure 2.3*b*). Vertical sections (*isopleths*) show the stability of phases over a wide temperature range but at a constant composition of one component or a constant ratio of two components. It is also common to see the liquidus surface from an isometric projection, which shows either isothermal contour lines or temperature troughs with arrows indicating the direction of decreasing temperature (Figure 2.3*d*).

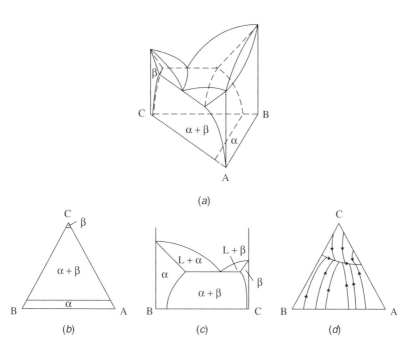

Figure 2.3   *(a)* Isometric projection (solid diagram) for a hypothetical ternary system; *(b)* isotherm (horizontal section); *(c)* binary-phase subset; *(d)* liquidus surface.

2.1.1. Univariant Equilibria in Binary Systems

Let's consider the simple case of a metal in equilibrium with its oxide and oxygen gas with no other species present. To represent the equilibrium, one may write:

$$x\text{M (s)} + O_2(\text{g}) = M_xO_2(\text{s}) \qquad K = \frac{a_{M_xO_2}}{a_M^x p_{O_2}} \qquad (2.6)$$

From the phase rule, this is seen to be a univariant equilibrium (there is 1 degree of freedom). If x is constant and the pressure is 1 atm, the metal and its oxide may be considered to be in their standard states, so their activities, $a_{M_xO_2}$ and a_M^x, respectively, are unity; hence,

$$K = \frac{1}{p_{O_2}} \qquad (2.7)$$

Substitution of Eq. 2.7 into the Nernst equilibrium expression of Eq. 2.3 yields

$$\Delta G = RT \ln(p_{O_2}) \qquad (2.8)$$

If one now defines $pO = -\log(p_{O_2})$, it is seen that the oxygen affinity increases with increasing pO as Eq. 2.8 becomes

$$-\Delta G = 2.303 RT \cdot pO \qquad (2.9)$$

Note the similarity between the definition of pO and that of the pH of aqueous solutions. In general, the potential of any quantity is equal to the negative of the common (or Briggian) logarithm of that quantity. The pO scale is used to predict the course of redox reactions, as a metal will reduce the oxides of other metals whose pO values are sufficiently smaller than its own. For example, CaO, MgO, and Al_2O_3 are all very thermodynamically stable, as evidenced by their high pO values (greater than 47). Accordingly, metallic calcium, magnesium, and aluminum are employed as reducing agents in industrial metallurgical processes to reduce the oxides of metals with smaller pO values, such as chromium oxide (Cr_2O_3), whose pO value is 30. Knowledge of pO values can also aid us in choosing a suitable reaction vessel (e.g., crucibles for molten materials). All other factors (melting point, cost, etc.) being equivalent, one would chose a metal crucible whose oxide has a pO value lower than the molten oxide it is intended to contain. In this way, no redox reaction will proceed between the molten oxide and the crucible. Similarly, to contain a molten metal, one would choose an oxide crucible whose pO value is higher than that of the molten metal. Figure 2.4 shows the oxygen affinity of various metals, the pO values of which may be obtained via knowledge of the free energy of formation and Eq. 2.9.

If we have measured the partial pressure of oxygen in equilibrium with a metal and its oxide, we may also use Eq. 2.8 or 2.9 to calculate the free energy of formation for the metal oxide at any given temperature. Of course, it is possible

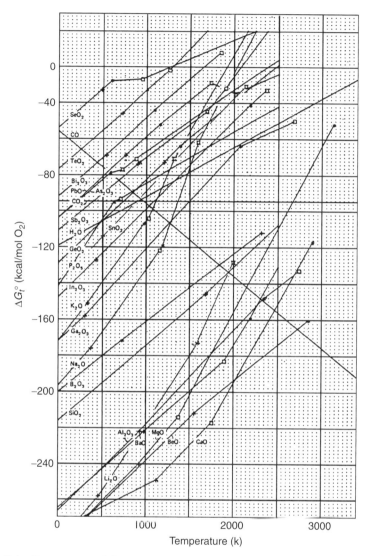

Figure 2.4 Free energy of formation, as a function of temperature, for various main group binary oxides. (From Reed, 1971. Copyright © The MIT Press. Reproduced with permission.)

to work in reverse and obtain the equilibrium constant and, in turn, the oxygen partial pressure, from graphing tabulated free-energy changes as a function of temperature. The advantages of this graphical approach were first pointed out by the Imperial College chemistry professor Harold Johan Thomas Ellingham (1897–1975) in 1944. In addition to metal–oxide systems, univariant equilibria in other metal–vapor binary systems including metal–sulfide, metal–nitride, and metal–halide systems, may be handled in a similar fashion. For each of these, an equilibrium reaction and the corresponding equilibrium constant expression, completely analogous to Eqs. 2.6 and 2.7, may be written. The interested reader is referred to the text *Physical Chemistry of Metals* (Darken and Gurry, 1953).

2.1.2. Thermochemical Calculations

Provided that all the thermochemical data are available (often they are not!), the thermodynamic driving force for any reaction involving solids can be quantified in the usual manner using Hess's law. This states that the overall *change* in a thermodynamic property for a chemical reaction (entropy, enthalpy of formation, free energy of formation) is the same regardless of the number of steps. That is, if a chemical equation can be written as a sum of two or more equations, the change in the overall reaction is the sum of the changes for the individual equations. This law was deduced in 1840 by the Swiss-born Russian chemist Germain Henri Hess (1802–1850). The principle is illustrated with enthalpy in Example 2.1.

Example 2.1   Given the following standard molar enthalpies of formation:

$$Mg(s) + 2Cr(s) + 2O_2(g) = MgCr_2O_4(s) \qquad \Delta H = -1759.2 \text{ kJ/mol}$$
$$Si(s) + O_2(g) = SiO_2(s) \qquad \Delta H = -909.07 \text{ kJ/mol}$$
$$2Cr(s) + \tfrac{3}{2}O_2(g) = Cr_2O_3 \qquad \Delta H = -1124.9 \text{ kJ/mol}$$
$$Mg(s) + Si(s) + \tfrac{3}{2}O_2(g) = MgSiO_3(s) \qquad \Delta H = -3089.38 \text{ kJ/mol}$$

calculate ΔH for the reaction

$$MgCr_2O_4(s) + SiO_2(s) = Cr_2O_3(s) + MgSiO_3(s)$$

SOLUTION   We must write the reaction of interest as a sum of the reactions for which we have thermochemical data. If a reaction is reversed, the enthalpy changes sign.

$$SiO_2(s) = Si(s) + O_2(g) \qquad \Delta H = +909.07 \text{ kJ/mol}$$
$$MgCr_2O_4(s) = Mg(s) + 2Cr(s) + 2O_2(g) \qquad \Delta H = +1759.2 \text{ kJ/mol}$$
$$2Cr(s) + \tfrac{3}{2}O_2(g) = Cr_2O_3 \qquad \Delta H = -1124.9 \text{ kJ/mol}$$
$$Mg(s) + Si(s) + \tfrac{3}{2}O_2(g) = MgSiO_3(s) \qquad \Delta H = -3089.38 \text{ kJ/mol}$$

$$\overline{MgCr_2O_4(s) + SiO_2(s) = Cr_2O_3(s) + MgSiO_3(s) \qquad \Delta H = -1546.01 \text{ kJ/mol}}$$

A large body of empirical thermochemical data was collected in the late 1940s by the metallurgical industry for many different metal oxides, chlorides, sulfides, and nitrides. The standard free energy of formation for these substances is known as a function of temperature and is plotted in this manner (ΔG_f versus T) in what are termed *Ellingham diagrams* (see, e.g., Ellingham, 1944; Darken and Gurry,

1953). Beginning in the 1950s, when many elements came under consideration as rocket propellant ingredients, substantial effort went into calculating or collecting empirical thermochemical data for a large number of other substances. Much of the early work in this area culminated in the production of the first JANAF (Joint Army Navy Air Force) tables. Other compilations of free-energy data for many binary compounds may be found in Reed's work (Reed, 1971). Such information is valuable in a variety of situations, as illustrated in Example 2.2. When used in conjunction with Hess's law, the data make possible prediction of the spontaneity for an even greater number of reactions.

Example 2.2   A solder alloy composed of 89% by weight bismuth and 11% silver (Bi–11Ag), forms a *dross*, or surface film, when molten (melting point = ~260°C) in air, which prevents the solder from wetting the metals to be joined. The dross varies from yellow to blue in color. It is found that a small addition of germanium to the alloy temporarily prevents this from occurring (Lalena et al., 2002). Rationalize this observation based on free energy of formation data for the metal oxides given in Table 2.1.

TABLE 2.1   Free-Energy Data

Bi_2O_3		GeO_2		Ag_2O	
T(K)	ΔG_f (kcal/mol)	T(K)	ΔG_f (kcal/mol)	T(K)	δG_f (kcal/mol)
0	−92	0	−129	0	−14
544	−69	1210	−73	480	0
1098	−44	1389	−64		
1852	−12	2500	−19		

Source: Read (1971).

SOLUTION   Linear interpolation yields values for ΔG_f (260°C) of −69, −104, and +1.55 kcal/mol for Bi_2O_3, GeO_2, and Ag_2O, respectively. The dross on the surface of the undoped Bi–11Ag alloy must be Bi_2O_3. Silver will not oxidize at this temperature, due to the positive ΔG_f. The reader may also want to use Eq. 2.8 for each oxide in order to observe the trend in pO values. In the Ge-doped alloy, the germanium would be expected to reduce the Bi_2O_3 chemically, since it has a higher oxygen affinity than bismuth (more negative ΔG_f). Because germanium oxide is volatile, this mechanism works only as long as some germanium remains present.

By the mid-1950s it was also recognized that the coupling of computer technology with empirical thermochemical data could be applied to many problems in the metals sector, primarily the steel industry. The field has grown significantly since its conception and now finds wide application in the semiconductor and ceramics fields as well. The principle behind *computational thermochemistry* is that algorithms for the minimization of the total Gibbs energy can be used to model phase equilibria in underinvestigated systems. When applied specifically for the production of phase diagrams, it is termed the *CALPHAD* (calculation of phase diagrams) *method*, a field pioneered by Larry Kaufman (b. 1931) and Mats Hillert (b. 1924).

Certainly, the experimentally determined phase boundaries of a well-investigated system can be modeled with mathematical expressions. However, the real advantage of the CALPHAD technique lies in its ability to *extrapolate* the stability ranges of known phases, as new components are added, to reasonably represent the uninvestigated higher-order system. With extrapolation, one essentially calculates the equilibrium phase boundaries in a $(n + m)$-component system by modeling the Gibbs energies of the liquid, solid solutions, and sublattice phases taking part in the equilibria of an n-component system, as the mth component(s) is added, and then minimizing the *total* Gibbs energy of the $(n + m)$-component system. For example, the stability range for a binary oxide with composition $A_xB_yO_z$ can be extrapolated as a ternary solid solution phase forms with the addition of a third metal, C. In the ternary phase, it is expected that C substitutes on one or more of the cation sublattices [e.g., $(A,C)_xB_yO_z$, $A_x(B,C)_yO_z$, or $(A,C)_x(B,C)_yO_z$]. Unfortunately, extrapolation of thermochemical data is of no value in finding hitherto unknown unique structures or phases that may exist in a high-order system but which have no analog in any of the subsystems.

Here, we only present the simplest thermodynamic expressions used in the CALPHAD method for the major phase classes observed in multicomponent systems: namely, disordered miscible and immiscible phases and ordered sublattice phases. The reader is referred to specialized textbooks for further discussion. The Gibbs energies for disordered two-component solid and liquid solution phases are most easily represented by the regular solution model (Eq. 2.10) or one of its variants:

$$G = x_A{}^\circ G_A + x_B{}^\circ G_B + x_A x_B \Omega + RT(x_A \ln x_A + x_B \ln x_B) \qquad (2.10)$$

In this expression, ${}^\circ G_A$ is the standard molar Gibbs energy (${}^\circ$ signifies the value at $p = 1$ atm) of the pure component A, and x_A is the mole fraction of component A. The last two terms on the right-hand side of Eq. 2.10, combined, constitute the Gibbs energy of mixing, ΔG_{mix}. If ΔG_{mix} is negative, a system is able to lower its total Gibbs energy by forming a solution, the extent of which will depend on the magnitude of ΔG_{mix}. When ΔG_{mix} has a positive sign, the components are immiscible. No solution formation will occur because the system is in a lower energy state as a two-phase mixture. Hence, Eq. 2.10 may also be used to determine the total Gibbs energy for immiscible or partially miscible systems (mechanical mixtures). The Gibbs energy of the mechanical mixture serves as a reference state for the properties of a solution, in which there is *chemical mixing* between components on an atomic or molecular level. The regular solution model was introduced by the American chemist Joel Henry Hildebrand (1881–1983) in 1929 (Hildebrand, 1929).

By contrast, the Gibbs energy of an ordered sublattice phase is modeled using what is known as the *compound energy formalism*. For the simplest sublattice phase, in which there is only a single component (z) of fixed stoichiometry on one sublattice (v) and two randomly mixed components (i, j) in a second sublattice (u) [e.g., (Ga,In)Sb], the compound energy formalism yields

$$G = \sum_i y_i \, {}^{\circ}G_{i:z} + RT \sum_i y_i \ln y_i + \sum_i \sum_{j>i} y_i y_j \sum_v L_{ij}^v (y_i - y_j)^v \qquad (2.11)$$

In this expression, y_i is the fractional site occupation for component i (the number of atoms of component i on the sublattice divided by the total number of sites on that sublattice), L_{ij}^v is an interaction energy parameter for mixing between components i and j on the sublattice, and ${}^{\circ}G_{i:z}$ is the Gibbs energy of the compound when the sublattice u is completely occupied by i.

In passing, we note that classical thermodynamics can also be modified to treat small finite systems. This field has been amply called *nanothermodynamics* (Hill, 2001), and it allows the calculation of thermodynamic properties for nano-sized systems. However, additional terms must be added to account for effects that are negligible in macroscopic systems, particularly surface energy (Section 2.3). Interestingly, it is found that properties (e.g., chemical potential) that are intensive in the bulk state are size dependent for small finite systems. Another important way in which macroscopic and nanoscale systems differ is that the thermodynamic properties of the latter are sensitive to their environment. The American chemist and biophysicist Terrell Hill was the first to investigate the thermodynamics of small systems extensively, while he was on sabbatical at Cambridge University (Hill, 1962, 1963, 1964). The difference between the applicability of thermochemical approaches and first-principles approaches (Section 1.2), in the context of nanoscale solids, should be emphasized. Nanothermodynamics is appropriate for determining energy changes associated with reactions in real systems comprised of nano-sized particles, which could contain as few as 10^3 atoms. First-principles simulations would be appropriate for calculating the total energy of a macroscopic solid, using a representative collection of atoms that reasonably accounts for all the long-range interactions, which, with today's computing technology, is on the order of 10^3 atoms.

It will be instructive to look now a little more closely at internal energy. The *first law of thermodynamics* is formulated in differential form as

$$dU = dq + dw \qquad (2.12)$$

where dq and dw are, respectively, the heat and work exchanged between the system and the surroundings. The first quantity on the right-hand side of Eq. 2.12 is $T \, dS$ and, for compressional work *against* a hydrostatic pressure, the second quantity is $-P \, dV$. Making these substitutions, we have

$$dU = T \, dS - P \, dV \qquad (2.13)$$

In his early work, J. W. Gibbs refined this relation by adding a term containing the sum of the free energies of each pure component comprising the system:

$$dU = T \, dS - P \, dV + \sum_i \mu_i \, dN_i \tag{2.14}$$

where μ_i is the chemical potential (free energy per mole) of the ith component and N_i is the number of moles. When Eq. 2.13 or 2.14 is integrated, the result is a value for ΔU. Absolute values of the internal energy are not defined by the first law of thermodynamics, only changes in U. Thus, thermodynamics does not require us to consider any molecular interpretation for what the internal energy of a macroscopic substance really consists of. Fortunately, the combination of statistical mechanics and first-principles (ab initio) methods is of great value in this regard. We will return to this topic shortly.

2.2. STRUCTURAL ENERGETICS

In Section 2.1, we remarked that classical thermodynamics does not offer us a means of determining absolute values of thermodynamic state functions. Fortunately, first-principles (FP), or ab initio, methods based on the density-functional theory (DFT) provide a way of calculating thermodynamic properties at 0 K, where one can normally neglect zero-point vibrations. At finite temperatures, vibrational contributions must be added to the zero-kelvin DFT results. To understand how ab initio *thermodynamics* (not to be confused with the term *computational thermochemistry* used in Section 2.1) is possible, we first need to discuss the statistical mechanical interpretation of absolute internal energy, so that we can relate it to concepts from ab initio methods.

Simply put, *absolute internal energy is the total potential and kinetic energy of the particles of a substance*. The contributions to the internal energy of a solid include (1) the kinetic energy of the atoms or ions comprising the substance (obviously of lesser importance in solids than in liquids and gases, but the atoms of a solid can still translate and vibrate), (2) the potential energy arising from interactions *between* these particles (a function of internuclear distances), and (3) the electronic energy. Having stated this, let us first examine the simplest of these contributions—the potential energy arising from the interactions between atoms—for the simplest possible case, a highly ionic solid.

2.2.1. Electrostatic Energy

Ionic solids minimize their internal energy by optimizing the electrostatic interactions. Hence, a discussion of the electrostatics conveniently parallels a thermodynamic treatment. The lowest-energy structure is normally that which minimizes all the electrostatic repulsions between ions and maximizes all the attractive interactions, both short and long range. The balance between these competing requirements results in ionic solids being highly symmetric structures with maximized coordination numbers and volumes. The potential energy per mole of compound associated with the particular geometric arrangement of ions forming the structure is referred to as the *lattice energy*, symbolized as $U_{lattice}$. It is equivalent to the heat of formation from 1 mol of its ionic constituents in the gas phase. Calculation of the total electrostatic energy involves summation of *long-range* attractions between oppositely charged ions and repulsions between like-charged ions, extending over the entire crystal, until a convergent mathematical series is obtained. In 1918, Erwin Rudolf Madelung (1881–1972) was the first to carry out this type of summation for the electrostatic energy of NaCl (Madelung, 1918).

The final expression for the long-range force on any one ion is

$$U_{ion} = \frac{1}{4\pi\varepsilon_0} \left(-\frac{Mq_+q_-e^2}{r} \right)$$ (2.15)

in which $4\pi\varepsilon_0$ is the permittivity of free space (1.11265×10^{-10} C^2 J^{-1} m^{-1}), e the electron charge (1.6022×10^{-19} C), q the ion charge, and M the Madelung constant. The Madelung constant is dependent only on the geometric arrangement of ions and the distance that r is defined in terms: nearest neighbor, unit cell parameter, and so on (beware of different conventions when using Madelung constants from the literature!). The constant has the same value for all compounds within any given structure type. If r is in meters, the units of Eq. 2.15 will be in joules per cation.

Johnson and Templeton calculated values of M for several structure types using the Bertaut method (Johnson and Templeton, 1962). Their results are partially reproduced in Table 2.2, where the second column lists the Madelung constant based on the shortest interatomic distance in the structure. The third column gives *reduced* Madelung constants based on the *average* shortest distance. For less symmetric structures (i.e., when there are several nearest neighbors at slightly different distances, as in the ZnS polymorphs), the reduced Madelung constant is the more significant value (Lalena and Cleary, 2005)

To obtain the complete expression for the lattice energy of an ionic crystal, we must (1) add the term representing the short-range repulsive forces, (2) include Avogadro's number, N (6.022×10^{23} mol^{-1}), and (3) make provisions for ensuring that we do not overcount pairs of interactions. In so doing, the final expression for the lattice energy of an ionic crystal containing $2N$ ions was given by M. Born and A. Landé (Born and Landé, 1918) as

$$U_{lattice} = \frac{N}{4\pi\varepsilon_0} \left(-\frac{Mq_+q_-e^2}{2} + \frac{B}{r^n} \right)$$ (2.16)

This equation gives the lattice energy in joules per mole. We can avoid having to determine a value for the parameter B by using the equilibrium interatomic distance as the value of r for which U is a minimum. This gives $dU/dr = 0$

TABLE 2.2   Madelung Constants for Several Structure Types

Compound	$M(R_0)$	$M\langle R\rangle^a$
Al_2O_3 (corundum)	24.242	1.68
$CaCl_2$	4.730	1.601
CaF_2(fluorite)	5.03879	1.68
$CaTiO_3$ (perovskite)	24.7550	—
CsCl	1.76268	1.76
Cu_2O	4.44249	1.48
La_2O_3	24.179	1.63
$MgAl_2O_4$ (spinel)	31.475	—
MgF_2	4.762	1.60
NaCl (rock salt)	1.74756	1.75
SiO_2 (quartz)	17.609	1.47
TiO_2 (anatase)	19.0691	1.60
TiO_2 (rutile)	19.0803	1.60
V_2O_5	44.32	1.49
ZnO	5.99413	1.65
ZnS (zinc blende)	6.55222	1.638
ZnS (wurtzite)	6.56292	1.641

Source: Johnson and Templeton (1962).

[a] $M\langle R\rangle = M(R_0)\langle R\rangle/R_0$, where R_0 is the shortest interatomic distance and $\langle R\rangle$ is the average shortest distance.

and the following expression for U:

$$U_{\text{lattice}} = \frac{N}{4\pi\varepsilon_0}\left(-\frac{Mq_+q_-e^2}{r}\right)\left(1 - \frac{1}{n}\right) \qquad (2.17)$$

If the value of n is not known, an approximate value may be obtained from Table 2.3. In cases where there are significant contributions from covalent bonding, Eqs. 2.16 and 2.17 will not reflect the true binding energy of the crystal. Nevertheless, they are still useful in comparing relative energies for different compounds with the same structure.

TABLE 2.3   Values of the Born Exponent

Cation–Anion Electron Configurations	Example	n^a
$1s^2-1s^2$	LiH	5
$1s^22s^2p^6-1s^22s^2p^6$	NaF, MgO	7
$[Ne]3s^2p^6-[Ne]3s^2p^6$	KCl, CaS	9
$[Ar]3d^{10}4s^2p^6-[Ar]3d^{10}4s^2p^6$	RbBr, AgBr	10
$[Kr]4d^{10}5s^2p^6-[Kr]4d^{10}5s^2p^6$	CsI	12

[a] For mixed-ion types, use the average (e.g., for NaCl, $n = 8$).

When an extremely stable ionic lattice (highly negative U_{lattice}) is formed from, say, a gas and a liquid, the enthalpy, H, or heat content of the lattice is lower than that of the substances from which it is formed. That is, heat is evolved (negative ΔH). In fact, such chemical reactions are enthalpy driven, since going from disordered reactants to a highly ordered lattice produces a negative entropy change and a positive value for $-T\Delta S$ in the free-energy expression: $\Delta G = \Delta H - T\Delta S$. This is an example of how the enthalpy change and entropy change can oppose one another. Most chemical reactions at low temperatures are, in fact, exothermic and have small entropy contributions (either slightly negative or slightly positive in sign). Hence, most low-temperature reactions are enthalpy driven. Similarly, reactions in which both the enthalpy of formation and entropy favor spontaneity, but in which the enthalpy term far outweighs the entropy contribution, can be considered enthalpy driven.

By contrast, if we vaporize a solid at a very high temperature (energy must be absorbed to break up the lattice), this process is certainly entropy driven, due to the negative $-T\Delta S$ term (positive ΔS) and highly positive ΔH (endothermic) term. Thus, reactions occurring at high temperatures accompanied by an entropy increase (e.g., gas evolution, order–disorder phase transformation, alloy formation) can be entropy driven. This is also true for moderately high-temperature processes with only small enthalpy changes, where the lattice energy advantage is not enough to drive the reaction.

2.2.2. Electronic Energy

The same energetic principles apply to covalent bonding. In fact, covalent, ionic, and van der Waals forces are all just different manifestations of the Coulomb force. For example, in covalent substances, the nucleus–electron attractive forces are maximized, and the nuclear–nuclear repulsions are minimized because the electrons, which are located between the nuclei in overlapping atomic orbitals, shield the two nuclei from each other. Thus, the strength of the covalent bond is attributable to the simultaneous attraction of both electrons to both nuclei. Maximizing these cohesive forces, or bond energies, between covalently bonded atoms leads to a large enthalpy of formation (ΔH_f). Caution must be exercised in using tabulated bond energies to approximate actual enthalpy changes, however. In some cases, the difference between the total bond energy of the products and the total bond energy of the reactants may give, at best, only a very crude estimate for the enthalpy of formation. The simple bond energy approximation is very different from a real statistical mechanical treatment.

The primary difference between covalent and ionic bonding is that with covalent bonding, we must invoke quantum mechanics. In molecular orbital (MO) theory, molecules are most stable when the bonding MOs or, at most, bonding plus nonbonding MOs, are each filled with two electrons (of opposite spin) and all the antibonding MOs are empty. This forms the quantum mechanical basis of the *octet rule* for compounds of the p-block elements and the *18-electron rule* for d-block elements. Similarly, in the Heitler-London (valence bond) treatment of solids, coordination polyhedra are most stable when a pair of electrons with opposing spins occupies each two-center bonding site. This occurs in silicon, for example, which is isostructural with diamond.

There may be more than one way the octet rule can be satisfied, however. As for diamond itself, sp^3 hybridization of the carbon atom's valence orbitals is less thermodynamically favorable than sp^2 hybridization, in direct analogy with the comparison between alkanes and alkenes. In graphite, as in ethylene, the octet rule is still satisfied. Bonds between two sp^2 carbon atoms, and between an sp^2 carbon and an sp^3 carbon, are somewhat stronger than bonds between two sp^3 carbons. The single C–C bonds in graphite are intermediate in length between that of a normal single C–C bond and that of a double carbon–carbon bond. Hence, the bonds between carbon atoms within the graphite sheets are even stronger than the C–C bonds in diamond, despite the fact that diamond is the hardest substance known and graphite is used as a lubricant! Still, once formed, diamond is extremely kinetically stable under ambient conditions. By contrast, the thermodynamic stability of multiple bonds decreases in moving down the fourteenth column of the periodic table since the enlarged inner cores result in longer bond lengths and weaker π overlap. Consequently, silicon has a much lower tendency than carbon to form unsaturated compounds containing multiple bonds (although some such molecular organometallic compounds are known) and, in turn, a lower tendency to resort to hybridization, so it has no graphite like allotrope; the tetrahedral network of sigma bonds is much preferred.

Chemists find the valence bond approach very appealing for formalizing their intuitive ideas about molecular (and crystal) geometry and Lewis's model of shared electron pairs. Armed with the valence bond model and its extensions, like the hybridization schemes for the valence atomic orbitals, one can rationalize a great many molecular and crystal structures. However, the most accurate picture of the electronic wave functions in extended solids, at least for those where the internuclear distances are less than some critical value, is the Bloch, or band, model. This is equivalent to molecular orbital theory for molecules. It is the *only* satisfactory form of the electronic wave function in materials exhibiting metallic conductivity. In the Bloch model, the valence electrons are considered delocalized in crystal orbitals extending over the entire crystal, just as, in MO theory, they are delocalized in molecular orbitals over entire molecules. One of the first approaches to calculating electronic energy levels in solids was the LCAO (linear combination of atomic orbitals), or tight-binding, method, based on the Hartree–Fock (independent electron) approximation (Lalena and Cleary, 2005).

2.2.3. Total Energy

The LCAO tight-binding (Hartree–Fock) method and its successor, the density-functional theory (DFT), were used originally to solve electronic structure problems. More recently, both have been applied to the calculation of *total energy*, which includes contributions due to core–core, core–electron, and electron–electron interactions. By varying the coordinates of the core, the dependence of the total energy on the core coordinates can be examined. This is referred to as the *potential surface* (also called *energy landscape*) of the system. These types of calculations are called *first-principles* (ab initio), or *quantum, molecular dynamics (MD) simulations*. They are more amenable to larger system sizes than classical (non-quantum mechanical) molecular dynamics and Monte Carlo simulations, which simply treat atoms in a many-particle system as hard spheres and provide no information about the electronic structure. In quantum MD simulations, the atomic cores are still treated as classical particles. The electrostatic interactions between cores are often approximated as a sum over pair potentials, including those Coulomb interactions between pairs of ions separated by long distances, which is very similar to the expression for the Madelung energy. However, the (classical) force on the core due to the electron system is also included, while the electronic structure problem itself is solved quantum mechanically. Alternatively, by keeping the nuclear positions fixed, the quantum molecular dynamics method can be restricted to just the electronic structure component of the total energy.

The concept of total energy is very similar, but not identical, to internal energy. Both are the sum of the potential energy and kinetic energy contributions in a system. However, internal energy is an extensive thermodynamic state function for a *macroscopic* system ($\sim 10^{23}$ atoms), irrespective of atomic numbers and atomic geometry. Furthermore, absolute values of internal energy are not defined by laws of thermodynamics, only *changes* in U that accompany chemical reactions or physical (phase) transformations of macroscopic substances. In density-functional theory, the total energy is decomposed into three contributions: the electron kinetic energy, the interactions between the cores, and a term called the *exchange-correlation energy* that captures all the many-body interactions. Note the similarity with the statistical mechanical interpretation for absolute internal energy given earlier. In fact, the total energy can be argued to accurately represent a sort of *specific* absolute internal energy, making it an intensive property. Total energy is an intrinsic physical property that may be calculated for a single molecule, or, using a finite-sized representative collection of atoms (usually on the order of 10^3), for a macroscopic solid. This is somewhat akin to the way that the Madelung energy for an ionic crystal is calculated, but total energy includes electronic contributions as well.

Other thermodynamic functions, in addition to internal energy, can also be calculated from first principles. For example, at a finite temperature, the Helmholtz free energy, A, of a phase containing N_i atoms of the ith component, N_j atoms of the jth, and so on, is equal to the DFT total energy at zero kelvin (neglecting zero-point vibrations) plus the vibrational contribution (Reuter and Scheffler, 2001):

$$A(T, V, N_i, \ldots, N_j) = E^{\text{tot}}(V, N_i, \ldots, N_j) + A^{\text{vib}}(T, V, N_i, \ldots, N_j) \quad (2.18)$$

The Helmholtz free energy, in turn, is related to the Gibbs free energy, G, via

$$G(T, P, N_i, \ldots, N_j) = A(T, P, N_i, \ldots, N_j) + pV(T, P, N_i, \ldots, N_j) \quad (2.19)$$

The DFT total energy at zero kelvin is thus a reference system for determining the free energy, corresponding to a simple Einstein solid with all the atoms vibrating around lattice points independently with the same frequency. The vibrational contribution, $A^{\text{vib}}(T, V, N_i, \ldots, N_j)$, contains the vibrational energy (including the zero-point energy), E^{vib}, and entropy, S^{vib}. Each of these components can be calculated from the partition function for an N-component system, Z, and using the relation $A^{\text{vib}} = E^{\text{vib}} - TS^{\text{vib}}$. The reader is referred to Ashcroft and Mermin (1981) for the partition function expression, as well as the vibrational and entropy defined in terms of Z.

As with thermochemical techniques, the most stable structure can be predicted from DFT total energy calculations by the principle of energy minimization: *global* energy minimization. Metastable structures that are kinetically stable can also be predicted by locating *local* minima in the potential surface surrounded by barriers of sufficiently high energy. It becomes increasingly difficult to predict the structure, stability, and properties of solids definitely, as the system size increases. This is because approximations have to be employed for the energy calculations on large systems with multiple minima. In general, the systems readily treated by these methods are limited to relatively small numbers of atoms, say a few hundred.

To make accurate total energy calculations, one must investigate the entire energy landscape, having an approximate crystal structure in mind, which is used as a starting point for optimizing the geometry to obtain the lowest-energy structure. Investigating anything less than the entire energy landscape runs the risk of not revealing the true minimum. The structure predictions can sometimes be surprisingly different from what one would expect based on chemical intuition. One such case involves the group IVB nitrides, which adopt the spinel structure at high pressures and temperatures.

There are two possible cation distributions in spinel: the "normal" spinel, AB_2X_4, in which there is no alloying distribution of A or B cations on either the four-coordinate site (A) or the eight-coordinate site (B), and the "inverse" spinel, $B(AB)X_4$, with equal amounts of A and B cations randomly distributed over the octahedral sites. Since it is well known that silicon tetrahedral bonds are stronger than germanium tetrahedral bonds, and that the Si^{4+} ion is smaller than the Ge^{4+} ion, initial DFT investigations were restricted to the possible "normal" configurations. It was predicted that this nitride should prefer to adopt the structure $SiGe_2N_4$, in which the four-coordinate sites are exclusively occupied by silicon and the eight-coordinate sites by germanium, and that $GeSi_2N_4$ was unstable or metastable to decomposition into the binary nitrides Si_3N_4 and Ge_3N_4. However, the reverse cation site preference, $(Si_xGe_{1-x})_3N_4$, was subsequently observed by x-ray powder diffractometry experiments in polycrystalline samples with $x \approx 0.6$ (Soignard et al., 2001). These silicon and germanium site preferences were later confirmed by density-functional total energy calculations (Dong et al., 2003).

The geometry observed apparently provides for a much more symmetrical distribution of bond lengths around the four-coordinate nitrogen atom (sp^3 hybridized). In fact, the "normal" spinel, $GeSi_2N_4$, where the tetrahedral sites are

Walter Kohn (b. 1923) received a Ph.D. in physics from Harvard University in 1948. In the early to mid-1960s, Kohn played the lead role in developing density-functional theory, which has become the predominant method for determining the electronic structures and total energies of molecules and solids. For this work he was awarded the 1998 Nobel Prize in Chemistry. Kohn also made seminal contributions to superconductivity, semiconductor physics, and surface physics. Kohn started his career as an instructor at Harvard and then moved to Carnegie Mellon (1950–1960), the University of California–San Diego (1960–1979), and finally, the University of California–Santa Barbara (1979), where he became emeritus in 1991. At UCSB, Kohn was the founding director of the world-renowned Institute for Theoretical Physics. Kohn was elected to the U.S. National Academy of Sciences in 1969, and he is a member of the Royal Society of London and the Bavarian Society.

occupied exclusively by silicon and the octahedral sites are occupied exclusively by germanium, is predicted to be the only stable crystalline phase in the Si_3N_4–Ge_3N_4 system, although kinetically stable metastable solid solutions may exist for Ge- and Si-rich compositions (Wong et al., 2003).

2.3. GRAIN BOUNDARY ENERGY AND SURFACE ENERGY CONTRIBUTIONS

Determinations of the energies of solids, and energy changes accompanying reactions and transformations of solids, must inevitably consider grain boundaries (solid–solid interfaces) and surfaces. Surface and grain boundary free energies are thermodynamic quantities that profoundly affect observed crystal morphology and behavior. Atoms at a grain boundary or surface are in markedly different environments from the atoms in the interior of a bulk sample. Since surface atoms are not as strongly bound as bulk atoms, they vibrate with lower frequency and larger amplitude. Their energy is higher than that of the bulk atoms and they have more freedom of movement. Consequently, surface atoms melt and sublime at temperatures lower than that required for bulk melting. In fact, there is often a roughly linear relationship between surface energy and the heat of sublimation, as well as many other physical properties. Similarly, atoms at homophase and heterophase grain boundaries are in regions of structural incoherency.

2.3.1. Surface Energy and Its Effect on the Equilibrium Crystal Shape

Thermodynamically, a solid surface is sufficiently characterized by two parameters: *surface energy* (a scalar), with units of energy per area, and *surface stress* (a second-rank tensor), also with units of energy per area. For a liquid, these two are the same, but they can be very different for solids. We do not discuss the latter in this book, nor shall we distinguish between surface energy and *surface tension*, which is defined as the reversible work done in creating unit area of new surface. In one-component systems, surface energy and surface tension are numerically equal.

The total energy, and the total free energy, of a crystal is the sum of the energy of the bulk of the crystal and the excess energy due to the simple presence of the surface. Surface atoms have fewer neighboring atoms, and hence bonds, compared to bulk atoms. Because there is less stabilization provided by chemical bonding, the surface has a higher energy per unit area relative to the bulk. Hence, smaller crystals, with their larger surface area/volume ratio, are less thermodynamically stable than large crystals. For this reason, most of the small crystals growing in a solution disappear slowly, as the remaining ones grow larger by consuming them. This process, known as *Ostwald ripening*, is discussed in Section 4.2.1. It is observed in the growth of single crystals in solution under ambient conditions, as well as in the coarsening of polycrystalline samples grown in subcritical aqueous phase and by solidification processes.

Not only does the surface energy affect crystal size, but it influences crystal shape as well. In the nineteenth century, Josiah W. Gibbs and Pierre Curie independently concluded that the macroscopic shape of a crystal in equilibrium will be such that the total surface energy is minimized (Gibbs, 1878; Curie, 1885). The surface energy is equal to the product of the *specific* surface energy and the surface area: In differential form, the total Gibbs free energy for the crystal is thus

$$dG_{\text{total}} = dG_{\text{bulk}} + \gamma_s \, dA \qquad (2.20)$$

where γ_s is the surface free energy per unit area A or specific surface free energy. Although integrating Eq. 2.20 yields *changes* in the free energies, estimates for absolute values may be obtained from first principles. For example, rearranging Eq. 2.20 to solve for γ_s and using the relation that G_{bulk} is the sum of all the

$N_i\mu_i$'s, the most stable surface composition and geometry of a metal oxide, MO, is the one that minimizes γ_s:

$$\gamma_s(T, P) = \frac{1}{A}\left[G(T, P, N_M, N_O) - N_M\mu_M(T, P) - N_O\mu_O(T, P)\right] \quad (2.21)$$

In Eq. 2.21, N_M is the number of metal atoms and N_O is the number of oxygen atoms. Similarly, μ_M and μ_O are the chemical potentials of a metal atom and an oxygen atom, respectively. Equation 2.21 has been used to determine the lowest-energy surface structure, or termination, of RuO_2 (110) surfaces (Reuter and Scheffler, 2001).

Although the equilibrium form of a crystal should possess a minimum surface free energy for a given volume, the minimum surface energy of a crystal, unlike with a liquid, does not correspond to that of a sphere (i.e., the geometry with minimum surface area). Surface energy is a function of the crystal face orientation, or (*hkl*) indices, as was pointed out in Section 1.2.1. Many high [Miller] index planes and, hence, high surface energy planes, are needed to approximate a sphere as a closed form. Surface energy is proportional to the distance of the surface from the center of the crystal because of the dependence on the interatomic potentials or, more specifically, the interactions both short and long range, between the surface atoms and their neighboring atoms. This was demonstrated early in the twentieth century from calculations on the energies of different sodium chloride faces by Madelung, Born, Stern, and Yamada (Madelung, 1918, 1919; Born and Stern, 1919; Yamada, 1923, 1924), based simply on electrostatic attractions and repulsions. They found that the surface energy of the (110) plane of rock salt is about 2.7 times larger than that of the (100) plane.

2.3.1.1. Wulff Construction   The Russian crystallographer Georgii Viktorovich Vulf (1863–1925), who used the name Georg Wulff in his German publications, discovered that for every given volume there is a unique convex body whose boundary consists of planar faces, such that this boundary surface has a lower energy than the boundary surface of any other piecewise smooth body of the same volume (Wulff, 1901). A polar plot is first constructed of $\gamma(\theta)$. Next, a Cartesian coordinate system is defined with an abscissa along $\theta = 0$ and an ordinate along $\theta = \pi$. For an arbitrary angle, a radial line of length γ is drawn, which ends on the polar plot. At this geometric point, a perpendicular is drawn. It can be shown that this perpendicular is tangent to a rescaled crystal shape. Varying the angle generates a family of tangents that, when superposed, creates an envelope of the rescaled crystal shape. The crystal morphology is the shape that gives the minimum volume from this construction. The Wulff construction can be used for finding the equilibrium morphologies of macroscopic crystals and large clusters (i.e., those containing hundreds or thousands of atoms), provided that the surface energies are known. The procedure for finding this *equilibrium crystal shape* (ECS) can be outlined as follows:

1. For all crystal faces, draw vectors normal from the center of the equilibrium crystal shape. The lengths of these vectors are proportional to the

surface energies. Knowledge of the complete surface energy function is thus a necessary condition for determination of the ECS.

2. Draw lines (planes in two dimensions) normal to these vectors at the points where these vectors intersect the polar plot of the surface energy.

3. The resulting closed polyhedron, formed by the intersections of these planes normal, is the equilibrium (lowest-energy) form. More precisely, if one takes the convex (inner) envelope of those crystal planes that are orthogonal to the vectors passing through the origin at the points where these vectors intersect the surface energy plot, the crystal form with the lowest surface energy may be determined.

Such a polar plot of the surface energy as a function of the direction of the surface normal will have at least the same symmetry as the point group of the crystal since surfaces related by symmetry will have the same surface energy. Consequently, for cubic crystals, one quadrant is sufficient to specify the entire plot. The Wulff plots of crystals have sharp cusps (local minima) corresponding to the faces with lower surface energy. These faces are the ones the crystal uses to minimize its total surface energy. The ratio of the surface energy of any crystal face to the distance of that face from the center of the equilibrium crystal shape (the Wulff point), which is the point at which the crystal begins growing, is constant. Hence, the lowest-energy faces are closest to the center of a crystal.

As might be expected, this is easiest to visualize in two dimensions, but the principles apply directly to three dimensions as well. For sodium chloride, at low temperatures, a plot qualitatively similar to Figure 2.5a is obtained. In this two-dimensional section [taken through the (100) plane], the center square corresponds to the Wulff construction described above and it is the absolute minimum energy. The {100} facets are lower in energy than the {110} or {111} facets, the latter not shown in this 2D section. Accordingly, in three dimensions, a cube represents the lowest-energy crystal form for sodium chloride at low temperatures.

The edges separating facets may be considered phase boundaries and may be of two types: first-order, or discontinuous, transitions (sharp), and second-order, or continuous, transitions (rounded). High-index orientations that do not appear in the ECS are unstable phases that tend to facet into thermodynamically stable low-index orientations. This is a form of spinodal decomposition (Cahn, 1962). Thermodynamic instability is found when the surface energy per unit area is severely anisotropic (i.e., when the surface energy differences between adjacent orientations are very large). Numerous facets (pyramidlike or steplike structures) of low surface energy (low index) will develop on these high-energy surfaces if kinetically feasible.

With the monatomic face-centered-cubic (FCC) metals, a qualitative two-dimensional Wulff construction taken through the (111) plane and a truncated octahedron equilibrium crystal shape similar to those shown in Figure 2.5b are expected. The symmetry of a cubic crystal tells us that we will have six (100) vectors and eight (111) vectors. This 2D section is therefore sufficient to allow

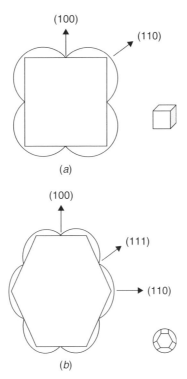

Figure 2.5 *(a)* Two-dimensional section [made through the (100) plane] of a polar plot of the surface energies of sodium chloride at low temperatures. The energy of the (110) corresponds to a maximum and that of (100) corresponds to a minimum. As explained in the text, the square (a cube in three dimensions) corresponds to the crystal form of lowest energy. *(b)* two-dimensional section [made through the (111) plane] of a polar plot of surface energy for an FCC monatomic metal oriented along <1$\bar{1}$0>. The equilibrium crystal shape is show to the right.

us to predict the three-dimensional ECS shown. The larger hexagonal facets have (111) orientation, and the smaller square facets have (100) orientation. This ECS is predicted if one only considers nearest-neighbor interactions in calculating the surface energies. The {110} facets, for example, do not appear because if one assumes that the minimal radial distance to every face in the Wulff construction is proportional to its surface energy, the ratio of the surface energies of the {100} and {110} facets should obey the constraint $\gamma_{100}/\gamma_{110} \leq 1/\sqrt{2}$. However, if longer-ranged potentials are used, further faceting occurs, giving the Wulff polyhedron a more rounded shape. The {110} surface can be low enough in energy that narrow {110} facets comprise part of the ECS, lying at the intersections between hexagons [i.e., the (111) and (11$\bar{1}$)]. The edges between these {110} and {111} facets can further be modestly rounded due to a limited number of higher-index orientations being marginally stable.

The flat (100) sodium chloride surface, corresponding to the cusp in the Wulff plot, is *singular*. Surfaces with inclinations close to those of singular surfaces (i.e., only slightly different in orientation) have energies lying on the sides of the cusps. These are *vicinal surfaces* that contain two-dimensional terraces (TLK model; Section 1.4.1). Such defects result in higher-energy surface configurations than those of singular surfaces. It may be possible, however, that either surface relaxation or interactions between neighboring steps can stabilize vicinal surfaces against faceting, as described above. Such factors could then play a crucial role in determining the crystal morphology and the equilibrium crystal shape. Frenken and Stoltze 1999 found from effective medium theory using empirical first- and second-nearest-neighbor interaction potentials that, at zero kelvin the entire range of orientations between (100) and (111) is unstable with respect to faceting into linear combinations of (111) and (100) orientations. At finite temperatures, their calculated excess vibrational entropy was explained as being responsible for stabilizing the free energy of these vicinal surfaces against faceting (Frenken and Stoltze, 1999). However, the accuracy of omitting higher-order interactions has been questioned and a more recent employment of density-functional theory (DFT) on several vicinal surfaces distributed over the [110] and [100] zones of FCC lead (Pb), which are also predicted to be unstable with respect to faceting into (111) and (100) orientations, has implicated the general importance of surface relaxation (Section 2.3.1.3) in stabilizing vicinal surfaces (Yu et al., 2006). Relaxation lowers the surface energy anisotropy: 5.6% lower surface energy for Pb(111), 13.6% lower for Pb(110), which enables new orientations to be stable and appear in the ECS for lead. However, the jury is still out, as surface relaxations in noble and transition metals are smaller than for lead. Hence, its effect on governing the ECS may be correspondingly less significant for these metals.

2.3.1.2. Approximating Surface Energy Using a Hard-Sphere Broken-Bond Model

*Wulff's theorem provides a phenomenological method for calculating the ECS provided that the surface energies of the various faces are known. We have already stated that as in the case for the binding energy of a bulk crystal (Section 2.2), surface energy is described via interatomic potentials. The details of such computations are beyond the scope of this book. In this section, however, we introduce an early method, based on the TLK model, of approximating the relative surface energies of different faces using a hard-sphere broken-bond picture. This is possible because surface energy originates from the excess energy due to broken bonds (atoms on the surface have fewer neighbors compared to those in the bulk). Consequently, the higher the number of broken bonds in forming a surface, the higher the surface energy since those atoms are stabilized less by chemical bonding. Consider the monatomic simple cubic structure in Figure 2.6. Each sphere in this figure represents an atom. Adjacent spheres, separated by the shortest possible distance, are the nearest-neighboring atoms of the same type (the dark spheres in the upper right figure). The bonds between nearest neighbors possess bond strength ϕ_1. Bonds to the second-nearest-neighbor atoms, also of the same type (lightly shaded spheres in the upper right figure), have bond strength ϕ_2 (where $\phi_2 < \phi_1$).

Atoms in the bulk simple cubic crystal have the six nearest neighbors and 12 second-nearest neighbors shown in the upper right figure for a total binding energy of $6\phi_1 + 12\phi_2$. Each bond is shared between two atoms. Hence, the mean sublimation energy of the crystal is one-half that value, or

$$3\phi_1 + 6\phi_2 \tag{2.22}$$

Per unit volume of the crystal, the mean sublimation energy of a simple cubic structure with lattice parameter a is Eq. 2.22 divided by a^3. Compare this with adatoms adsorbed on a smooth face, which have one nearest neighbor and four second-nearest neighbors. Their binding energy is thus significantly lower $(\phi_1 + 4\phi_2)$.

Terrace (face) sites in the interior of a step have five nearest neighbors and eight second-nearest neighbors (binding energy $= 5\phi_1 + 8\phi_2$). That is, they have one fewer nearest neighbor (there is a missing atom directly above it) and four fewer second-nearest neighbors than a bulk atom, due to the absence of an overlying atomic plane. The result is an excess energy, over that of the atoms in the bulk, of

$$e_t = \phi_1 + 4\phi_2 \tag{2.23}$$

This equation is divided by $2a^2$ to obtain the excess energy per unit area. Similarly, ledge sites on a completed step have four nearest neighbors and six second-nearest neighbors (binding energy $= 4\phi_1 + 6\phi_2$) This is one fewer nearest neighbor and two fewer second-nearest neighbors than terrace atoms have. Thus, a ledge site has an excess energy over terrace atoms of

$$e_l = \phi_1 + 2\phi_2 \tag{2.24}$$

To obtain the excess ledge site energy per unit length, we would divide Eq. 2.24 by $2a$. Finally, at a kink site, there are three nearest neighbors and six second-nearest neighbors (binding energy $= 3\phi_1 + 6\phi_2$). This is one fewer nearest neighbor and the same number of second-nearest neighbors as atoms in ledge sites. Hence, the excess energy, relative to the ledge site, of a kink site in this case is

$$e_k = \phi_1 \tag{2.25}$$

This number must be divided by 2 to obtain the excess energy per atom. Note that the binding energy of a kink site is the same as the mean sublimation energy of the crystal. This means that removing an atom from a kink site requires the same amount of energy as removing an atom from the bulk. Kink sites are thus crucial to the crystal growth process, as they provide energetically favored binding sites to the growth units.

For the surface in Figure 2.6, it can be shown that the total excess surface energy is the sum of a contribution by the singular terraces and a contribution by the ledges, given by

$$\gamma = e_t \cos\theta + \frac{e_l}{a}\sin\theta \tag{2.26}$$

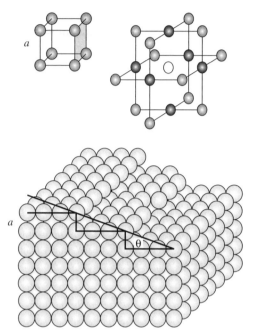

Figure 2.6 The six nearest neighbors (dark) and the 12 next-nearest neighbors (light) surrounding an atom (unshaded) in the bulk of a simple cubic lattice. The next-nearest neighbors are located at the vertices of three perpendicular planes passing through the center sphere, while the nearest neighbors lay at the midpoints of the plane edges or, alternatively, at the vertices of an octahedron surrounding the center sphere. Surface atoms are surrounded by fewer neighboring atoms, as illustrated in the bottom figure.

This function can be plotted on a polar diagram and used to predict the shape of the surface energy plot cusps in the Wulff construction. The results are semi-quantitative but useful for finding the relative anisotropic surface energy, in that for cubic crystals, minima are found at low-index (111), (110), and (100) orientations. The interested reader is referred to Venables (2000) and Howe (1997) for details.

2.3.1.3. Surface Reconstruction and Relaxation   Two very important ways by which a crystal can lower its surface energy are *surface reconstruction* and *surface relaxation*. These processes occur at metal, semiconductor, ceramic, and glass surfaces. In general, the driving forces for relaxation and reconstruction are more-or-less electronic in origin. Hence, we can examine these effects in the context of the free electron model or the LCAO (Bloch) model. However, as early as 1919, Madelung had predicted from simple electrostatic considerations how a contraction (relaxation) of surface region interionic distances in the direction normal to the surface plane should occur in ionic solids with the rock salt structure (Madelung, 1919). With metals, the free electron model offers an easy-to-understand qualitative explanation. At a metal surface, there tends to be a smoothing out of the surface free electron charge density, which lowers the electron energy. This creates a surface dipole layer that repels the topmost layer of positive ion cores. If these atoms were constrained to remain structurally coherent with interior atoms, the surface would be under tensile surface stress (Howe, 1997). The surface can lower its energy and at the same time relieve the surface stress by contracting uniformly, which simply decreases the interlayer spacing between the first (surface) and second atomic layers. This is termed *surface relaxation*, and it does not alter the symmetry of the surface but can lower the surface energy anisotropy, which may enable new orientations to become stable in the equilibrium crystal shape.

By contrast, surface reconstruction is a surface symmetry-lowering process. With reconstruction, the surface unit cell dimensions differ from those of the *projected* crystal unit cell. The energy change associated with surface relaxation and reconstruction (as much as 10%) is usually much larger than that associated with surface roughening, especially with ceramics and semiconductors.

Let us now switch to the LCAO (Bloch) theory in rationalizing reconstruction at ceramic and semiconductor surfaces. In the LCAO approach, valence electrons are housed in extended crystal orbitals just like conduction band electrons. Nevertheless, the valence band is narrow for highly ionic substances (e.g., MgO) and the electron wave functions can be considered much more localized, or atomiclike. Similarly, highly directional, or angular, bonding (e.g., the diamond structure of silicon) also implies a more localized picture for the electron wave functions. In these cases, the valence-bond (Heitler–London) picture becomes applicable. With this model, it is readily seen that dangling, or broken (one electron containing) bonds at the surface will result in the presence of one unpaired electron per atomic orbital, a highly unstable configuration. Consequently, the individual bonds undergo a high elastic strain to reconstruct into a structure with fewer dangling bonds (e.g., dimerization).

One such well-known surface reconstruction is the 2×1 reconstruction of the Si(001) face. In the bulk of a silicon crystal, each silicon atom is tetrahedrally coordinated by four other atoms, that is, each atom bonds with four neighboring atoms. Viewed from the (001) plane, two of these bonds reach to the level below and two reach to the next level above. If the crystal is cleaved along a (001) plane, each surface atom is left with two broken (dangling) bonds, each containing one electron. At low temperatures, the surface lowers its energy by reconstructing so that the atoms form pairs, known as *dimers*. These dimer rows buckle so that alternating atoms are slightly raised and lowered from the plane. In this way, one of the dangling bonds belonging to each dimer is filled with electrons and pushed up while the other dangling bond is emptied and pulled down. This leaves the reconstructed surface with only two dangling bonds per dimer.

2.3.2. Grain Boundary Energy

For a polycrystal, the free energy is the sum of the free energies of the individual crystallites (lumped together as dG_{bulk}) and the grain boundaries:

$$dG_{total} = dG_{bulk} + \sum \gamma_{GB} \, dA \tag{2.27}$$

where γ_{GB} is the grain boundary energy per unit area. Like solid–vapor interfacial (surface) energy, grain boundary (solid–solid interfacial) energy is highly dependent on orientation. Reliable grain boundary energies are difficult to obtain. As one might imagine, *measuring* grain boundary energies is difficult and tedious. One method for determining the *average* grain boundary energy involves annealing a sample under vacuum and measuring (via scanning electron microscopy or optical interferometry) the dihedral angle that forms where a grain boundary intersects a surface. This *thermal-grooving technique* yields the grain boundary energy via (Howe, 1997)

$$\gamma_{GB} = 2\gamma_{SV} \cos \frac{\theta_{SV}}{2} \tag{2.28}$$

where γ_{SV} is the surface energy for the solid–vapor and θ_{SV} is the dihedral angle. Analytical calculation of grain boundary energy by atomistic modeling is no simpler, since this requires accurate knowledge of the structure. The determination of grain boundary structure (including misorientation) requires careful sample preparation and high-resolution instruments. Nonetheless, some experimental work has been performed, and it is possible to make some generalized statements.

First, low-Σ boundaries tend to have relatively lower grain boundary energy, on average. The entropy term is undoubtedly the dominant contribution to the free energy in these cases. Grain boundary energy tends to be small for low-angle boundaries, and it has been found experimentally that as the misorientation angle, θ, exceeds $15°$, the grain boundary energy typically begins to level off, eventually becoming independent of the angle. However, one must be very cautious when attempting to correlate the three parameters, Σ, θ, and γ. Increases in Σ do often correspond to increases in θ, but not *all* high-angle boundaries are high Σ also. For example, the high angle coherent twin boundary ($\theta = 60°$) is a "low sigma" $\Sigma 3$ structure. Furthermore, the atoms at the interface of a high-angle coherent twin boundary are coherent, which results in a very low energy boundary. Similarly, although many low-Σ boundaries tend to have relatively low energies, the energy does not always show a simple relationship to Σ.

Polycrystals will contain a distribution of grain boundaries that minimize the total free energy due to the solid–solid interfaces. Determination of grain boundary energy is but one aspect of the broader, and extremely challenging, area of materials science and engineering concerned with microstructure evolution. The ultimate goal of simulating grain growth is, of course, the prediction of the final microstructure, including texture (distribution of crystallite orientations). Modeling allows investigators to examine the effects on microstructure of varying single parameters. It is important to be able to achieve specific textures because the macroscopic anisotropy to the properties of a polycrystalline substance will depend on the crystallite orientation distribution. Only materials crystallizing in the cubic crystal class or those with a completely random crystallite orientation distribution (regardless of the crystal class) have isotropic properties.

Grain boundaries are metastable configurations and therefore exhibit dynamical behavior in response to external forces (e.g., thermal, mechanical). When heat-treated, a polycrystal attempts to minimize its free energy by evolving toward a single crystal through a grain growth process. In this phenomenon, atoms move from the side of the grain boundary with a high free energy to the low-energy side. The free energy of the system can thus be reduced as the low-energy crystallites consume the high-energy crystallites. On annealing, polycrystalline grains with low dislocation densities will grow by consuming grains with high dislocation densities. Similarly, at curved boundaries atoms are more likely to diffuse from the convex side to the concave side in order to flatten the interface. In this way, the interfacial area and energy are reduced.

Materials processing is generally aimed at achieving a texture and/or grain boundary structure that will maximize a specific property of interest. Approaches taken with polycrystalline metals and alloys usually involve deformation (e.g., cold rolling) and annealing, or a combination thermomechanical method. These processes are discussed more fully in Chapter 7. Polycrystalline metals deform by mechanisms involving slip and, where slip is restricted, *rotation* of the individual grains. Both processes, of course, must satisfy the condition that the interfaces along which the grains are connected remain intact during deformation. As the extent of deformation increases, larger grains may break up into smaller grains of different orientations, giving rise to an orientation spread. Similar changes accompany hot working and annealing. In these procedures, the free energy of a metal decreases due to rearrangement of dislocations into lower-energy configurations of decreased dislocation density. This is termed *recovery*. However, in competition with this is *recrystallization*, in which the resulting grain structure and texture depend on the spatial distribution and orientation of the recrystallization nuclei. The competition between these two processes and their effect on texture and grain boundary structure is a hotly debated subject in metallurgy.

2.4. MASS TRANSPORT AND NONEQUILIBRIUM THERMODYNAMICS

Nothing in Section 2.3 concerning the thermodynamics and energetics of solid-state reactions, was different in principle from gas- or solution-phase systems. Indeed, thermodynamics is universal. The significance and respect of thermodynamics within the scientific community is well established. Albert Einstein once said (Klein, 1967):

> [A law] is more impressive the greater the simplicity of its premises, the more
> different are the kinds of things it relates, and the more extended its
> range of applicability. Therefore, the deep impression which classical
> thermodynamics made on me. It is the only physical theory of uni-
> versal content which I am convinced, that within the framework of
> applicability of it basic concepts will never be overthrown.

In fact, Einstein considered J. W. Gibbs, the architect of classical (equilibrium) thermodynamics, one of the greatest scientific intellects of modern time (Clark, 1971). However, although it deals with processes in which systems exchange energy or matter, classical thermodynamics is concerned directly only with equilibrium states. The criterion for equilibrium is that the state functions and hence chemical and physical properties are uniform and constant in time throughout the system. Crystal growth and, more generally, chemical reactions in solids proceed by atomic diffusion, or mass transport, in which the local concentrations and component activities change with time.

Other examples of transport properties include electrical and thermal conductivity. Transport of a physical quantity along a determined direction due to a gradient is an irreversible process by which a system transitions from a nonequilibrium state to an equilibrium state (e.g., compositional or thermal homogeneity). Therefore, it is outside the realm of equilibrium thermodynamics. (For this reason, equilibrium thermodynamics is more appropriately termed *thermostatics*.) Transport processes must be studied by irreversible thermodynamics.

In this book we are concerned only with mass transport, or diffusion, in solids. *Self-diffusion* refers to atoms diffusing among others of the same type (e.g., in pure metals). *Interdiffusion* is the diffusion of two dissimilar substances (a diffusion couple) into one another. *Impurity diffusion* refers to the transport of dilute solute atoms in a host solvent. In solids, diffusion is several orders of magnitude slower than in liquids or gases. Nonetheless, diffusional processes are important to study because they are basic to our understanding of how solid–liquid, solid–vapor, and solid–solid reactions proceed, as well as [solid–solid] phase transformations in single-phase materials.

Diffusion is a concentration gradient–induced process that follows *Fick's laws of diffusion*—macroscopic, or continuum-level, empirical relations—derived by the German physiologist Adolf Eugen Fick (1829–1901) in 1855 (Fick, 1855). *Fick's first law* is written as

$$J_i = -D_i \nabla n_i \tag{2.29}$$

where D is known as the *diffusivity* or *diffusion coefficient* (units of cm^2/s), $\mathbf{J}_i$ the *net* diffusional flux (the number of particles of species i crossing a unit area), and ∇n_i the concentration gradient of species i, which provides the *thermodynamic driving force*. It should be noted that ∇n_i might refer to a concentration gradient involving solute atoms, vacancies, or interstitials. The diffusional flux is actually a first-rank tensor, or vector; there exist three equations representing the diffusional fluxes along three principal axes. Accordingly, the diffusion coefficient is a second-rank tensor.

Equation 2.29 states that the flux is proportional to the concentration gradient (the higher the value for D_i, the larger $\mathbf{J}_i$ for the same ∇n_i) and that flow will cease only when the concentration is uniform. It was later proposed by Einstein (Einstein, 1905) that the force acting on a diffusing atom or ion is, in fact, the negative gradient of the chemical potential, which is dependent not only on concentration but on temperature and pressure as well. However, for our purposes, we consider T and P to be uniform, in which case the concentration gradient will determine the flow.

If the net rate, which is given by the difference between the forward and reverse reaction rates, is small compared to the two rates in opposite directions (i.e., if the system is not far from equilibrium), the forward and reverse processes occur with equal probability. This is the principle of microscopic reversibility, a concept of *classical irreversible thermodynamics*. A more general statement of this principle is that the mechanical equations of motion of the individual particles are invariant under the reversal of the time from t to $-t$. It implies that the nonequilibrium system can be divided up into volume elements, or subsystems, for which the state functions will have the same values as they would have in equilibrium. As a consequence of this local equilibrium, a linear relationship between the diffusion coefficient and chemical potential gradient in Eq. 2.29 may be assumed. Local equilibrium was, in fact, presumed to prevail by the German physical chemist Carl Wagner (1901–1977) in his early work relating the diffusivity of the rate-determining species with the reaction rate constant (Section 2.5), years before classical irreversible thermodynamics was generally accepted. It is now known that nonlinear response occurs in systems far from equilibrium, a phenomenon studied with *extended irreversible thermodynamics*.

Fick's first law represents steady-state diffusion. The concentration profile (the concentration as a function of location) is assumed constant with respect to time. In general, however, concentration profiles do change with time. To describe these non-steady-state diffusion processes use is made of *Fick's second law*, which is derived from the first law by combining it with the *continuity equation* $\left[\partial n_i/\partial t = -(J_{in} - J_{out}) = -\nabla J_i\right]$

$$\frac{\partial n_i}{\partial t} = \nabla(D_i \nabla n_i) \tag{2.30}$$

If the material is sufficiently homogeneous, D_i can be considered constant (independent of concentration), so that Eq. 2.30 reduces to

$$\frac{\partial n_i}{\partial t} = D_i \nabla^2 n_i \tag{2.31}$$

Fick's second law, which is also known simply as the *diffusion equation*, indicates that nonuniform gradients tend to become uniform.

The equations above imply that in the case of multiple diffusing species, each species will have its own intrinsic diffusion coefficient. Interestingly, when two dissimilar substances are joined together (forming a diffusion couple) and allowed to homogenize by interdiffusion, an apparent bulk flow occurs as a result of the differing intrinsic diffusion coefficients. Specifically, the side of the couple with the fastest diffuser shortens and the side with the slowest diffuser lengthens. Smigelskas and Kirkendall demonstrated that fine molybdenum wire could be used as a marker in the Cu–Zn system to provide an alternative description of interdiffusional processes between substances with dissimilar diffusion coefficients (Smigelskas and Kirkendall, 1947). A brass (Cu–Zn) bar, wound with molybdenum wire, was plated with copper metal. The specimen was annealed in a series of steps in which the movements of the molybdenum wires were recorded. The inert markers had moved from the interface toward the brass end of the specimen, which contained the fastest diffuser—zinc. This is now called the *Kirkendall effect*. A similar marker experiment had actually been performed by Hartley a year earlier while studying the diffusion of acetone in cellulose acetate (Hartley, 1946), but most metallurgists were not familiar with this work (Darken and Gurry, 1953).

Lawrence Stamper Darken (1909–1978) subsequently showed (Darken, 1948) how, in such a marker experiment, values for the intrinsic diffusion coefficients (e.g., D_{Cu} and D_{Zn}) could be obtained from a measurement of the marker velocity and a single diffusion coefficient, called the *interdiffusion coefficient* (e.g., $D = N_{Cu}D_{Zn} + N_{Zn}D_{Cu}$, where N_i are the molar fractions of species i), representative of the interdiffusion of the two species into one another. This quantity, sometimes called the *mutual* or *chemical diffusion coefficient*, is a more useful quantity than the more fundamental intrinsic diffusion coefficients from the standpoint of obtaining analytical solutions to real engineering diffusion problems. Interdiffusion, for example, is of obvious importance to the study of the chemical reaction kinetics. Indeed, studies have shown that interdiffusion is the rate-controlling step in the reaction between two solids.

2.5. CHEMICAL REACTION AND PHASE TRANSFORMATION KINETICS IN SOLIDS

Kinetics is the field of study dealing with the time rates of change and the mechanisms of reactions and transformations, topics we now take up. Chemical reactions that occur between different components in a single phase (gases or solutions) are called *homogeneous reactions*. In general, such chemical reactions are reversible. That is, reactants combine to produce products, which, in turn, combine to give back the reactants. It is possible to formulate the time dependence of either the forward or the reverse reactions in terms of the time rate of change of the concentration of the reactants or products. The overall reaction rate is the difference between the rates of the forward and reverse reactions. For example, consider the reaction between substances A and B to yield substance C:

$$a\mathrm{A(g)} + b\mathrm{B(g)} \rightleftharpoons c\mathrm{C(g)} \qquad (2.32)$$

The velocity of the forward reaction is

$$v_f = -\frac{dA}{dt} = -\frac{dB}{dt} = \frac{dC}{dt} \qquad (2.33)$$

By analogy, the velocity of the reverse reaction is

$$v_r = -\frac{dC}{dt} = \frac{dA}{dt} = \frac{dB}{dt} \qquad (2.34)$$

The net reaction has an overall rate given by

$$r = v_f - v_r \qquad (2.35)$$

At thermodynamic equilibrium, the forward and reverse rates are equal and no net reaction occurs.

The atomic processes that are occurring (under conditions of equilibrium or non equilibrium) may be described by statistical mechanics. Since we are assuming gaseous- or liquid-phase reactions, collision theory applies. In other words, the molecules must collide for a reaction to occur. Hence, the rate of a reaction is proportional to the number of collisions per second. This number, in turn, is proportional to the concentrations of the species combining. Normally, chemical equations, like the one given above, are stoichiometric statements. The coefficients in the equation give the number of moles of reactants and products. However, if (and only if) the chemical equation is also valid in terms of what the molecules are doing, the reaction is said to be an *elementary reaction*. In this case we can write the rate laws for the forward and reverse reactions as $v_f = k_f[A]^a[B]^b$ and $v_r = k_r[C]^c$, respectively, where k_f and k_r are rate constants and the exponents are equal to the coefficients in the balanced chemical equation. The net reaction rate, r, for an elementary reaction represented by Eq. 2.32 is thus

$$r = k_f[A]^a[B]^b - k_r[C]^c \qquad (2.36)$$

A rate law is a statement of the *law of mass action*, first formulated by the Norwegian chemist Peter Waage (1833–1900) and his brother-in-law, Cato Guldberg (1836–1902), in 1864 (Waage and Guldberg, 1864), building on the earlier work of French chemist Claude Berthollet (1748–1822). Unfortunately, it is dangerous to infer a rate law from a chemical equation unless it is known to be elementary. For a stoichiometric reaction involving a series of elementary reactions (i.e., a *mechanism*), the exponents in the rate law *may* be equal to the coefficients in the balanced chemical equation, but this need not be the case. There is no simple connection between a chemical equation for a stoichiometric reaction and the order of the reaction; the exponents must be determined by experiment.

If the reaction in Eq. 2.32 is elementary, the following condition holds at equilibrium:

$$v_f = v_r \qquad k_f[A]^a[B]^b = k_r[C]^c \qquad (2.37)$$

Hence, the product of the concentrations of the reaction products divided by the product of the concentrations of the starting materials (each raised to a power given by the coefficients in the chemical equation) equals a certain numerical value, K_c, characteristic of the reaction. For Eq. 2.32 this gives

$$K_c = \frac{k_f}{k_r} = \frac{[C]^c}{[A]^a [B]^b} \tag{2.38}$$

where the concentrations of the species used in the equilibrium constant expression *must* be the equilibrium values. Note that Eq 2.38 is valid regardless of whether the reaction is elementary or a complex multistep reaction. In other words, the thermodynamic equilibrium constant expression depends only on the stoichiometry of the reaction, not its mechanism. Known as the *mass action equilibrium principle*, this was first derived by Waage and Guldberg. Paradoxically, although their procedure always gives the correct result, the kinetic derivation is not generally valid. This is because the forward and reverse rate laws are not necessarily given by the denominator and numerator on the right-hand side of Eq. 2.38. Josiah Willard Gibbs and, shortly afterward, Jacobus Henricus van't Hoff (1852–1911), independently showed the satisfactory thermodynamic basis for the equilibrium constant expression.

The mass action law assumes that the reaction medium is homogeneous. In *heterogeneous reactions* (involving different substances in multiple phases), the densities and effective concentrations of pure condensed phases (liquids or solids) are constant. The concentrations of such species are set to unity in the equilibrium constant expression for such reactions. For example, given the following decomposition,

$$CaCO_3(s) \rightleftharpoons CaO(s) + CO_2(g)$$

the thermodynamic equilibrium constant expression is simply

$$K_c = [CO_2]$$

As with homogeneous reactions, rate laws for heterogeneous reactions are kinetic statements and must be determined experimentally. The exponents (called *orders*) of a rate law depend on the reaction mechanism.

The basic processes involved in solid-state chemical reactions are (1) interdiffusion of reacting species toward the interface, (2) nucleation of crystalline substances, and (3) growth of the product phase(s). When a chemical reaction proceeds through a series of steps, the slowest of these is the rate-limiting step. Overall, the rate-limiting step in solid–solid reactions is generally diffusion, due to the macroscopic lengths in the densely packed media through which species must migrate to reach the reaction interface. However, this is not the case with *every* solid-state reaction, and even when it is, nucleation still plays an important role in governing crystal morphology. For example, thin-film reactions are considered to be in the interface rate-controlled regime with the growth determined by the processes occurring at the reaction front. Similarly, for some bulk-phase transformations, including crystallization from solvents, solidification from melts, and crystallization from amorphous solids, either nucleation or the subsequent crystal growth process is the rate-limiting step. The observed crystalline morphology and/or microstructure are under kinetic control; the morphology that appears is the one with the maximum growth rate. Hence, in this section there are two limiting cases that we will be considering. The first case is the type controlled by surface or interface kinetics. These reactions may follow various types of rate laws, including linear, logarithmic, cubic, and fourth power. In the second limiting case, the rate-limiting factor is the diffusion of some particular species (reactant), across the phase boundary, to the other reactant. These reactions are said to be under *diffusion control* and a parabolic rate law is often observed, as first reported by Tammann (1920) and later explained theoretically by Wagner (1936). We end this chapter with a brief discussion of diffusion in amorphous solids and a few words about reaction mechanisms.

2.5.1. Surface and Interface-Controlled Reactions in Crystals

The growth rates of thin films on solid substrates, as well as the initial stages of bulk reactions of solids, are considered to be in the interface-controlled kinetic regime. In the latter case, nucleation is initially the overriding factor, before further reaction becomes rate-limited by counterdiffusion of the species involved. Nucleation is also the rate-limiting step in forming crystalline products from amorphous intermediates, melts, or solvents. Even when the rate-limiting step of crystal growth is diffusion, nucleation strongly influences crystal morphology. There are some useful analogies that we can draw upon from homogeneous reactions in understanding certain aspects of nucleation. For example, without knowledge of the exact rate law for a particular reaction, collision theory inspires us to anticipate that a homogeneous reaction should generally proceed faster if the concentrations of the reactants are higher. After all, the more molecules present in a given volume, the more likely they will combine. In a similar fashion, with heterogeneous reactions between solids the rate is proportional to the surface area of the particles. This is why finely divided metal powders are pyrophoric. It is also the reason that solid-state reactions proceed faster if the reacting species are fine-grained. High surface areas ensure that the maximum number of atoms or molecules of each substance are in contact.

Let us first consider homogeneous nucleation from the vapor phase. Classical nucleation theory supposes that at any given time a wide distribution of nuclei (clusters), ranging in size from one molecule and up, will exist as adatoms collide with and bind to the surface. Since the nuclei have an equilibrium vapor pressure, there is some critical cluster size and external pressure (*above* the growing cluster), below which the nuclei are unstable and disappear due to the evaporative loss (desorption) of atoms. At equilibrium, adsorption and desorption occur at the same rates and no net crystal growth is observed; crystal growth is a nonequilibrium process. Crystal growth can be categorized into different models, based on the need for nucleation. In the Volmer–Weber–Kossel–Stranski model of crystal growth, two-dimensional nuclei must first be formed on a flat surface (Volmer and Weber, 1926; Kossel, 1927, Stranski and Krastanov, 1938). This is discussed further in Section 3.3. Once formed, adatoms preferentially attach at the ledges (edges) of the nuclei boundaries, for reasons discussed earlier (Section 2.3.1.2.). However, a new two-dimensional nucleus is required for formation of the next layer. The crystal growth rate is thus dependent on the nucleation rate, which, in turn, is dependent on the degree of supersaturation.

By comparison, in the Burton–Cabrera–Frank model, nucleation sites are provided by screw dislocations that intersect the crystal surface, which supply an inexhaustible source of the necessary monolayer steps. This allows for crystal growth at much lower levels of supersaturation. In this model, the edge from a screw dislocation grows laterally and rotates about itself, resulting in an Archimedean growth spiral of steps (Figure 1.5). Once a moving step has reached the initial position of the directly underlying step, the spiral has made a complete turn. This rotation increases the height of the spiral, causing the macroscopic face to advance. In a seminal paper which assumed that mass transfer is governed by a set of adatom diffusion equations on each step, Burton et al. (1951) showed that the rate of growth of a face is proportional to the dimensions of the screw dislocation (the distance between its steps) and to the residence time (lifetime before desorption) of an attaching adatom on the crystal surface.

Kinetic models for crystal growth usually quantify the *interface velocity*, which is a direct measure of the crystal growth rate. The rate of stable cluster formation on a surface is proportional to the equilibrium surface concentration of clusters (of critical radius r_c), n_{rc}, and the frequency of addition of adatoms to these clusters, v_a (Howe, 1997):

$$I_c \propto n_{rc} v_a \tag{2.39}$$

The equilibrium cluster concentration is given by

$$n_{rc} = n_a \exp\left(-\frac{\Delta G_c}{kT}\right) \tag{2.40}$$

where n_a is the number of adatoms per unit area of surface and ΔG_c is the free energy of formation of clusters with radius r_c. Sophisticated experiments are required for the determination of nucleation rates. It will be instructive for us to model crystal growth rate in terms of macroscopic observables when the surface reactivity is indeed the rate-limiting process.

This is a comparatively simple task for a bimolecular reaction using a mechanism first suggested by Irving Langmuir (1881–1957) and developed further by Sir Cyril Norman Hinshelwood (1897–1967) for heterogeneous catalysis involving two adsorbed species on a solid surface (Langmuir, 1921; Hinshelwood, 1926). This is discussed more in Section 3.2. Incidentally, both Langmuir and Hinshelwood were subsequently awarded the in Nobel Prize Chemistry for their research in surface chemistry (Langmuir in 1932) and into the mechanisms of chemical reactions (Hinshelwood in 1956). For now, we follow the treatment presented in most undergraduate physical chemistry textbooks (see, e.g., Laidler and Meiser, 1982). Consider the deposition of a hypothetical AB compound on a solid substrate from A and B vapor. When an A atom is adsorbed, it occupies a distinct surface site, in competition with the B atoms. The rate of surface adsorption of A is therefore proportional to the partial pressure of A, p_A, and to the fraction of available (unoccupied) surface sites given by $(1 - \theta_A - \theta_B)$. We can express this mathematically as

$$v_a^A = \frac{d\theta_A}{dt} = k_a p_A (1 - \theta_A - \theta_B) \tag{2.41}$$

Similarly, the rate of desorption, or evaporation, of A is proportional to the number of sites occupied by that species:

$$v_d^A = -\frac{d\theta_A}{dt} = k_d \theta_A \tag{2.42}$$

The B adsorption and desorption rates are

$$v_a^B = \frac{d\theta_B}{dt} = k_a p_B (1 - \theta_B - \theta_A) \tag{2.43}$$

$$v_d^B = -\frac{d\theta_B}{dt} = k_d \theta_B \tag{2.44}$$

At equilibrium, the adsorption and desorption rates of a particular species are equal. Hence, equating Eqs. 2.41 and 2.42 yields

$$\frac{\theta_A}{1 - \theta_A - \theta_B} = \frac{k_a}{k_d} p_A = K_A p_A \tag{2.45}$$

where K_A is equal to k_a/k_d and is known as the *adsorption coefficient* for A. Similarly, equating Eqs. 2.43 and 2.44 leads to

$$\frac{\theta_B}{1 - \theta_A - \theta_B} = \frac{k_a}{k_d} p_B = K_B p_B \tag{2.46}$$

Equations 2.45 and 2.46 are two simultaneous equations that can be solved to give the fractions of surface sites covered by A and B:

$$\theta_B = \frac{K_B p_B}{1 + K_A p_A + K_B p_B} \tag{2.47}$$

$$\theta_A = \frac{K_A p_A}{1 + K_A p_A + K_B p_B} \tag{2.48}$$

In general, equations such as Eqs. 2.47 and 2.48, which relate the amount of a substance adsorbed on a surface to the concentration in the gas phase at a fixed temperature, are known as *adsorption isotherms*. Now, the rate of the bimolecular reaction is proportional to the probability that the two species are adsorbed on neighboring sites, which, in turn, is proportional to the surface fractions covered by each of the species. Hence, we have

$$r = k_r \theta_A \theta_B \tag{2.49}$$

where k_r is the rate constant. Combining Eqs. 2.47, 2.48, and 2.49 gives

$$r = \frac{k_r K_A K_B p_A p_B}{(1 + K_A p_A + K_B p_B)^2} \tag{2.50}$$

The preceding treatment is, undoubtedly, an oversimplification. For example, many diatomic molecules dissociate upon adsorption (e.g., H_2, SiH, GeH). Each atom from the dissociated molecule then occupies its own distinct surface site and this naturally changes the rate law expression. When these types of details are accounted for, the Langmuir–Hinshelwood mechanism has been very successful at explaining the growth rates of a number of thin-film chemical vapor deposition (CVD) processes. However, more important, our treatment served to illustrate how crystal growth from the vapor phase can be related to macroscopic observables: namely, the partial pressures of the reacting species.

We've been discussing crystalline film growth from vapors. The exact structure of a surface also has important consequences on a solid's chemical reactivity. This is prominently displayed, for example, during the etching of anisotropic single crystals (e.g., Si, GaAs). It is found that different etch rates occur in different crystallographic directions. This anisotropic etching is due not to differing diffusion rates of etching reactants to and from the surface, but by different reaction rates on different crystallographic planes (diffusion-controlled etching can mask the effects of anisotropic surface kinetics, however, and result in isotropic etching). Similarly, certain silicon-rich reconstructed SiC surfaces are known to be much more reactive to hydrogen and oxygen (by three orders of magnitude) than pure silicon. This is because the reconstructed surface has additional planes of Si atoms on top of the Si-terminated unreconstructed surface. A larger number of silicon atoms are present per unit volume of surface than in pure silicon.

What about solid–liquid and solid–solid reactions and phase transformations? Homogeneous nucleation from an undercooled melt frequently results in a thermodynamically metastable bulk phase with a low kinetic barrier, in accordance with *Ostwald's step rule* (Ostwald, 1897). The homogeneous nucleation rate from an undercooled melt is related to the crystal-melt interfacial energy. However, direct measurement of γ (e.g., contact angle) is difficult. Turnbull found that the surface or interfacial energy is roughly proportional to the latent heat of fusion and inversely proportional to the negative two-thirds power of the density (Turnbull, 1950), where the proportionality constant (the Turnbull coefficient) is approximately 0.45 for metals and 0.32 for nonmetals. Molecular dynamics simulations have been used to show the molecular origin of Turnbull's rule for close-packed metals interacting with continuous potentials, such as the

Lennard-Jones and inverse-power repulsive potentials (Davidchack and Laird, 2005). Using these simulations, they were further able to show how both the magnitude and anisotropy in the surface or interfacial energy scale with potential.

While the magnitude of the interfacial energy is important in controlling the rate kinetics of crystal growth from a melt, the anisotropy plays a critical role in determining the exact crystal morphology. The anisotropy in γ gives rise to an equilibrium shape known as the *Wulff shape*, which was discussed in Section 2.3.1. Under isothermal growth conditions, such as when long-range transport of energy (i.e., heat) and/or mass (i.e., solute) are very fast, a crystal will nucleate in the equilibrium Wulff shape and subsequently grow asymptotically toward morphologies containing different corners and facets (Sekerka, 2005). Nonisothermal temperature fields and nonuniform compositional fields, however, may give rise to intricate crystal morphologies and even morphological instabilities leading to highly convoluted surfaces, as in dendrites. This happens in the planar interface-destabilizing event known as *constitutional undercooling*, which arises from variations in liquid composition. When the temperature gradient across an interface is positive (i.e., the solid is below the freezing temperature, the interface is at the freezing temperature, and the liquid is above the freezing temperature), a planar solidification front is most stable. However, as a solidification front advances, solute is redistributed at the interface. Commonly, solute is rejected into the liquid, where it accumulates into the solute boundary layer. Depending on the temperature gradient, such liquid may be undercooled below its freezing temperature, even though it is hotter than liquid at the front. Constitutional undercooling is difficult to avoid except with very slow growth rates.

For solid–solid reactions, nucleation is obviously facilitated in such processes if there is a structural similarity between the products and one or more of the starting materials in order to reduce the reconstruction necessary for continued growth. Thus, *epitactic reactions* (those controlled by the surface or interface structure) and *topotactic reactions* (those controlled by the bulk crystal structure) generally proceed under milder conditions. For those cases where reconstruction or relaxation is required, there is evidence that suggests an initially incoherent product phase forms, or nucleates, at the interface between two reacting solids, which subsequently becomes more coherent as the product phase grows. For example, transmission electron microscopy studies have found that, in the reaction between solid MgO (rock salt structure) and sapphire (Al_2O_3), the grains of the $MgAl_2O_4$ thin-film product phase tilt after initial formation to accommodate the lattice misfit between the three phases (Sieber et al., 1996). Similarly, powder x-ray diffraction experiments on quenched disordered Li_2TiO_3–Li_3NbO_4 solid solutions reveal the appearance of broad supercell reflections upon annealing, which narrow with increased annealing time. This indicates that the disordered phase transforms into the cation-ordered phase by a nucleation and growth mechanism. The reader should be aware that these types of kinetic experiments are very difficult to perform and reproduce accurately.

2.5.2. Diffusion-Controlled Chemical Reactions in Crystals

As stated earlier, heterogeneous reactions may be rate limited, or rate controlled, in two distinct ways. The first (e.g., as in thin films) is by surface or interface reactivity. The second (e.g., as in bulk solid reactions) is by volume diffusion, which we now discuss. The German physical chemist Carl Wagner, along with physicist Walter Schottky, studied the thermodynamics of point defect formation in solids and later proposed that diffusion (mass transport) proceeds by point defect–mediated mechanisms (Wagner and Schottky, 1931). In these processes, species migrate through a crystalline substance via a particular type of point defect known as a *vacancy*, which may be of two types, a lattice site vacancy or interstitial site vacancy. The vacancies may be created in several ways, including:

1. The substitution of aliovalent impurities on a sublattice.
2. Thermal displacement of an ion from a sublattice site to the crystal surface (*Schottky defect*) or to an interstitial site (*Frenkel defect*).
3. In transition metal compounds where the transition metal is stable in more than one oxidation state, by nonstoichiometry. For example, in CaO and FeO (both of which possess the rock salt structure) Schottky defects are thermally formed. Additionally, however, FeO (wüstite) exhibits mixed valence and, accordingly, has Fe^{2+} and Fe^{3+} ions and vacancies distributed over the cation sublattice.

Diffusion of atoms or ions in crystalline solids can occur by at least three possible mechanisms, as shown schematically in Figure 2.7. In some solids, transport proceeds primarily by the *vacancy mechanism*, in which an atom jumps into an adjacent, energetically equivalent vacant lattice site. The vacancy mechanism is generally much slower than the interstitial mechanism (discussed below). Nonetheless, it is thought to be responsible for *self-diffusion* in all pure metals and for most substitutional alloys (Shewmon, 1989).

When thermal oscillations become large enough, an atom already in an interstitial site, or one that migrates there, can move among energetically equivalent vacant interstitial sites without permanently displacing the solvent (lattice) atoms. Termed the *direct interstitial mechanism*, this is known to occur in interstitial alloys and in some substitutional alloys when the substitutional atoms spend a large fraction of their time on interstitial sites. Although it is found empirically that only solute atoms with a radius less than 0.59 that of the solvent atom dissolve interstitially, diffusion studies have shown that solute atoms with radii up to 0.85 that of the solvent can spend enough time in interstitial sites for the interstitial mechanism to dominate solute transport (Shewmon, 1989).

Atoms larger than this would produce excessively large structural distortions if they were to diffuse by the direct interstitial mechanism. Hence, in these cases diffusion tends to occur by what is known as the *interstitialcy mechanism*. In this process, the large atom that initially moves into an interstitial position displaces one of its nearest neighbors into an interstitial position and takes the displaced

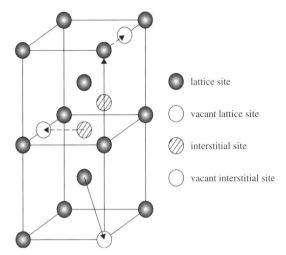

lattice site

vacant lattice site

interstitial site

vacant interstitial site

Figure 2.7 Diffusion of atoms or ions in crystalline solids can occur by at least three possible mechanisms illustrated here. In the vacancy mechanism (bottom arrow), an atom in a lattice site jumps to an adjacent vacant lattice site. In the interstitial mechanism (middle arrow), an interstitial atom jumps to an adjacent vacant interstitial site. In the intersitialcy mechanism (top two arrows), an interstitial atom pushes an atom residing in a lattice site into an adjacent vacant interstitial site and occupies the displaced atom's former site. (After Lalena and Cleary, 2005. Copyright © John Wiley & Sons, Inc. Reproduced with permission.)

atom's place in the lattice. This is the dominant diffusion process for the silver ion in AgBr, where an octahedral site Ag^+ ion moves into a tetrahedral site, then displaces a neighboring Ag^+ ion into a tetrahedral site and takes the displaced Ag^+ ion's formal position in the lattice.

To treat solid–solid reactions, Wagner introduced the concepts of local equilibrium and counterdiffusion of cations between the solids. The latter concept forms the basis for Darken's subsequent introduction of the interdiffusion coefficient, which was discussed in Section 2.4. To maintain a state of local equilibrium, the exchange fluxes across the interface must be large compared to the net transport of matter across the boundary. This is analogous to the criterion that the forward and reverse reaction rates be the same, or nearly so, for a reversible reaction to be considered at thermodynamic equilibrium.

Wagner's model is best illustrated by way of example. For this purpose we reproduce the detailed kinetic analysis of the $CaO–FeO–Ca_2Fe_2O_5$ system by Fukuyama (Fukuyama et al., 2002). In the reaction between FeO and CaO, the product phase $Ca_2Fe_2O_5$ is formed initially at the CaO–FeO interface, where Fe^{3+} is highly stabilized. Subsequent growth of this product phase occurs via diffusion of Ca^{2+} in FeO (away from the $FeO–Ca_2Fe_2O_5$ interface), diffusion of Fe^{2+} in FeO (toward the $FeO–Ca_2Fe_2O_5$ interface), and diffusion of Fe^{3+} through $Ca_2Fe_2O_5$ (toward CaO). The iron vacancies created by the formation of

Carl Wilhelm Wagner (1901–1977) received his Ph.D. in physical chemistry in 1924 from the University of Leipzig. In 1929, he coauthored the famous book *Thermodynamik* with W. Schottky and H. Ulich. A year later, Wagner and Schottky studied the thermodynamics of point defect formation in solids. They then went on to propose the point defect–mediated mechanism to mass transport in solids, and Wagner extended the analysis to electronic defects. Subsequently, Wagner introduced the concept of local equilibrium, an oxidation rate theory, and the concept of counter diffusion of cations in the reaction between solids to form a high-order (e.g., tertiary) phase. Wagner's work forms the basis of what has become known as *defect chemistry*. Because of the immense importance of this area of study to the field, Wagner is considered by many to be the "Father of Solid-State Chemistry." In 1933 he spent one year at the University of Hamburg and then became professor of physical chemistry at the Technische Hochschule Darmstadt, where he stayed until 1945 and proposed the concept of counter diffusion of cations in solid state reactions and the mechanism of ion conduction. From 1945 to 1949, Wagner was a scientific advisor at Fort Bliss, Texas. From 1949 to 1958, he was a Professor of Metallurgy at the Massachusetts Institute of Technology. In 1958, he moved back to Germany, assuming the position of Director of the Max Planck Institute of Physical Chemistry in Göttingen, retiring in 1966. Wagner was a member of several national academies and was elected to the U.S. National Academy of Sciences in 1967.

Source: M. Martin, Life and achievements of Carl Wagner, 100th birthday, *Solid State Ionics*, **2002**, *15–17*, 152–153.

$Ca_2Fe_2O_5$ diffuse away from the $CaO–Ca_2Fe_2O_5$ interface toward the $FeO–Ca_2Fe_2O_5$ interface. This is illustrated in Figure 2.8. Although the product is formed at the $CaO–Ca_2Fe_2O_5$ interface, it decomposes simultaneously at the

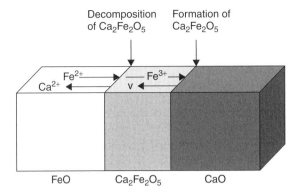

Figure 2.8   Atomic migration across the $FeO-Ca_2Fe_2O_5-CaO$ interface.

$FeO-Ca_2Fe_2O_5$ interface due to dissolution in the FeO phase (Fukuyama et al., 2002). The overall chemical reaction may be expressed as

$$2CaO(s) + 2FeO(s) + \frac{1}{2}O_2(g) \rightleftharpoons Ca_2Fe_2O_5(s)$$

The growth rate of $Ca_2Fe_2O_5$ is given by the difference between the forward (formation) and reverse (dissolution) rates:

$$r = \frac{dx}{dt} = j_{Fe^{3+}}V_m - j_{Ca^{2+}}V_m \tag{2.51}$$

where j_i is the flux of component i, V_m is the molar volume of $Ca_2Fe_2O_5$, and dx/dt is the interface distance per unit time. The two terms on the right-hand side are given by:

$$j_{Fe^{3+}}V_m = \frac{V_m}{x}k_1 \tag{2.52}$$

$$j_{Ca^{2+}}V_m = \frac{V_m}{x}k_2 \tag{2.53}$$

Substitution of Eqs. 2.52 and 2.53 into Eq. 2.51 and integration yields

$$x^2 = 2(V_m k_1 - V_m k_2)t \tag{2.54}$$

Thus, the $Ca_2Fe_2O_5$ product grows according to a parabolic rate law (as observed experimentally), implying a diffusion-controlled process. Wagner derived the relationship between the reaction rate constant and the diffusivity of a rate-determining species (Wagner, 1936). Assuming that the intrinsic diffusivity of

Fe^{3+} is much greater than that of Ca^{2+} (i.e., Fe^{3+} is the rate-determining species), k_1 may be expressed as

$$k_1 = (RT)^{-1} \int_{\mu_{Fe}(FeO)}^{\mu_{Fe}(CaO)} D_{Fe^{3+}} c_{Fe^{3+}} \, d\mu_{Fe} \tag{2.55}$$

where $\mu_{Fe}(FeO)$ and $\mu_{Fe}(CaO)$ are the chemical potentials of Fe at the CaO–$Ca_2Fe_2O_5$ and FeO–$Ca_2Fe_2O_5$ interfaces, respectively, and $c_{Fe^{3+}}$ is the concentration (mol/m^3) of Fe^{3+} in the $Ca_2Fe_2O_5$ phase. The second rate constant, k_2, can be related to $j_{Ca^{2+}}$ via Fick's first law (Eq. 2.29) in the FeO solid solution range as

$$k_2 = j_{Ca^{2+}} x = -D \frac{dc_{Ca^{2+}}}{dx} x \tag{2.56}$$

where D is the interdiffusivity given by $D = N_{FeO} D_{Ca^{2+}} + N_{CaO} D_{Fe^{2+}}$, in which N_{FeO} and N_{CaO} are the molar fractions of FeO and CaO. Two other well-known applications of Wagner's parabolic growth model are the high-temperature oxidation of metals and spinel oxide (AB_2O_4) growth from AO and B_2O_3 after the spinel product reaches a certain volume.

Finally, we should point out that grain boundaries also contribute to the chemical properties exhibited by a material. The kinetic rational for this is the fact that atomic diffusion normally occurs more rapidly through the grain boundary than it does through the grains. This is because the interface is usually less dense than the grains. For example, a metal will usually oxidize or corrode more quickly at the grain boundaries, a condition known as *intergranular corrosion*. The oxidation rate may be very dependent on the grain boundary structure. However, because the grain boundary volume is a small fraction of the total volume of a sample, intragranular diffusion is usually the dominant mass transport process, except for very small grain sizes or at low temperatures.

2.5.3. Atomic Diffusion in Amorphous Substances

What happens when we consider microscopic, or atomistic, descriptions of diffusion without the convenient framework of a lattice? Amorphous solids display no long-range translation order characteristic of crystals or long-range orientational order characteristic of quasicrystals. However, there are different types of amorphous solids, including (1) polymers, which contain regular "crystalline" portions–(the polymer chains themselves) and noncrystalline portions in between; (2) network glasses, or vitreous solids, obtained from melts (i.e., an amorphous substance that exhibits a glass transition; it softens and flows upon heating), which possess short-range order but not long-range order; and (3) nonglassy amorphous solids (i.e., an amorphous substance that does not exhibit a glass transition).

All of these different types of amorphous materials posses short-range order, usually in the form of local coordination environments around individual atoms (Voronoi polyhedra). For example, the local tetrahedral coordination of silicon atoms in SiO_2 glass constitutes short range order. Point defects can be defined in amorphous solids as anomalies in the valence (charge)/Voronoi volume ratio. Nevertheless, most studies of diffusion in amorphous solids seem to indicate a more direct mechanism, not involving point defects. Here we consider only the most rudimentary explanation for diffusion in amorphous materials.

The fundamental mechanisms of mass transport in fluids (gases and liquids) and amorphous solids are still not entirely clear. Atomic dynamics in condensed matter is influenced by the *cage effect*, where diffusion is coupled to structural relaxations. Each atom finds itself in a cage formed by its immediate neighbors, which influences its diffusive motion. Kinetic expressions describing diffusion mechanisms are based on not-so-very-realistic hard-sphere models that are insensitive to long-range atomic interactions. Recently, Mikhail Dzugutov, at the Royal Institute of Technology in Sweden, has proposed a universal scaling law for atomic diffusion in condensed matter, which, by definition, is applicable to crystalline and amorphous solids as well as liquids (Dzugutov, 1996). This theory is based on the argument that the frequency of local structural relaxations is proportional to the excess entropy, S, the number of accessible configurations per atom (or, equivalently, the difference between the system's thermodynamic entropy and that of the equivalent ideal gas). In an equilibrium system, this number is reduced by a factor e^S. Hence, the diffusion constant, D, and e^S must be related by a universal linear relationship. Dzugutov's results, obtained from molecular dynamics simulations, imply that atomic diffusion can be accounted for by the frequency of binary collisions and the excess entropy. It was found that throughout the range of S corresponding to the liquid domain, the results were applicable to other types of condensed matter (e.g. quasicrystals and ionic conductors) and could be described by

$$D = 0.049e^s \tag{2.57}$$

where S is restricted to a two-particle approximation.

2.5.4. Reaction Mechanisms in Solid-State Chemistry

Although knowledge of the mechanism of a reaction—the pathway, or series of elementary steps making up the reaction—is not essential to determining the overall rate law, the exponents (called *orders*) of those rate laws depend on the reaction mechanism. Moreover, knowledge of reaction mechanisms promotes optimization of synthetic parameters for a particular system as well as strengthening our predictive power for comparable systems. The former is important not only for speeding up reactions yielding thermodynamically favored products but for obtaining kinetically stable metastable phases. Indeed, synthetic organic chemistry is built on techniques that enable preparation of desired products (by and far, molecules) in high yield through their high kinetic selectivity. Specifically, kinetic control allows one to optimize the yield of each intermediate step. A long-time goal of materials chemists has been to develop similar kinetic control in synthesizing new solids.

For solid-state chemical reactions, the basic mechanism is known: (1) inter-diffusion of reactants to the interface; (2) nucleation of crystalline intermediates and (3) growth of the product phase(s). For bulk reactions in solids, because of the densely-packed macroscopic lengths over which the species must migrate to reach the interface, diffusion is the overall rate-limiting process. Consequently, each of the aforementioned processes may be observed simultaneously at various rates, depending on the surfaces reacting (Fukuto et al., 1994). An additional complicating factor arising from the dominance of interdiffusion is the eventual nucleation and growth of all thermodynamically stable phases. The desired product yield thus suffers to the extent that such multiphase products are obtained. Of most importance, however, is that the complex set of circumstances just described has prevented good understanding of the nucleation and growth processes. This is unfortunate because it is control of the nucleation step that must be obtained to gain kinetic control.

Investigation of the kinetics in solid-state reactions have been simplified in some experiments by utilizing markers with atomic-resolution transmission electron microscopy and specific crystallographic orientations. In certain cases, this method has allowed elucidation of the reaction mechanism. An often-cited example is the formation of $MgAl_2O_4$ spinel. A recent study investigated the solid-state formation of $MgAl_2O_4$ thin films from a (001)-oriented MgO substrate and Al_2O_3 vapor as well as between a deposited MgO film and a ($1\bar{1}02$) oriented Al_2O_3 substrate (Sieber et al., 1996). In this system, the orientation (texture) and morphology of the $MgAl_2O_4$ thin film product is different for the two reactions, which has been explained by different interface-controlled reaction mechanisms.

On MgO(001) substrates, the reaction starts with the epitaxial formation of spinel product since the anion FCC sublattices of each phase are similar. As reaction proceeds, polycrystalline product is obtained, owing to a slight (4.1%) lattice misfit between MgO and $MgAl_2O_4$, which gives rise to orientational degrees of freedom for the spinel nuclei. Prior to coalescence, the nuclei rotate and tilt, resulting in low-angle grain boundaries instead of an exact cube-to-cube orientation.

With deposited MgO film and ($1\bar{1}02$)-oriented Al_2O_3 substrates, the difference between the hexagonal close-packed (HCP) oxygen sublattice of Al_2O_3 and the FCC oxygen sublattice of $MgAl_2O_4$ requires reconstructive steps during the growth of the spinel, which depends on the coherency of the interface. Before reaction, the interface is incoherent due to the chosen crystallographic orientations. However, after initial growth, the $MgAl_2O_4$ grains tilt by $5.8°$ around their [010] axis, producing coherent interfaces. Although the detailed mechanism of this tilt is not clear, the coherent reaction front facilitates reconstruction of the HCP oxygen sublattice of Al_2O_3 into the FCC oxygen sublattice of $MgAl_2O_4$.

In the last decade, progress has also been made with using superlattices as templates, or structure-directing agents, to kinetically control solid-state reactions. This is accomplished by allowing interdiffusion to reach completion before the occurrence of heterogeneous nucleation, thus trapping the system in the metastable state with the lowest nucleation barrier. David Johnson's group at the University of Oregon has demonstrated success by several methods in different systems. One way was the synthesis of the ternary compound $Cu_xMo_6Se_8$, even though the binary compounds (e.g., $MoSe_2$) are more thermodynamically stable (Fister et al., 1994). This was achieved by utilizing multilayer heterostructures (superlattices) of composition $Cu_2Mo_6Se_8$, where the interdiffusion distances are so small that interfacial nucleation of the binary compounds can be avoided. They have also prepared metastable amorphous Nb_5Se_4 (Fukuto et al., 1994), before the onset of homogeneous crystallization, by using a superlattice comprised of niobium and selenium multilayers with a critical thickness below which the kinetically stable intermediate is formed. Finally, by studying the composition dependence in the nucleation energy of InSe and recalling the empirical rule that the easiest compound to nucleate is the most stable phase closest in composition to the glassy phase, they were able to show that it is possible to use the composition of an amorphous intermediate to change nucleation energies (Oyelaran et al., 1996). Despite the important progress that has been made, much more work is needed before the widespread use of kinetic control will be possible in the preparation of bulk materials.

REFERENCES

Ashcroft, N. W.; Mermin, N. D. *Solid State Physics*, CBS Publishing, Tokyo, **1981**.

Born, M.; Landé, A. *Sitzungsber. Deutsch. Akad. Wiss. Berlin*. **1918**, *45*, 1048.

Born, M.; Stern, O. Sitz. *Preuss. Akad. Wiss.* **1919**, 901.

Burton, W. K.; Cabrera, N.; Frank, F. C. *Philos. Trans. R. Soc. London A.* **1951**, *243*, 299.

Cahn, J. W. *Acta Metall.* **1962**, *10*, 179.

Clark, R. W. *Einstein: The Life and Times*, Avon Books, New York, **1971**. p. 754.

Curie, P. *Bull. Soc. Fr. Minéral.* **1885**, *8*, 145–150.

Darken, L. *Trans. AIME.* **1948**, *175*, 184.

Darken, L. S, Gurry, R. W. *Physical Chemistry of Metals*, McGraw-Hill, New York, **1953**.

Davidchack, R. L.; Laird, B.B. *Phys. Rev. Lett.* **2005**, *94*, 086102.

Dong, L.; Deshippe, J.; Sankey, O.; Soignard, E., McMillan, E.F. *Phys. Rev.B* **2003**, *67*, 94104.

Dzugutov, M. *Nature*. **1996**, *381*, 137.

Einstein, A. *Ann. Phys.* **1905**, *17*, 549.

Ellinghmam, H. J. T. *J. Soc. Chem. Ind.* **1944**, *63*, 125.

Fick, A. Pogg. *Ann. der Physik and Chemie*. **1855**, *94*, 59.

Fister, L.; Johnson, D. C.; Brown, R. *J. Am. Chem. Soc.* **1994**, *116*, 629.

Frenken, J. W. M.; Stoltze, P. *Phys. Rev. Lett.* **1999**, *82*(17), 3500.

Fukuto, M.; Hornbostel, M. D.; Johnson, D. C. *J. Am. Chem. Soc.* **1994**, *116*, 9136.

Fukuyama, H.; Hossain, M. K.; Nagata, K. *Met. Mater. Trans. B.* **2002**, *33*, 257.

Gibbs, J. W., *On the Equilibrium of Heterogeneous Substances*, **1878**. In *Collected Works*, Vol. *1*, New York: Longmans, Green, 1898, pp. 55–353.

Hartley, G.S. *Trans. Faraday Soc*. **1946**, *42*, 6. Green Longmans, New York; 1898, pp. 55–353.

Hildebrand, J. H. *J. Am. Chem. Soc*. **1929**, *51*, 66.

Hill, T. L. *J. Chem. Phys*. **1962**, *36*, 3182.

Hill, T. L. *Thermodynamics of Small Systems*, Part I, W.A. Benjamin, New York, **1963**.

Hill, T. L. *Thermodynamics of Small Systems*, Part II, W.A. Benjamin, New York, **1964**.

Hill, T. L. *Nano Lett*. **2001**, *1*, 273.

Hinshelwood, C. N. *Kinetics of Chemical Change*, Clarendon Press, Oxford, **1926**.

Howe, J. M. *Interfaces in Materials*, Wiley, New York, **1997**.

Johnson, Q. C.; Templeton, D. H. *J. Chem. Phys*. **1962**, *34*, 2004.

Klein, M. J. *Science*. **1967**, *157*, 509.

Kossel, W. *Nachr. Ges. Wiss. Goettingen Math. Phys. Kl*. **1927**, 135.

Laidler, K. J.; Meiser, J. H. *Physical Chemistry*, Benjamin-Cummings, Menlo Park, CA, **1982**.

Lalena, J. N.; Cleary, D. A. *Principles of Inorganic Materials Design*, Wiley, Hoboken, NJ, **2005**, pp. 335–368.

Lalena, J. N.; Dean, N. F.; Weiser, M. W. J. *Electron. Mater*. **2002**, *31*, 1244.

Langmuir, I. *Trans. Faraday Soc*. **1921**, *17*, 621.

Madelung, E. *Phys,, Z*. **1918**, *19*, 524.

Madelung, E. *Phys. Z*. **1919**, *20*, 494.

Ostwald, W. *Z. Phys. Chem*. **1897**, *22*, 289.

Oyelaran, O.; Novet, T.; Johnson, C. D.; Johnson, D. C. *J. Am. Chem. Soc*. **1996**, *118*, 2422.

Reed, T. B. *Free Energy of Formation of Binary Compounds: An Atlas of Charts for High-Temperature Chemical Calculations*, MIT Press, Cambridge, MA, **1971**.

Reuter, K., Scheffler, M. *Phys. Rev. B*. **2001**, *65*, 35406.

Runyan, W. R.; Bean, K. E. *Semiconductor Integrated Circuit Processing Technology*, Addison-Wesley, Reading, MA, **1990**.

Sekerka, R. F. *Cryst. Res. Technol*. **2005**, *40*(4–5), 291–306.

Shewmon, P. *Diffusion in Solids*, The Minerals, Metals, and Materials Society, Warrendale, PA, **1989**.

Sieber, H.; Hesse, D.; Pan, X.; Senz, S.; Heydenreich, J. *Z. Anorg. Allg. Chem*. **1996**, *622*, 1658.

Smigelskas, A.; Kirkendall, E. *Trans. AIME*. **1947**, *171*, 130.

Soignard, E.; Somayazulu, M.; Dong, J.; Sankey, O. F.; McMillan, P. F. *Solid State Common*. **2001**, *120*, 237.

Stranski, I. N.; Krastanov, L. Sitzungsber. *Akad.Wiss. Wien. Math. Nat. Kl IIb*. **1938**, 797.

Tammann, G. *Z. Anorg. Allg. Chem*. **1920**, *111*, 78.

TurnBull, D. *J. Chem. Phys*. **1950**, *18*, 198.

Venables, J. A. *Introduction to Surface and Thin Film Processes*, Cambridge University Press, Cambridge, **2000**.

Volmer, M.; Weber, A. *Z. Phys. Chem*. **1926**, *119*, 277.

Waage, P.; Guldberg, C. M. *Forhandl. Vidensk. Selsk. Christiana*. **1864**, 35.

Wagner, C. Z. *Phys. Chem. B.* **1936**, *34*, 309.

Wagner, C.; Schottky, W. *Theorie dergeordneten Mischphasen II (Diffusionsvorgänge)*, Bodenstein-Festband, **1931**, p.177.

Wulff, G.Z. *Kristallogr*. **1901**, *34*, 449.

Yamada, M. *Phys.Z.* **1923**, *24*, 364.

Yamada, M. *Phys.Z.* **1924**, *25*, 52.

Yu, D. K.; Bonzel, H. P.; Scheffler, M. *New, J. Phys*. **2006**, *8*, 65.

3 Solid–Vapor Reactions

In this chapter we examine the interactions that occur between a solid and a vapor. These interactions can lead to intercalation, physical vapor deposition, or chemical vapor deposition. The latter two interactions, along with molecular beam epitaxy, which will be discussed briefly, are used to produce thin films on substrates. The production of thin films has wide applications encompassing integrated-circuit manufacturing, optical devices, chemical sensors, and ionic conductors to name a few. We first review the properties of gases with respect to the important relationships needed when considering film growth. Next, we review the isotherms typically used when considering the adsorption of gases on solid surfaces. After reviewing gases and adsorption, we examine intercalation, physical vapor deposition, chemical vapor deposition, and molecular beam epitaxy. In this chapter we refer to the vapor-phase species as molecules even if the actual species consists of monatomic atoms such as O(g) or Ar(g).

3.1. VAPOR-PHASE FUNDAMENTALS

A discussion of vapor-phase fundamentals begins with the basic gas laws, which apply to any vapor-phase deposition technique. These techniques employ gases at low pressure (less than 1 atm) and therefore are well described by basic laws such as the ideal gas law and the kinetic gas theory, which are presented in undergraduate physical chemistry. For the purposes of vapor deposition, the critical gas parameters include (1) concentration, (2) velocity distribution, (3) flux, and (4) mean free path. The *concentration* of gas particles in a low-pressure gas, less than 1 atm, is given by the *ideal gas law*,

$$PV = nRT \tag{3.1}$$

which can be rewritten as

$$[\text{gas}] = \frac{P}{RT} \tag{3.2}$$

where [gas] is the molarity of the gas, moles of gas per liter. From this we see that the concentration of a gas is linearly proportional to the pressure of the gas at a given temperature.

Inorganic Materials Synthesis and Fabrication, By John N. Lalena, David. A. Cleary, Everett E. Carpenter, and Nancy F. Dean
Copyright © 2008 John Wiley & Sons, Inc.

Gas particles, at a given temperature and pressure, do not all have the same velocities. Instead, the velocities are described by the famous *Maxwell velocity distribution*, $g(v_x)$,

$$g(v_x) = \left(\frac{m}{2\pi k_B T} \right)^{1/2} e^{-mv_x^2/2k_B T} \tag{3.3}$$

This results in a Gaussian distribution centered on $v_x = 0$, shown in Figure 3.1a. In most vapor deposition techniques, it is the speed rather than the velocity that we

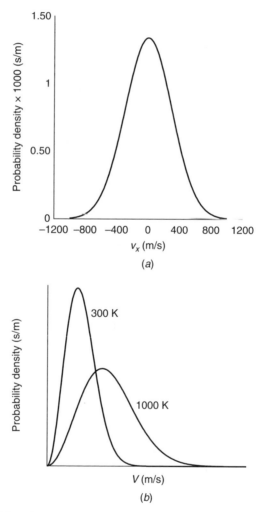

(a)

(b)

Figure 3.1 (*a*) Plot of equation (3.3) for $N_2(g)$ at 300 K; (*b*) plot of Maxwell speed distribution for $N_2(g)$ at 300 and 1000 K.

are interested in. The speed distribution is shown in Figure 3.1b. The important feature of this plot is that as the temperature of a gas sample is increased, a wider distribution of speeds is present. In addition, a larger percentage of the gas molecules is traveling at higher speeds.

Flux is the rate at which gas-phase molecules impinge on a surface. Intuitively, the pressure of the gas will be a factor in establishing this rate. The mass of the assemblage of gas molecules and the temperature will also affect the collision frequency. Using the Maxwell velocity distribution along one direction and treating the gas as an ideal gas leads to the following expression for the collision frequency (Chang, 2000):

$$J = \frac{P}{(2\pi m k_B T)^{1/2}} \tag{3.4}$$

where P is the pressure of the gas, T the temperature, k_B is Boltzmann's constant, and m is the mass of the gas molecule. The units for J, a flux, are area^{-1} time^{-1}, and in SI units would be m^{-2} s^{-1}. At a pressure of 0.1 torr and a temperature of 298 K, nitrogen molecules will strike a 1-cm^2 surface at the rate of 3.88×10^{19} s^{-1}. If we use the accepted value for the area of a nitrogen molecule, 0.162 nm^2, this corresponds to a monolayer consisting of 6.17×10^{14} molecules/cm^2. If all the molecules hitting the surface stick to the surface, a nitrogen film will grow at approximately 6.3×10^4 monolayers/s. If a nitrogen monolayer has a thickness of 5 Å, this would result in a film growth of 30 µm/s. Such a rough calculation highlights one of the challenges in vapor-phase film deposition: the presence of residual gas. At a background pressure of 1×10^{-7} torr, the same calculation results in a nitrogen film growth rate of 3×10^{-5} µm/s. An alternative method for illustrating the importance of the background pressure is to consider how long it would take to form a monolayer. In the case above, the background pressure of 0.1 torr would produce a monolayer in 16 µs. The background pressure of 1×10^{-7} torr would produce a monolayer in 16 s if the assumptions, molecular size and sticking probability, are valid.

The *mean free path* of a gas molecule is the average distance it will travel before colliding with another gas-phase molecule. Conceptually, this will be equal to the velocity divided by the frequency of collisions. The frequency of collisions between like molecules, Z_{11}, is given as

$$Z_{11} = \frac{1}{\sqrt{2}} \pi d_1^2 \left(\frac{8RT}{\pi M_1}\right)^{1/2} \left(\frac{P_1 N_A}{RT}\right)^2 \tag{3.5}$$

and the frequency of collision between unlike molecules, Z_{12}, is given as

$$Z_{12} = \pi(r_1 + r_2)^2 \left[\frac{8RT}{\pi}\left(\frac{1}{M_1} + \frac{1}{M_2}\right)\right]^{1/2} \frac{N_1}{V} \frac{N_2}{V} \tag{3.6}$$

where r_1 and r_2 the radii of the two molecules, M_1 and M_2 the molecular weights, R the universal gas constant, T the absolute temperature, N_A is Avogadro's number, N_1 the number of molecules of type 1, N_2 the number of molecules of type 2, and V the volume of the sample. The equation for the mean free path, λ, is

$$\lambda = \frac{1}{\sqrt{2}\,\pi d^2}\frac{RT}{PN_A} \tag{3.7}$$

where the symbols have their usual meaning and d is the diameter of the gas-phase molecule. This has the correct intuitive form given that mean free path increases as the pressure, P, decreases.

Having defined the mean free path, we next define the *Knudsen number*, Kn. The Knudsen number is defined as the ratio of the mean free path to a characteristic distance:

$$Kn = \frac{\lambda}{d} \tag{3.8}$$

The Knudsen number is significant in vapor-phase processes because for large values, say Kn > 10, it means that the molecules will undergo very few collisions with each other over the characteristic distance. This type of flow, called *molecular flow*, will be important when we discuss molecular beam epitaxy. Conversely, a small Knudsen number, Kn < 0.1, means that the vapor-phase molecules will undergo many collisions prior to traveling the characteristic distance, and this type of flow is called *viscous flow*. This type of flow is important in film deposition, where troughs must be filled in completely (Figure 3.2b). In such a case, it is important that the molecules arrive at the surface from all angles to ensure complete coverage and leave no voids in the troughs. If the angular range of approaching molecules is narrow and the angle is not 0° with respect to the surface normal, *shadowing* and incomplete filing will occur (Figure 3.2c). Shadowing occurs with low pressure where molecular paths are uninterrupted. As a material begins to deposit on a surface, it will have fixed orientation with respect to an incoming beam of molecules. Growth will occur faster on the side facing the incoming beam and a shadow will form on the other side. Although this can be a detriment, it can also form the basis for a new technique such as glancing angle physical vapor deposition, which we discuss shortly. When the Knudsen number is approximately equal to 1, the flow is called *Knudsen flow* or *transition flow*. At very low pressures, the mean free path can approach the distance between the substrate and the source material. In molecular beam epitaxy, the gas pressure in the source beam is kept low enough to meet this criterion. However, with a very low pressure comes a low deposition rate. For oxygen, O_2, at 25°C, a pressure of 2×10^{-4} torr is needed to achieve a mean free path of 24.3 cm.

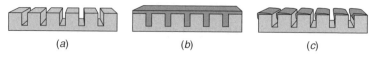

(a)   (b)   (c)

Figure 3.2 Schematic representation of film formation in crevices. Shadowing is shown in (c).

3.2. VAPOR ABSORPTION AND ADSORPTION

We begin by defining two important terms: absorption and adsorption. The International Union of Pure and Applied Chemistry offers precise definitions for absorption and adsorption (IUPAC, 1972). *Absorption* is used to describe a process where a component is transferred from one phase to another. Hydrogen gas can be absorbed by $LaNi_5$ (Jurczyk, 2003). *Adsorption* is used to describe the increased or decreased concentration of a component at an interface. Water molecules will adsorb to an aluminum oxide surface (Al-Abadleh and Grassian, 2003). The concentration of water molecules on the aluminum oxide surface will be greater than the vapor phase, shown schematically in Figure 3.3.

Absorption, in the context of a vapor–solid interface, means that vapor molecules penetrate the solid to occupy vacancies or interstices within the interior of the solid. Adsorption, in the same context, means that vapor molecules stick to the surface of a solid but do not migrate to the interior of the solid. This distinction is easier to apply to a single-crystal surface where there is a clear delineation of surface and interior atoms. In a high-surface-area solid such as an aerogel, which can have a surface area as high as 3000 m^2/g, such distinctions become problematic. Both adsorption and absorption begin with adsorption, although in some cases an adsorbed molecule may be in such a state for only a brief time before moving to the interior.

Because any surface deposition process begins with adsorption, we begin our discussion there. Molecules adsorb on the surface of a solid for a variety of

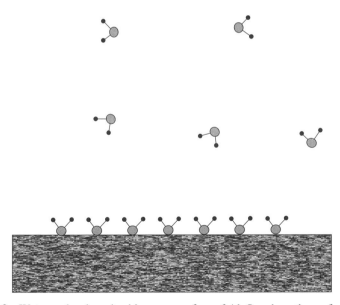

Figure 3.3   Water molecules adsorbing on a surface of Al_2O_3 where the surface concentration of water is higher than the bulk vapor concentration.

reasons, and the extent of adsorption depends on several factors, some of which can be controlled by the synthetic chemist. The primary reason that a molecule adsorbs to the surface of a solid is the exothermic nature of chemical bond formation between the substrate atoms and the vapor-phase molecules. Adsorbed molecules can bond to a surface atom via a covalent bond or London (van der Waals) forces. Those that adsorb due to covalent bonding with the surface are referred to as *chemisorbed*, and those that adsorb due to London forces are referred to *physisorbed*. The strength of the bonding interaction will, in part, determine the mobility of the adsorbed atom. This mobility will affect the subsequent fate of the adsorbed atom with respect to intercalation, chemical vapor deposition, or physical vapor deposition.

The nature of surface adsorption has traditionally be characterized with the construction of a variety of isotherms, the two most common being the Langmuir isotherm (Langmuir, 1918) and the Brunauer–Emmett–Teller (BET) isotherm (Brunauer et al., 1938). Most physical chemistry undergraduate texts present a detailed explanation of at least these two isotherms (McQuarrie and Simon, 1997; Levine, 2002). An *isotherm* is a mathematical relationship relating how much of a surface is covered with adsorbed molecules as a function of the pressure of the gas consisting of those molecules. As the name implies, the relationship is for a given temperature. Intuitively, one would expect the surface coverage to increase with increasing pressure, but the microscopic details of how the rates of adsorption and desorption will affect the surface coverage must also be taken into account.

The *Langmuir isotherm*, regarded as the starting point for any discussion of isotherms because of its simplicity and applicability, was derived by Irving Langmuir in 1918. Langmuir made four assumptions:

1. The adsorbed molecules do not interact with each other.
2. The energy released when a molecule binds to the surface is independent of how much of the surface is covered with adsorbed molecules.
3. A fixed number of sites are available for molecular adsorption.
4. Only one layer of adsorbed molecules will form.

A gas, X(g), at a given pressure is assumed to be in equilibrium with adsorbed molecule, X(ads):

$$X(g) + \text{surface} \xrightarrow{k_1} X(ads) : \text{surface}$$

$$X(ads) : \text{surface} \xrightarrow{k_{-1}} X(g) + \text{surface}$$

where k_1 is the rate constant for adsorption and k_{-1} is the rate constant for desorption. The fixed number of sites available for molecular adsorption is N_0. The fraction of these sites occupied at equilibrium is traditionally labeled θ. The rate of desorption will be equal to

$$\text{rate} = k_{-1}\theta N_0 \tag{3.9}$$

and the rate of adsorption will be

$$\text{rate} = k_1 P (1 - \theta) N_0 \tag{3.10}$$

where P is the pressure of the gas. The adsorption rate is linear in P because it will depend on the collision frequency of the gas with the surface, and this collision frequency is linear in pressure, as shown in Eq. 3.4. Equating these rates at equilibrium and solving for θ results in

$$\theta = \frac{aP}{1 + aP} \tag{3.11}$$

where $a = k_1/k_{-1}$. This is the desired result, a functional form relating surface coverage, θ, to equilibrium pressure, P. The temperature dependence of θ is contained in the temperature dependence of k_1 and k_{-1}.

The Langmuir isotherm provides important physical insights. First, at low pressure, the surface coverage will be linear in pressure. Second, at high pressure, all the adsorption sites will be filled and no additional molecules can be adsorbed. Experimentally, the Langmuir isotherm is investigated by a series of well-controlled steps. First, the surface to be investigated is cleaned of adsorbed species, typically by heating it. Second, a known amount of gas in introduced into the sample chamber holding the sample (Figure 3.4a). Because gas pressures are low, the ideal gas law (Eq. 3.1) can be used to calculate the number of moles of gas introduced given P_0, V, and T. The equilibrium P then allows one to calculate the amount of gas adsorbed (Figure 3.4b). A series of gas injections of increasing amount of gas allows one to produce a plot of adsorbed molecules versus equilibrium pressure. This is an isotherm. Using the ideal gas law, these adsorbed molecules, even though they are not in the gas phase, can be reported as the equivalent volume, v, of gas at $0°C$ and 1 atm, standard temperature and standard pressure (STP) in the gas world. Often, then, it is not the number of

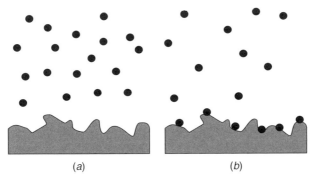

| (a) | (b) |

Figure 3.4   (a) Clean surface and gas prior to adsorption; (b) equilibrium between adsorbed gas and vapor-phase gas.

moles of gas adsorbed that is plotted in an isotherm, but rather, this equivalent volume.

Most of the assumptions in deriving the Langmuir isotherm are not accurate. It works best for cases of chemisorption where only one layer is formed. A straightforward check on the validity of the Langmuir isotherm is to plot the inverse of the surface coverage, θ, versus the inverse of the equilibrium pressure:

$$\frac{1}{\theta} = \frac{1}{aP} + 1 \tag{3.12}$$

An alternative and equivalent verification can be sought by using the equivalent volume, v, mentioned above instead of the surface coverage. The surface coverage equals the ratio of the equivalent volume to the equivalent volume of a complete monolayer,

$$\theta = \frac{v(P)}{v(P \to \infty)} \tag{3.13}$$

This results in a verification equation of

$$\frac{1}{v(P)} = \frac{1}{aPv(P \to \infty)} + \frac{1}{v(P \to \infty)} \tag{3.14}$$

Isotherms that follow the Langmuir isotherm (i.e, form only a monolayer) are termed *type I isotherms*.

The *BET isotherm* (Adamson, 1982) addresses some of the shortcomings of the Langmuir isotherm: most important, the restriction of one adsorbed layer. The resulting isotherm is

$$\frac{P}{(P^* - P)v} = \frac{1}{v_{mon}b} + \frac{b-1}{v_{mon}b} \frac{P}{P^*} \tag{3.15}$$

where b is a constant at a given temperature, v_{mon} the equivalent volume of a monolayer of adsorbed gas, and P^* the equilibrium vapor pressure of the liquid phase of the adsorbed gas at the experimental temperature. A plot of $P/(P^* - P)v$ versus P/P^* is constructed with the linear portion giving b and, more important, v_{mon}. When multiple adsorbed layers are formed, the isotherm is called a *type II isotherm*. The BET isotherm is an example of a type II isotherm. Type I and type II isotherms have characteristic shapes as shown in Figure 3.5, where the equivalent volume of adsorbed gas is plotted versus the equilibrium vapor pressure. Type I, (Figure 3.5a) reaches a plateau when the first monolayer is completed and no subsequent monolayers form. Type II (Figure 3.5b), where the BET isotherm is typically applied, shows a inflection point where the first monolayer is completed and subsequent layers are beginning to form. In both of these isotherms, the equilibrium pressure, P, is typically much less than the equilibrium vapor pressure of the liquefied gas, P^*.

Irving Langmuir (1881–1957) received his Ph.D. under Walther Nernst in 1906 at Göttingen University. He taught chemistry briefly at Stevens Institute of Technology in New Jersey, then joined the Research Laboratory at General Electric in 1909, eventually becoming associate director. Langmuir investigated the properties of adsorbed films, electric discharges in high vacuum, and other vacuum phenomena. He was awarded the 1932 Nobel Prize in Chemistry for his discoveries and investigations in the field of surface chemistry. Langmuir was the first nonacademic chemist to receive the prize. He defined chemisorption, introduced adsorption isotherms, and studied heterogeneous catalysis. It is not widely acknowledged in textbooks, but Langmuir contributed to atomic theory as well, having proposed the octet rule independent of Gilbert Lewis. He also introduced the terms *valence, isoelectronic, isomer*, and *isobar*. Langmuir served as president of the American Chemical Society in 1929, and an ACS journal of surfaces and colloids, *Langmuir*, is named in his honor. Langmuir was elected to the U.S. National Academy of Sciences in 1918.

Both isotherms contain a term representing the volume of a monolayer of adsorbed gas that, again using the ideal gas law, can be converted to the number of molecules needed to complete a monolayer. This has several practical consequences. First, if the size of the molecule is known, the surface area of the solid can be determined. Several gases are used routinely for this purpose: nitrogen, argon, krypton, and oxygen. The accepted surface areas of these molecules are (Sing, et al., 1985):

$$
\begin{array}{llll}
\text{nitrogen:} & 16.2 \ \text{Å}^2 & \text{krypton:} & 19.5 \ \text{Å}^2 \\
\text{argon:} & 13.8 \ \text{Å}^2 & \text{oxygen} & 14.1 \ \text{Å}^2
\end{array}
$$

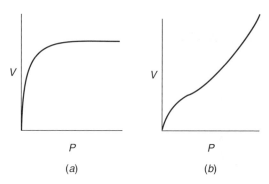

Figure 3.5   (*a*) Type I and (*b*) type II isotherms.

Alternatively, if the surface area is know, an estimate of molecule size is possible. In our discussion of physical and chemical vapor deposition we make use of the Langmuir isotherm because of the low pressures involved.

As already stated, the strength of the bonding between the substrate and the adsorbed species distinguishes physical adsorption from chemical adsorption. The strength of this bonding, termed the *enthalpy of adsorption*, can be determined using

$$\left(\frac{\partial \ln P}{\partial T}\right)_{\theta} = -\frac{\Delta \overline{H_a}}{RT^2} \tag{3.16}$$

where $\Delta \overline{H_a}$ is the differential molar enthalpy of adsorption, P the pressure of the gas in equilibrium with the solid surface, T the temperature, and R the gas constant. The enthalpy of adsorption for a physisorbed species will typically be less than 20 kJ/mol (exothermic), whereas that of a chemisorbed species will typically be greater than 200 kJ/mol (exothermic) (Atkins, 1978). To extract ΔH_a from Eq. 3.15, a series of isotherms at different temperatures is collected. The coldest corresponds to the upper trace in Figure 3.6, and the warmest to the lower trace. The P and T data points at constant θ are indicated with the tie line. It is interesting to consider that adsorption is almost always an exothermic process because of the formation of a chemical bond, regardless of whether covalent or van der Waals in character, yet the processes we discuss presently, particularly chemical vapor deposition, typically use surfaces with very high temperatures where surface adsorption is not favored by Le Châtelier's principle.

3.3. FILM FORMATION BASICS

Film growth, by whatever technique is used, can typically be described in terms of three main events: (1) nucleation, (2) crystal growth, and (3) grain growth. *Nucleation* is the process of the vapor-phase molecules or atoms coming together to form a condensed unit. The theory of nucleation and growth has been developed extensively over many years. When a nucleus is very small, the energy

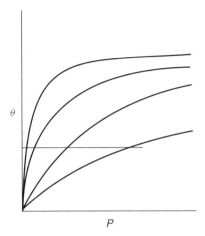

Figure 3.6   Series of isotherms needed to determine ΔH_a.

required to grow larger is dominated by surface energy requirements, and the resulting growth has a positive Gibbs free energy. Once nuclei pass beyond a certain size, the surface effects become less dominant and the condensed phase continues to grow. Surface energy and nucleation were discussed in Section 2.3.

   The theory of nucleation and growth of films on substrates is complex because of the interactions between nucleation sites, adsorbed atoms (adatoms), defects, and soon. Ratsch and Venables (2003) have reviewed nucleation and growth models on flat surfaces, both perfect and with defects. The authors highlight the wide range of time scales and dimensions that must be treated when modeling nucleation and growth of thin films. Individual atomic motion may occur on a scale of angstroms and picoseconds, yet films are often produced on a scale of micrometers and hours. Film growth, on an intermediate scale, is characterized by three main growth types. In the *Frank–van der Merwe growth type*, each layer is completed before the next layer is started. In the *Volmer–Weber growth type*, islands of the deposited material several layers thick form and eventually coalesce. The *Stranski–Krastanov model* is a mix of the first two. Initially, complete layers grow sequentially (Frank–van der Merwe growth), but then islands begin to appear (Volmer–Weber growth). These three models are illustrated in Figure 3.7. In the Frank–van der Merwe growth mechanism, the assumption is that the adsorbed atoms have a stronger attraction for the substrate than they do for one another. Conversely, Volmer–Weber growth results when the attractions between the adsorbed atoms are stronger than the attractions between the adsorbed atoms and the substrate. It is not attraction, of course, which drives these mechanisms, but rather, chemical bonding. Hence, even if one cannot predict which mechanism will be observed, the mechanism observed can be rationalized in terms of the types of chemical bonding possible between adsorbed molecules and the substrate.

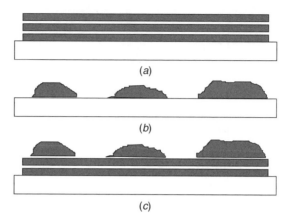

(a)

(b)

(c)

Figure 3.7 Three models of film growth: (a) Frank–van der Merwe, layer by layer; (b) Volmer–Weber, islands; (c) Stranski–Krastanov, combination of layer by layer and islands.

3.4. VAPOR-PHASE INTERCALATION

As mentioned at the beginning of this chapter, intercalation results when adsorbed molecules migrate into the bulk of a solid. Intercalation in the solid-state chemistry world is viewed as the reaction between a mobile phase, liquid or vapor, and a solid phase. The chemistry of liquid–solid intercalation is discussed in Chapter 4. In this chapter we focus on vapor–solid intercalation. Vapor–solid intercalation means that the mobile phase arrives at the solid in the vapor phase, although the stable phase of the mobile phase may be a liquid or solid at room temperature and ambient pressure. Intercalation implies a chemical interaction between the host, the solid, and the guest, the migrating molecule. This interaction could be redox, acid–base, or ion exchange, to name a few. In addition, intercalation is also used to describe *reversible* insertion reactions where some degree of order is present within the inserted species (Figure 3.8.). The intercalated compound is rarely crystalline but often retains some degree of periodicity along one direction of the material. In Figure 3.8 the intercalated organic layer is likely to be disordered, yet periodicity is retained in the stacking direction. The

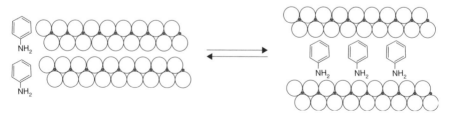

Figure 3.8 Intercalation of aniline into a layered solid.

general topic of intercalation chemistry has been reviewed in several monographs (Whittingham and Jacobson, 1982; McKelvy and Glaunsinger, 1990).

Often, a liquid source is used to accomplish a vapor–solid intercalation. For example, Luca and Thomson (2000) reported the vapor-phase intercalation of aniline (boiling point $= 184°C$) into halloysite, an aluminosilicate clay. In this work the authors pretreated the host material such that it would be able to oxidize the intercalated aniline monomers, resulting in intercalated polyaniline. The vapor-phase intercalation was accomplished by exposing the host to aniline vapors. The individual aniline molecules intercalated into the available voids within the host. The host then oxidized the aniline monomers to produce polyaniline (Figure 3.9). This is a clever approach to producing an intercalated polymer. Polymers, with their large molecular weights, large size, and negligible vapor pressure would be impossible to intercalate as polymers. Instead, the polymer is synthesized within the host lattice from the intercalated monomers.

Vapor-phase intercalation has its advantages and disadvantages. One advantage of vapor-phase intercalation is that the host is forced to retain its stoichiometry. In liquid-phase intercalation, the host can release atoms, or more likely, ions, to the liquid to which it is being exposed to. For example, when the layered semiconductor $Cd_2P_2S_6$ is soaked in an aqueous solution of potassium chloride and ethylenediaminetetraacidic acid (EDTA), cadmium ions leave the lattice, and potassium ions enter the lattice (Clement et al., 1986):

$$Cd_2P_2S_6(s) + KCl(aq) \rightarrow Cd_{1.50}P_2S_6K_{1.00}(H_2O)_2(s) + \frac{1}{2}CdCl_2(aq)$$

With vapor-phase intercalation, the host is much less likely to decompose as just described. The disadvantage of vapor-phase intercalation is that with far fewer molecules impinging on the surface of the host, reaction rates are much slower. The disadvantage can also be an advantage since the reduced reaction rate allows the intercalated material to remain more homogeneous with respect to the amount of material intercalated.

In the example with aniline, the aniline vapor was provided by the equilibrium vapor liquid aniline. Vapor-phase intercalation can be done with compounds that are gases at room temperature and ambient pressure. The most common gas used for intercalation reactions is ammonia. Ammonia intercalation can be accomplished by exposing a host to the vapor generated by a concentrated aqueous ammonia solution. This multi-component vapor containing $NH_3(g)$, $H_2O(g)$,

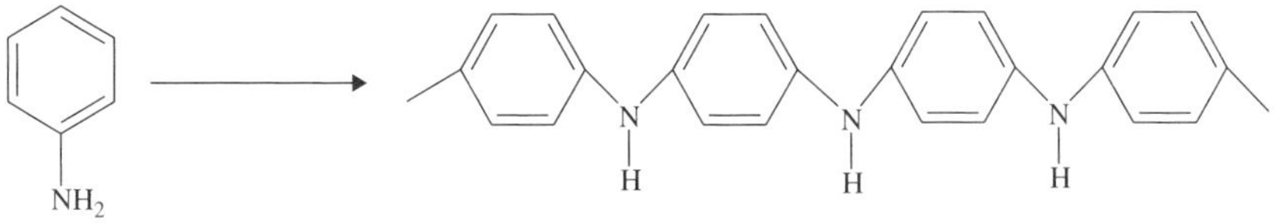

Figure 3.9 Polymerization of aniline to form polyaniline.

$O_2(g)$, and $CO_2(g)$ can complicate the interpretation of the resulting compound, so instead, a pure ammonia atmosphere is often used instead. This can be done under flowing conditions, where the concentration of ammonia is low, resulting in a slow intercalation, or with a sealed tube, where a higher concentration of vapor can be achieved. With a sealed tube, the pressure in the tube is a critical parameter. At room temperature, if the pressure is too high, (i.e., 10 atm), the ammonia vapor will condense and the complications of liquid-phase intercalation mentioned above must be considered. If the pressure is too low, complete intercalation will not be achieved. Normally, the ammonia vapor is condensed into a small sample tube, which is then sealed off. To prevent contact between the host and the condensed ammonia prior to complete vaporization of the condensed ammonia, a sidearm is used on the sample tube. As shown in Figure 3.10a, the ammonia gas is introduced into the entire sample vessel. The vacuum line is isolated, and the trapped ammonia is frozen in the sidearm, (Figure 3.10b) using liquid nitrogen. The resulting cryopumped vacuum is used to collapse the sample tube when it is heated with a glassblowing torch. When the sample tube returns to room temperature (Figure 3.10c), several atmospheres of ammonia gas are contained in the sample tube, with a pressure less than 10 atm if the initial pressure of ammonia in the vacuum line is chosen, or guessed, correctly.

A beautiful example of vapor-phase intercalation involves the reaction of lead(II) iodide with hydrazine (Ghorayeb et al., 1984). Lead(II) iodide is a layer structure (CdI_2 structure, P3m, no. 164) which forms large yellow transparent hexagonal crystals. When these crystals are exposed to hydrazine vapor, they turn colorless! Similar chemistry is observed with ammonia (Cleary, unpublished results). In the ammonia case, when the colorless intercalated crystals are removed from the ammonia atmosphere, the ammonia deintercalates spontaneously and the crystal returns to its yellow color:

$$PbI_2(s), \text{ yellow} + xNH_3(g) \rightleftharpoons PbI_2\{NH_3\}_x, \text{ colorless}$$

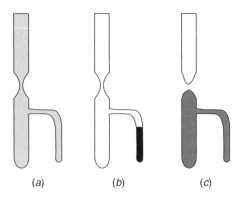

(a) (b) (c)

Figure 3.10   Three steps in preparing an ammonia intercalation reaction cell: (a) introduction of ammonia; (b) condensation into sidearm; (c) sealing of sample cell and vaporization of ammonia.

This rapid intercalation–deintercalation of ammonia can form the basis for a chemical sensing device. Thomas and Cleary (1996) used $Mn_2P_2S_6$, a layered material similar to PbI_2, as the host material for ammonia intercalation and showed a rapid change in the dielectric properties of the host upon exposure to ppm concentrations of ammonia.

Ammonia intercalation has been rationalized in terms of the small size and basic character of the molecule. The basic character appears to be more important given the large number of alkylamines that have been intercalated into a variety of hosts. The most dramatic example of this type of chemistry is shown by the intercalation of n-octyldecylamine into TaS_2 (Gamble et al., 1971). In this spectacular case, the van der Waals gap of the host TaS_2 will expand by over 50 Å to accommodate a bilayer of n-octyldecylamine. The intercalation reactions of alkyl amines continues to be investigated. Figueiredo and Oliveira (2005) have reported the intercalation of the vapors of n-butylamine, n-hexylamine, and n-octylamine into V_2O_5. One of the reasons there continues to be interest in the intercalation chemistry of alkyl amines is that these intercalating agents, because they can be so large, are useful for exfoliating layered host lattices down to atomic dimensions. This type of intercalation is generally done with the intercalating agent in the form of a pure liquid or solution (see Chapter 4).

Intercalation of molecular species such as ammonia, n-alkylamines, and aniline allows the synthetic chemist to investigate the resulting materials using a wide range of spectroscopic tools. Infrared spectroscopy is particularly important for establishing the existence of the starting intercalating agent in the final product. Raman spectroscopy has also been used, but sample decomposition and deintercalation become problems with this method. Optical spectroscopy, especially luminescence spectroscopy, has found application in the characterization of intercalated compounds (Cleary et al., 1986). Finally, x-ray diffraction, normally powder type, can be used to establish the swelling of a host lattice resulting from the intercalation of a molecule.

Vapor-phase intercalation–deintercalation has been used to describe the behavior of oxides at high temperature. Manca et al. (2000) published a study of the superconducting properties of $Y_1Ba_2Cu_3O_{7-x}$ where $x = 0.7$. The stoichiometric compound $Y_1Ba_2Cu_3O_6$ is semiconducting, and $Y_1Ba_2Cu_3O_7$ is superconducting. The $Y_1Ba_2Cu_3O_7$ compounds can be converted to $Y_1Ba_2Cu_3O_{7-x}$ by heating to several hundred degrees Celsius, resulting in the deintercalation of oxygen. The language of intercalation is used to describe this chemistry, although this represents somewhat extreme reaction conditions. Intercalation reactions are generally regarded as occurring at low temperature. More important, intercalation reactions are reversible. The example above is termed intercalation, as opposed to simple decomposition, because when the oxygen vapor pressure is raised, the stoichiometry returns to $Y_1Ba_2Cu_3O_7$:

$$Y_1Ba_2Cu_3O_7 \rightleftharpoons Y_1Ba_2Cu_3O_{7-2x} + xO_2(g).$$

Vapor-phase intercalation can also be done with a guest material that is a solid at room temperature. For example, Harley and McNeil (2004) report the

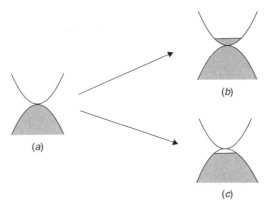

Figure 3.11 (*a*) Band structure of pure graphite showing filled valence band touching the empty conduction band; (*b*) resulting band structure after intercalation with a reducing agent (e.g., Cs); (*c*) band structure after intercalation with an oxidizing agent (e.g., I_2).

intercalation of cesium into single-walled carbon nanotubes. Cesium melts at 28°C, and above this temperature has enough vapor pressure to produce intercalating atoms. Harley and McNeil followed a common method used to prepare graphite intercalated with alkali metal atoms. The sample is maintained at several hundred degrees Celsius ($\approx 200°C$), and the sample temperature and alkali metal temperature are controlled to prevent simply coating the sample with condensed-phase alkali metal. Single-walled carbon nanotubes apparently have a redox intercalation chemistry similar to graphite, which was studied thoroughly in the 1970s and 1980s. What distinguishes graphite and single-wall carbon nanotubes from other intercalation hosts is their ability to either donate or accept electrons upon intercalation, which is dependent on the oxidation–reduction potential of the guest species. The schematic representation of the band structure of graphite (Figure 3.11) is used to account for this. Because the bandgap in pristine graphite is 0.0 eV, getting electrons into the valence band or out of the conduction band requires very little energy. Hence, the determining factor on the direction of electron flow in an intercalation reaction will be decided by the reduction potential of the guest:

$$Cs^+ + e^- \rightarrow Cs \qquad E° = -2.923 \text{ eV} \qquad \text{electrons into conduction band}$$
$$I_2(s) + 2e^- \rightarrow 2I^- \qquad E° = +0.535 \text{ eV} \qquad \text{electrons out of valence band}$$

3.5. PHYSICAL VAPOR DEPOSITION

In Section 3.4, we considered the intercalation of surface molecules into the interstices available within a host compound. We consider the case where the surface molecules remain on the surface to form a thin crystalline film. Thin-film formation is at the heart of integrated-circuit manufacturing. An example of

1. Twin-well Implants

2. Shallow Trench Isolation

3. Gate Structure

4. Lightly Doped Drain Implants

5. Sidewall Spacer

6. Source/Drain Implants

7. Contact Formation

8. Local Interconnect

9. Interlayer Dielectric to Via-1

10. First Metal Layer

11. Second ILD to Via-2

12. Second Metal Layer to Via-3

13. Metal-3 to Pad Etch

14. Parametric Testing

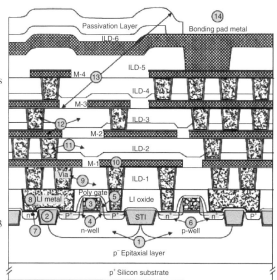

Figure 3.12   Modern integrated circuit. (From M. Quirk and J. Serda, *Semiconductor Manufacturing Technology*, Prentice-Hall, New Jersey, p. 220. Copyright © 2001. Reproduced with permission.)

a structure prepared by modern integrated-circuit manufacturing techniques is shown in Figure 3.12.

In physical vapor deposition, the task is to transfer a small amount of the source material to the substrate. Two basic techniques are used to accomplish this task: evaporation and sputtering. A wide variety of specific methods have been developed for this purpose. We examine a few of the evaporation techniques as well as sputtering. The simplest method is *resistive heating evaporation*. In this method, the source material is kept at a temperature greater than the substrate, and material sublimes or evaporates off the source material and condenses on the cooler substrate. This technique has several requirements. First, the source material must have a reasonable vapor pressure. Second, the composition of the vapor produced by heating the source material must be the desired chemical formula of the deposited material. Finally, the vapor must stick to the substrate. A more subtle problem is realized when one considers the geometry of the substrate relative to the evaporation source. Consider (Figure 3.13), a wafer 20 cm in diameter situated 50 cm directly over the source of the vapors, which we approximate as a small planar source. The growth rate (in cm/s) for this configuration is (Jaeger, 1993)

$$G = \frac{m}{\pi \rho r^2} \cos \phi \cos \theta \tag{3.17}$$

where m is the evaporation rate in g/s, ρ the density of the vapor, r the distance from the substrate to the source, ϕ the angle measured from the normal to the

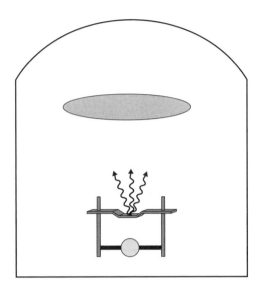

Figure 3.13   Resistive heater evaporator showing orientation of substrate to vapor flux.

source plane, and θ the angle between the vapor stream and the normal to the substrate surface. Under the conditions given, for the deposition of a 10-μm. film at the center of the substrate, the edge of the substrate would have a film thickness of 9 μm. This lack of uniformity in the evaporated film thickness due to geometrical considerations can be eliminated by a judicious choice of orientation of the substrate with respect to the source. If the substrate and source are kept equidistant for a point (i.e., if they both lie on a common sphere), the cosine terms in Eq. 3.17 become constants equal to $r/2r_0$, where r_0 is the radius of the sphere. In addition, the substrate can be spun to further homogenize the film thickness. Evaporation remains a common method for depositing thin films because of its simplicity. High deposition rates are possible, and the substrate does not need to be heated. This becomes an advantage when attempting to deposit a thin film on an organic-based polymer.

Because the source material will be heated to produce the vapor pressure, the process is conducted in a high vacuum to prevent chemical reaction of the source material. This represents an advantage for this method since the incorporation of impurities into the film is minimized with the use of such a vacuum. Several methods are used to heat the source material. The simplest is resistive heating. The source material is placed in a ceramic boat, which is then placed into a heater. Alternatively, the source material itself can be formed into a filament, heated, and the vapors produced that way. The temperature of the source materials is varied by varying the power to the resistive heater. Since this approach sends source material vapor in all directions, the source material and substrate are often placed inside a foil tent so that the cleanup is simpler.

For those materials that do not produce an appropriate vapor with resistive heating, other heating methods are employed, including electron beam and inductive heating. These techniques, which we discuss below, add complexity to the apparatus but also extend the range of source materials that can be used. A final method of physical vapor deposition that we discuss is sputtering.

Electron beam physical vapor deposition (EB-PVD) is accomplished by directing a narrow beam of high-powered ($\approx$20 kV, $\approx$500 mA) electrons at the source material. The energy released by these electrons when they are absorbed by the source material causes local melting in the source material. The entire source material is not melted, only the portion irradiated by the electron beam. This production of a small pocket of melted source material contained within solid source material is sometimes referred to as *skulling*. Hence, the vapor produced does not suffer contamination from the vessel containing the source material. Singh and Wolfe (2005) reviewed the use of electron beam–physical vapor deposition for the fabrication of nano- and macro-structured components.

The electron beam is typically arranged in one of two ways, as shown in Figure 3.14: the straight on approach, (*b*), or the bent approach, (*a*). Because the particles in an electron beam are charged, they can be deflected, steered, focused, and manipulated using magnetics and voltage plates. The film deposition rate in EB-PVD can range from 10 Å to 0.1 mm per minute. Because electrons are being deposited in the source material, it must be held in an electrical conductor. Copper is a favorite choice. As in resistive heating deposition, the entire process is contained within a vacuum system in which the pressure ranges from 10^{-7} torr at the gun to 10^{-2} torr where the source material is being vaporized. A sophisticated electron beam physical vapor deposition system can have multiple electron guns and multiple source materials. With a system of shutters, translators, and rotators, multiple layers can be deposited on a single substrate without venting the system to change source materials. In addition to resistive heating and electron beam heating, a source material can be heated by radio-frequency (RF) *induction heating*. The source material is placed in a crucible that is susceptible to RF heating, such as graphite (or quartz surrounded by a graphite susceptor), and the RF coil heats the crucible (or susceptor). In RF heating, as opposed to electron beam heating, more source material is heated, as in resistive heating.

The most common method for large scale physical vapor deposition is *sputtering*. Sputtering is fundamentally different from the three heating methods

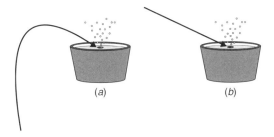

(*a*)   (*b*)

Figure 3.14 Two orientations of electron beam with respect to sample: (*a*) bent; (*b*) straight.

just described. In sputtering, the vapor is not generated by thermally inducing the source material to generate a vapor. Rather, pieces (albeit atomic pieces) are mechanically dislodged from the surface of the source material by momentum transfer with accelerated ions (Figure 3.15). The ions impinging on the source material are generated by an electrical discharge. Typically, argon atoms are used and are ionized in a glow discharge. One problem associated with sputtering is that large pieces, referred to as *droplets*, can be dislodged from the surface of the material being sputtered. To prevent these droplets from reaching the substrate, a technique known as *filtered vacuum arc deposition* is used (David et al., 2005). In this method the plasma is steered with a magnetic field to separate that desired vapor-phase material from the unwanted droplets.

Sputtering is used as a deposition technique. It can also be used in conjunction with a technique such as electron beam–physical vapor deposition. Prior to operating the electron gun, argon can be introduced into the vacuum chamber and ionized. The resulting ions can be used to sputter the *substrate*. This is an effective method of cleaning the substrate surface. The cleaner surface will accept the deposited film better. Another method for producing high-quality single-crystalline surfaces is to cleave a material under high vacuum. The easiest materials to do this with are layered compounds that readily exfoliate, such as graphite, mica, and transition metal disulpfides, such as those shown in Figure 3.16. This exfoliation, done under high vacuum, produces a clean, defect-free surface on which to deposit material. During the course of an experiment, additional surfaces can be produced by repeated exfoliation. If a substrate can tolerate it, a third method for cleaning is heating.

Once a vapor is produced, regardless of how it is produced, the resulting gas-phase molecules travel to and are deposited on the substrate. Normally,

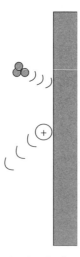

Figure 3.15 Sputtering shown as an ion impinging on a surface, transferring momentum to the substrate, and substrate material being ejected.

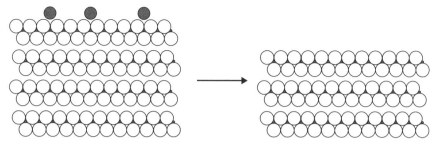

Figure 3.16 Edge-on view of the delamination of a layered material, resulting in a clean surface. Dark circles represent adsorbed impurities.

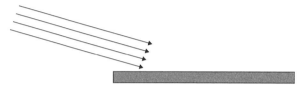

Figure 3.17 Glancing angle deposition where molecules arrive at a surface in a trajectory almost parallel to the surface.

one would like a uniform film thickness, and this presents challenges when one considers the geometrical arrangement of the flat substrate surface on the curvature of the gas-phase front, as already demonstrated using Eq. 3.17. In that case, judicious positioning of the substrate and source along with rotation of the substrate alleviate much of the nonuniformity in film thickness. An interesting alternative approach, designed specifically to avoid a perpendicular geometry between the incoming gas molecule trajectories and the surface of the substrate, is a type of physical vapor deposition called *glancing angle deposition*, GLAD, Figure 3.17 (Robbie, 1998). In this case, the substrate is significantly tilted (Figure 3.18) with respect to the incoming gas-phase molecules. This takes advantage of the *shadowing effect*, which one seeks to avoid in other methods. The result of glancing angle deposition leads to an interesting morphology on the surface as shown in Figure 3.19. These films are deliberately prepared to be anisotropic and porous. They have potential applications in areas such as optical coatings and chemical sensors.

3.6. CHEMICAL VAPOR DEPOSITION

So far we have consider two possibilities for the fate of a molecule adsorbed on a solid surface. It can intercalate into the solid or can become part of a physically deposited layer. In this section we consider a third possibility. Under the right conditions, a molecule on the surface of a solid can be induced to react chemically at that surface with the products of the reaction remaining on the surface. This is

Figure 3.18 Slanted morphology on a surface resulting from glancing angle deposition. (Courtesy of M. J. Brett. Copyright © 2006 M. J. Brett, University of Alberta. Reproduced with permission.)

Figure 3.19 Helical morphology on a surface resulting from glancing angle deposition. (Courtesy of M. J. Brett. Copyright © 2006 M. J. Brett, University of Alberta. Reproduced with permission.)

in contrast to a molecule coming to a surface, reacting, and having the products leave the surface. The former situation is referred to as *chemical vapor deposition*, and the latter is *heterogeneous catalysis:*

chemical vapor deposition: $A(g) \rightarrow A(surface) \rightarrow B(surface)$

heterogeneous catalysis: $A(g) \rightarrow A(surface) \rightarrow B(surface) \rightarrow B(g)$

In the case of chemical vapor deposition, the solid surface is referred to as the *substrate*, and the gas-phase molecule is referred to as the *precursor*. A wide range of issues concerning the film–substrate interface are important in chemical vapor deposition, including structural coincidence, chemical reactivity, thermal compatibility, and substrate morphology.

In this section we are concerned with epitaxial deposition. The word Greek *taxis* can mean "an arrangement" or "a positioning." The Greek preposition *epi* in this context means "upon." *Epitaxial*, then, means that the deposited layers are arranged on something: namely, the substrate or layers already deposited. The particular arrangement is crystalline. The term *epitaxial deposition* is reserved for crystalline deposition. *Epitaxial* is further refined to include homoepitaxial and heteroepitaxial. In *homoepitaxial deposition*, the deposited material is the same as the substrate; silicon on silicon and diamond on diamond are examples of homoepitaxial deposition. In *heteroepitaxial deposition* the deposited material is different from the substrate: diamond on silicon or GaN on sapphire.

Two important considerations in heteroepitaxial film growth are the mismatch in the lattice constants of the substrate versus the deposited film and the differences in thermal expansivity between the substrate and the epitaxial layer. Given the high deposition temperatures that are often used and the much lower operational temperatures of the devices being manufactured, most films must endure significant cooling, and hence compression (or tension), prior to use. Taking a cue from glassblowers who have faced this problem for decades, film makers have used intermediate layers, called *buffer layers*, to ease the difference in expansivity and lattice mismatch between the substrate and the film of interest. For example, to grow GaN on a silicon substrate, a buffer layer of AlN is deposited on the silicon prior to depositing the GaN (Zhang et al., 2005).

Chemical vapor transport is used to synthesize thin films of materials on a substrate. The film can be the same composition as the substrate or different. In order to proceed with chemical vapor transport, the constituent elements of the compound to be deposited as a thin film must be brought into the vapor phase. Given that many of the thin films of commercial importance involve elements with little or no practical vapor pressure, a lot of attention has been focused on preparing volatile compounds that contain the elements needed in thin-film preparations. Most chemical supply companies carry these compounds as stock items. The major classes of compounds include metal alkyl, metal carbonyl, metal alkoxide, metal β-diketonates, and organometallics. Examples of each are given in Table 3.1.

While the decomposition of these compounds under the right conditions produces exquisitely tailored materials, it is important to recall that most of these compounds are highly toxic, pyrophoric, and air and moisture sensitive. Therefore, special equipment is necessary to produce thin films by chemical vapor transport. The method itself appears quite simple. The appropriate vapors are brought into contact with a substrate maintained at a particular temperature. The vapors decompose, leaving behind the desired material. In Figure 3.20, the three vapors introduced into the reaction chamber decompose and produce the desired

TABLE 3.1   Volatile Compounds Used to Prepare Thin Films

Class	Example	Room-Temperature Phase
Metal alkyl	Dimethyl mercury, $(CH_3)_2Hg$	Liquid
Metal carbonyl	Nickel carbonyl, $Ni(CO)_4$	Liquid
Metal alkoxide	Tantalum(V) ethoxide, $Ta(OCH_2CH_3)_5$	Solid
β-Diketonates	Strontium hexafluoroacetylacetonate, $Sr(CF_3COCH_2COCF_3)$	Solid
Organometallic	Chromocene, $Cr(C_5H_5)_2$	Solid

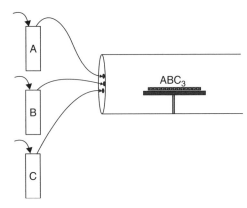

Figure 3.20   Three gases being introduced into a chemical vapor deposition chamber.

material. Many compounds used in chemical vapor deposition are liquids at room temperature or have melting points slightly above room temperature. To introduce the vapors of these liquids, a carrier gas is bubbled through a sample of the liquid-phase component. The carrier gas may also contain oxygen to control the oxidation state of the metal. This makes Figure 3.20 highly simplified. An actual CVD system needs temperature, pressure, and mass flow control and looks more like what is shown in Figure 3.21.

Like so many techniques that have widespread industrial applications, chemical vapor deposition has splintered into a myriad of techniques, each with advantages and disadvantages. An incomplete list of these techniques and their common abbreviations would include:

- AACVD: aerosol-assisted chemical vapor deposition
- PACVD or PECVD: plasma-assisted or plasma-enhanced chemical vapor deposition
- MOCVD: metal–organic chemical vapor deposition (also called OMVPE)
- LCVD: laser chemical vapor deposition

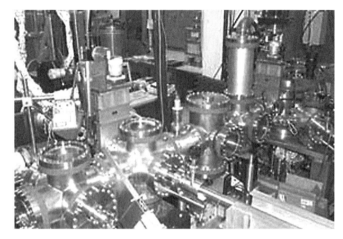

Figure 3.21 Chemical vapor deposition system. (Copyright © 2006 University of Wisconsin. Reproduced with permission.)

- HVPE: hydride vapor-phase epitaxy
- ALE or ALD: atomic layer epitaxy or atomic layer deposition

These techniques have crossovers in the sense that, for example, LCVD can be done with metal-organic precursors. Cheon and Zink (1997) deposited thin films of ZnS, CdS, and $Zn_xCd_{1-x}S$ onto quartz substrates. The source of the metal and the sulfur was the organometallic compound diethyl dithiocarbamate zinc or cadmium: $M(S_2CNEt_2)_2$. In this case a single precursor was used to produce a binary compound as a thin film.

Aerosol-assisted chemical vapor deposition is illustrated by work published by Siadati et al. (2004). An aerosol is a vapor suspension of finely divided particles. The particles can be solid, as in smoke, or liquid, as in fog. A toluene solution of $Zr(tfac)_4$, $Y(hfac)_3$, and $Ce(tmhd)_4$, converted to an aerosol, was used to deliver the metal in a carrier gas of O_2. The film deposited was CeO_2-doped Y_2O_3-stabilized zirconia. The ligands used in this work to produce volatile metal compounds are frequently used in chemical vapor deposition. Trifluoroacetylacetonate (tfac), hexafluoroacetylacetonate (hfac), and tetramethylheptanedionate (tmhd) are all β-diketonates, and their structures are shown in Figure 3.22. Like the diethyl dithiocarbamates mentioned above, these precursors could potentially be the source of both the metal and the oxygen in the chemically deposited film. However, to ensure that the metal remains at its highest oxidation state and to avoid a film with mixed valencies on the metal, oxygen is used as the carrier gas.

The use of plasmas is an interesting development in the field of chemical vapor deposition. We recall that a plasma is a vapor consisting of cations and electrons. It is most easily generated by establishing an electrical discharge through a low-pressure gas. A neon sign is an example of a plasma, specifically a glow discharge. When it comes to chemical vapor deposition, where the objective is to

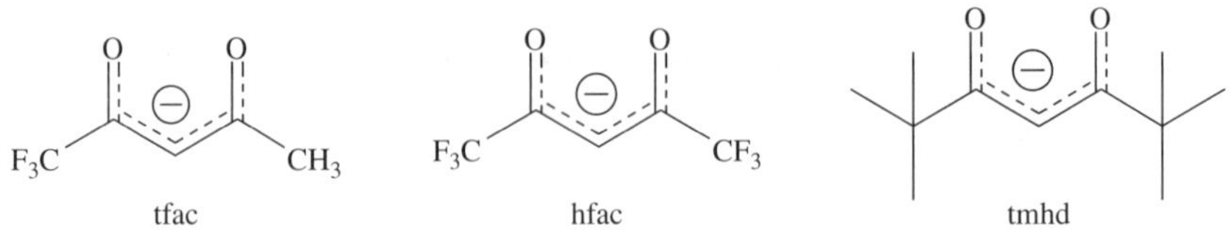

Figure 3.22 Chemical structures of three common ligands used to volatilize metal ions. The negatively charged oxygen atoms bind to the metal cations.

get a molecule to decompose on a surface, a plasma offers an alternative to high temperature for effecting this decomposition. Hence, PACVD would be in contrast to thermal chemical vapor deposition, although the abbreviation TCVD is not used since most of the techniques listed above use thermal decomposition. Cote et al. (1999) reviewed the use of plasma-assisted chemical vapor deposition of dielectric thin films for ultralarge-scale-integrated (ULSI) semiconductor circuits.

A plasma can be used to decompose a molecule that will not decompose at a reasonable elevated temperature, or it can be used to decompose a thermally unstable molecule but at a much lower temperature. The low decomposition temperature is important for a variety of reasons. First, it reduces the engineering requirements of a system if the system can be operated at a low temperature. More important, however, if high temperatures are not required for the decomposition of the precursor, the synthetic chemist has flexibility with respect to the deposition temperature. This allows for low-temperature deposition where the stoichiometry and structure of the film may vary from the high-temperature result. Other issues concerning plasma-deposited films include the higher density achieved with a plasma film and the ease of cleaning. Finally, and perhaps most important, in plasma CVD the substrate is not subjected to high temperatures. Hence, substrates that cannot tolerate high temperatures, such as polymers, can be used in plasma CVD, where substrate temperatures range from 100 to 500°C.

Chemical vapor deposition as a synthetic technique is used primarily to produce thin films of materials, although we shall see presently that bulk samples can be produced as well. The simplest example of the use of chemical vapor deposition for the production of single-crystalline thin films is that of silicon. Silicon films are deposited on silicon itself or other substrates, such as silicon dioxide or germanium, depending on the application of the particular layer in the device being fabricated. The source of the silicon is a volatile silicon compound such as SiH_4, $SiCl_4$, SiH_2Cl_2, or $SiHCl_3$. When these compounds are heated to high temperature in an inert atmosphere, they decompose to produce solid silicon and vapor phase by-products:

$$SiH_4(g) \rightarrow Si(s, cryst) + 2H_2(g)$$

By controlling the substrate temperature and the flow of the volatile compound to be decomposed, the rate of growth and quality of film can be controlled.

Metals are a frequent candidate for deposition by chemical vapor deposition techniques. Gladfelter (1993) reviewed the deposition of tungsten, copper, and aluminum, metals that are especially important in device manufacturing. He focuses on the issue of selectivity, where the deposition rate is a function of the different surfaces. As in the case shown in Figure 3.23*a*, deposition is not selective. The film forms equally well on the substrate and on islands of another material that have already been deposited. In the lower case, the deposited film forms selectivity on the islands and not on the substrate. This type of selectivity is often required in integrated-circuit fabrication. The brute-force approach is to use photolithography and etching to achieve the metal deposition shown in Figure 3.23*b*.

In addition to silicon and metals, a third important element being deposited as thin films is diamond (Celii and Butler, 1991; May, 2000). For many years, diamonds were synthesized by a high pressure/high temperature technique that produced bulk diamonds. More recently, the interest in diamonds has expanded to thin films. Diamond has a slew of properties that make it a desired material in thin-film form: hardness, thermal conductivity, optical transparency, chemical resistance, electrical insulation, and susceptibility to doping. Thin film diamond is prepared using chemical vapor deposition, and we examine the process in some detail as a prototypical chemical vapor example. Despite its importance and the intensity of research focused on diamond chemical vapor deposition, there remains uncertainty about the exact mechanism.

A mixture of hydrogen gas and a hydrocarbon, where the hydrocarbon is 1 to 2% of the mixture, at a pressure less than or equal to 1 atm is subjected to an activation source. This activation source could be a plasma produced with argon, a hot filament, or the heat of combustion produced in a torch using the hydrogen–hydrocarbon mixture as a fuel, to name only a few. This activation energy dissociates the hydrogen molecule into hydrogen atoms:

$$H_2(g) \rightarrow 2H^{\bullet}(g)$$

The hydrogen atoms abstract a hydrogen atom from the hydrocarbon to re-form molecular hydrogen:

$$C_xH_y(g) + H^{\bullet}(g) \rightarrow C_xH_{y-1}(g) + H_2(g)$$

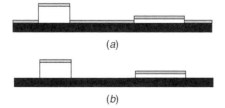

Figure 3.23   (*a*) Nonselective and (*b*) selective deposition.

Often, methane, CH_4, is used as the hydrocarbon. In addition to reacting with the hydrocarbon, the hydrogen also reacts with the surface where the diamond is being deposited. This surface is typically kept hot, 1000 to 1400 K. Presumably, the hydrogen in the hydrocarbon species plays a critical role in the deposition of diamond film. The stable phase of carbon at the pressure and temperature of the deposition is graphite. In a sense, the hydrogens act as protecting groups, preventing a planar carbon film, graphite, from forming. As each methyl radical displaces a surface hydrogen from the diamond (Figure 3.24) a new carbon layer is in progress, protected from converting to graphite with the hydrogens maintaining the tetrahedral sp^3 hybridization of the carbon atom. The importance of the hydrogens is further revealed by noting that if the substrate temperature gets too high, greater than 1400 K, diamond does not form. Presumably, at this high temperature, the hydrogens, chemically bonded to the diamond surface and preventing the formation of graphite, are thermally desorbed.

Silicon, diamond, and metal deposition are all examples of elemental deposition. Compounds, particularly oxides, are also deposited by chemical vapor deposition. Some of the important oxides deposited as thin films include SiO_2, $BaTiO_3$, $LiNbO_3$, $YBa_2Cu_3O_x$, indium-doped SnO_2, and $LiCoO_2$. These materials have properties such as superconductivity or lithium ionic conductivity that make their production as thin films a much-studied area of research. If the oxide is to be deposited on the bare metal (e.g., depositing SiO_2 onto Si), chemical vapor deposition is not really needed. Controlling the oxygen partial pressure and temperature of the substrate will produce the oxide film. Whether the film sticks to the substrate is another question! The production of SiO_2 films on Si is an advanced technology that the integrated-circuit industry has relied on for many years. Oxide films on metals have been used to produce beautiful colored coatings as a result of interference effects (Eerden et al., 2005).

The oxygen in the film can come from the precursor or be supplied by the carrier gas. Even if the precursor contains oxygen, as in a metal alkoxide [e.g., tantalum(V) ethoxide, $Ta(OCH_2CH_3)_5$], oxygen can be used in the carrier gas to prevent the metal from being reduced. This illustrates the added difficulty of depositing a compound as opposed to an element. Stoichiometry must be controlled when depositing a useful oxide film. In the case of Ta_2O_5, resulting

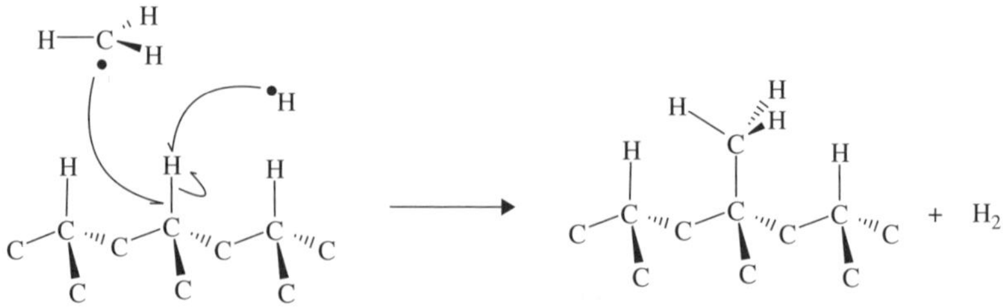

Figure 3.24 Possible mechanism for the growth of diamond film using methane as a precursor.

from the decomposition of $Ta(OCH_2CH_3)_5$, the stoichiometric compound Ta_2O_5 is desired. Ta_2O_5 is useful in the building of integrated circuits because of its high dielectric constant. In the case of indium tin oxide (ITO), a common transparent conductor, the desired film composition is 90% In_2O_3 and 10% SnO_2. Control of both precursors will be required to achieve this specific nonstoichiometric mixture.

An alternative to diamond for a hard thin film is silicon nitride, Si_3N_4. Silicon nitride is not only hard but is dense. This makes it difficult for ions to move through it. This is in contrast to many oxides, which present channels large enough for a sodium or potassium ion to move through. Silicon nitride, then, has long been used in the integrated-circuit business where a barrier is needed which prevents ionic movement. The film is produced by the oxidation of dichlorosilane, $SiCl_2H_2$, or silane, SiH_4, with ammonia, NH_3. As in the case of diamond CVD, the deposition of silicon nitride is often accomplished with plasma-assisted chemical vapor deposition.

Another important nitride deposited as a thin film is gallium nitride, GaN. People who work with gallium nitride are convinced that is the most important semiconductor next to silicon. It is a wide-bandgap material that emits in the visible portion of the electromagnetic radiation spectrum. It was the solution to the elusive blue light-emitting diode (Nakamura, 1997). Gallium nitride thin films can be prepared using a wide range of techniques, including HVPE (hydride vapor-phase epitaxy). It brings to the substrate surface two vapors: gallium chloride generated by reacting the gallium with hydrogen chloride gas, and ammonia, the hydride of nitrogen, hence the name *hydride vapor-phase epitaxy*:

$$Ga(s) + HCl(g) \quad \rightarrow \quad GaCl(g) + \tfrac{1}{2}H_2(g)$$
$$GaCl(g) + NH_3(g) \quad \rightarrow \quad GaN(s) + HCl(g) + H_2(g)$$

HVPE has become the favored method for the deposition of GaN because of its high deposition rate and low defect density (Lee and Auh, 2001). Recently, free-standing GaN samples have been prepared by depositing GaN on $LiAlO_2$ using HVPE and then etching away the $LiAlO_2$ (Jasinski et al., 2005). The samples were reported to be 50 mm in diameter and 350 μm thick. This is an application of chemical vapor deposition for the production of a bulk sample. Bulk GaN has also been realized by a technique labeled as *laser lift-off*. The laser doesn't go anywhere, but instead, is used to detach the GaN thick film from the substrate (Ambacher et al., 2001). Xu et al. (2003) reported growing GaN to a "film" thickness of 1 cm by the hydride vapor-phase epitaxy method.

Atomic layer deposition, also known as *atomic layer epitaxy* (Suntola and Antson, 1977), is a chemical vapor deposition technique capable of producing extremely thin uniform films (Ritala and Leskelä, 2001; Leskelä and Ritala, 2003). The method differs from conventional chemical vapor deposition in that the precursors, of which there are typically two, are not exposed to the substrate simultaneously. Rather, the first precursor is introduced into the reaction chamber, where it binds to the substrate at complete monolayer coverage: a Langmuir type I isotherm described in Section 3.2. The nonadsorbed precursor is pumped away. The second precursor is then admitted where it reacts with the adsorbed first precursor. Again, excess second precursor is added with the unreacted portion pumped away. An example of this type of chemistry is the deposition of Al_2O_3 on silicon (George et al., 1994). The surface of the silicon is first oxidized to provide surface hydroxyl groups. The first precursor, $Al(CH_3)_3(g)$, is adsorbed on the silicon surface, where it reacts with the hydroxl groups and forms an aluminum–oxygen bond and produces methane as a by-product:

$$Al(CH_3)_3(g) + H-O-Si(s) \rightarrow Al(CH_3)_2-O-Si(s) + CH_4(g)$$

After the nonadsorbed $Al(CH_3)_3(g)$ is removed, water vapor is introduced, causing the next reaction:

$$Al(CH_3)_2-O-Si(s) + H_2O(g) \rightarrow Al(OH)_2-O-Si(s) + 2CH_4(g)$$

Notice that after the second reaction, hydroxyl groups are again present, so that the process can be repeated, each time laying down a new layer of $Al_2O_3(s)$. With the sequential application of the two precursors, a compound is deposited as a thin film. Depositing a pure metal presents an interesting challenge. Lim et al. (2003) deposited several transition metals and the corresponding transition metal oxide using transition metal acetamidinate compounds for the metal precursor and hydrogen gas or water vapor as the second precursor. The precise details of the mechanism involving the hydrogen continue to be investigated, but it believed that the propensity for molecular hydrogen to dissociate into atomic hydrogen atoms upon transition metal absorption plays a key role. For example, the chemisorbed hydrogen atoms could be responsible for breaking the metal–nitrogen bond in the metal precursor:

$$MN_2CCH_3[CH(CH_3)_2]_2(ads) + 2H(ads) \rightarrow M(s) + H_2N_2CCH_3[CH(CH_3)_2]_2(g)$$

Two clues lead to such a proposed mechanism: (1) the hydrogen must precede the metal acetamidinate in order for deposition to occur, and (2) the NMR spectrum of the desorbed ligand shows resonances attributable to additional hydrogens relative to the starting material.

Because of the sequential nature of atomic layer deposition, it is a slow method for preparing thin films. The sequential nature, however, also produces a film of uniform thickness, referred to as a *conformal film*. This is important when the surface being coated is not atomically flat, but rather, has troughs and islands to be coated. Some of the most important technological materials, such as silicon and germanium, have not shown themselves to be amenable to the atomic layer deposition technique. This points to the need for continued research in the field of precursor synthesis.

The methods described so far in this section are designed to provide uniform coverage of the deposited material. If a pattern is needed in the deposited material, such as in the deposition of metal layers, as already mentioned, a mask must be used. Laser chemical vapor deposition, however, has the potential to produce structured deposition directly where chemical selectivity does not exist. Lasers have been used in chemical vapor deposition for several tasks. One task has been to vaporize a source material. This heating technique is called *laser ablation* and has been used to deposit diamond films (Park, 1998; Lade et al., 1999). A second task has been to remove material (Dennler et al., 2005; Boehme et al., 2006). In this task, it serves as the etching agent. The final task, and the most interesting, is to serve as the energy source necessary to accomplish a photochemical reaction as part of a chemical vapor deposition. This has the potential for depositing spatially confined product as shown schematically in Figure 3.25.

The uniform coverage (Figure 3.25a) is a result of uniform vapor composition at the surface and uniform substrate temperature. The spatially localized deposition (Figure 3.25b) is a result of the narrowly focused laser beam. The chemical vapor deposition occurs in the laser beam only where the required energy is present. This energy can be in the form of heat, providing the energy source for pyrolysis. It can also be in the form of photonic energy, where electronic transitions are effected in the precursors. Mulenko and Mygashko (2006) used the 488-nm line from an Ar^+ laser to photodecompose iron pentacarbonyl to iron tricarbonyl:

$$Fe(CO)_5 + h\nu \rightarrow Fe(CO)_3 + 2CO$$

The reactive nature of the $Fe(CO)_3$ radical with unreacted $Fe(CO)_5$ on the substrate (silicon) surface leads to the formation of a $Fe_2O_{3-x}(0 \leq x \leq 1)$-deposited film. Johansson et al. (1992) demonstrated that boron could be deposited from BCl_3 and H_2:

$$2BCl_3(g) + 3H_2(g) \rightarrow 2B(s) + 6HCl(g)$$

The particularly intriguing result of this laser chemical deposition was that the deposited boron formed as a free-standing, albeit narrow fiber. The shape of the

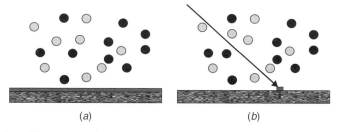

(a)   (b)

Figure 3.25 CVD versus photochemical laser deposition: (*a*) uniform film formation; (*b*) focused laser beam causes photochemical vapor deposition to occur only at the laser beam tip indicated by the arrow.

fiber could be manipulated by translating the substrate relative to the focal point of the laser beam. Although the authors used a 514.5-nm continuous-wave argon ion laser, they considered the laser effect to be a heating effect.

Laser chemical vapor deposition has been used to prepare films with an unfocused beam (Kimura, 2006) and free-standing fibers with a focused beam. A continuing challenge in this area of chemical vapor deposition is the production of lines that remain *on* the substrate surface. This is the operational mode that would be useful for replacing lithography as the method for a patterned deposition.

3.7. MOLECULAR BEAM EPITAXY

Physical vapor deposition and chemical vapor deposition are both techniques for producing thin films. Both rely on the transfer of mass from the vapor phase to a solid surface. A third technique, related to chemical vapor deposition but generally considered distinct from it, is molecular beam epitaxy (MBE) (Joyce and Joyce, 2004), in which a neutral beam of atoms is used to deposit a layer of adsorbed atoms. To deposit a compound, two molecular beams are used, depositing the constituent elements in the compound sequentially. Although this appears to make the deposition of any size film of any composition a simple matter (shown schematically below in Figure 3.26), the technical requirements for achieving this deposition are severe.

The heart of the MBE process is the Knudsen cell. In this cell, an element is heated to produce its vapor. Some elements require very little heating, such as the alkali metals or the gaseous elements themselves! Other elements, such as tantalum and tungsten, have famously minute vapor pressures at any convenient experimental temperature. Once a vapor is established in a Knudsen cell, a small opening allows a very tiny amount of the gas to escape (Figure 3.27). The dimensions of the hole and the wall thickness are designed so that the molecules leaving the cell all travel in the same direction and do not experience collisions with each other or the walls of the opening as they traverse it. The Knudsen cell provides a molecular beam, hence the name *molecular beam epitaxy*. Because of the conditions necessary to maintain a unidirectional molecular beam (i.e., low pressure), molecular beam epitaxy is necessarily a technique that suffers from a low deposition rate.

With multiple Knudsen cells containing different elements, molecular beam epitaxy can be used to deposit the elements required for thin-film compound formation. With its precise control over the composition of the beam, that is, a given beam with a given element is either on or off, molecular beam epitaxy can be used to produce very thin films with abrupt changes in composition (Figure 3.28a). This technical capability has led to the development of quantum wells and cascade lasers that require the type of compositional change shown in Figure 3.28a and would not operate as efficiently with a conventional stacking sequence shown in Figure 3.28b.

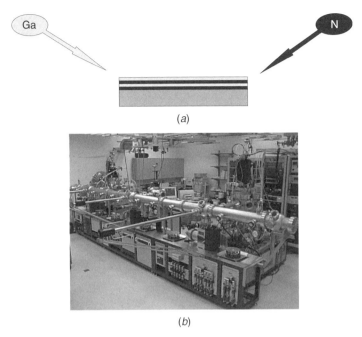

(a)

(b)

Figure 3.26 Molecular beam epitaxy: (*a*) simplified presentation of monatomic layers of gallium and nitrogen deposited sequentially to form a gallium nitride film; (*b*) molecular beam epitaxy system designed and built at Pacific Northwest National Laboratory, Richland, WA. (Reproduced with permission.)

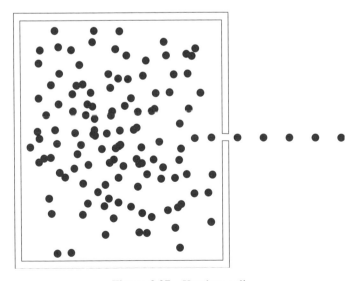

Figure 3.27   Knudsen cell.

(a)   (b)

Figure 3.28   (*a*) Thin layers with abrupt composition changes resulting from molecular beam epitaxial growth; (*b*) relative layer thickness as composition changes resulting from conventional layering procedures.

An additional advantage of molecular beam epitaxy over chemical vapor deposition is that lower substrate temperatures are used in molecular beam epitaxy. The high temperatures required to effect a chemical reaction in chemical vapor deposition are not needed in molecular beam epitaxy. Given the extremely thin nature of the films, atomic diffusion is kept to small distances, and hence the small diffusion coefficients do not seriously retard the overall reaction rate. The difficulty presented by small diffusion coefficients with respect to chemical reactions between bulk solids is discussed in Chapter 5.

One of the major difficulties, with respect to both cost and construction, is the vacuum system requirements for molecular beam epitaxy (Figure 3.26*b*). Molecular beam epitaxy requires a vacuum system capable of maintaining 10^{-11} torr. Despite the technical hurdles of this approach, it will produce the film needed when all other methods fail.

REFERENCES

Al-Abadleh, H. A.; Grassian, V. H. *Langmuir*. **2003**, *19*, 341–347.

Ambacher, O.; Kelly, M. K.; Miskys, C. R.; Hoppel, L.; Nebel, C.; Stutzmann, M. *Mater. Res. Soc. Symp. Proc.* **2001**, *617*, J1.7.1–J1.7.12.

Atkins, P. W. *Physical Chemistry*, 2nd ed. W. H. Freeman, San Francisco, CA, **1978**, p. 938.

Boehme, R.; Hirsch, D.; Zimmer, K. *Appl. Surf. Sci.* **2006**, *252*(13), 4763–4767.

Brunauer, S.; Emmett, P. H.; Teller, E. *J. Am. Chem. Soc.* **1938**, *60*, 309–193.

Celii, F. G.; Butler, J. E. *Annu. Rev. Phys. Chem.* **1991**, *42*, 643–684.

Chang, R. *Physical Chemistry for the Chemical and Biological Sciences*, University Science Books, Sausalito, CA, **2000**, pp. 64–65.

Cheon, J.; Zink J. I. *J. Am. Chem. Soc.* **1997**, *119*, 3838–3839.

Cleary, D. A.; Francis, A. H.; Lifshitz, E. *J. Lumin.* **1986**, *35*(3), 163–70.

Clément, R.; Garnier, O.; Jegoudez, J. *Inorg. Chem.* **1986**, *25*, 1404–1409.

Cote, D. R.; Nguyen, S. V.; Stamper, A. K.; Armbrust, D. S.; Tobben, D.; Conti, R. A.; Lee, G. Y., *IBM J. Res. Dev.* **1999**, 43.

David, T.; Goldsmith, S.; Boxman, R. L. *Vac. Technol. Coat.* **2005**, *6*, 40–45.

Dennler, G.; Lungenschmied, C.; Neugebauer, H.; Sariciftci, N. S.; Labouret, A. *J. Mater. Res.* **2005**, *20*(12), 3224–3233.

Eerden, M.; Tietema, R.; Krug, T.; Hovsepian, P. E. In *Proc. 48th Annual Technical Conference of the Society of Vacuum Coaters*, **2005**, 575–579.

Figueiredo, M. A.; Oliveira, H. *Mater. Res. Soc. Symp. Proc*. **2005**, *847*, 273–278.

Gamble, F. R.; Osiecki, J. H.; Cais, M.; Pisharody, R.; DiSalvo, F. J.; Geballe, T. H. *Science*. **1971**, *174*(4008), 493–497.

George, S. M.; Sneh, O.; Dillon, A. C.; Wise, M. L.; Ott, A. W.; Okada, L. A.; Way, J. D. *Appl. Surf. Sci*. **1994**, *82–83*, 460–467.

Ghorayeb, A. M.; Coleman, C. C.; Yoffe, A. D. *J. Phys. C Solid State Phys*. **1984**, *17*(27), L715–L719.

Gladfelter, W. L. *Chem. Mater*. **1993**, *5*, 1372–1388.

Harley, E. C. T.; McNeil, L. E. *J. Phys. Chem. Solids*. **2004**, *65*, 1711–1718.

IUPAC Committee on Colloid and Surface Chemistry. *Pure Appl. Chem*. **1972**, *31*(4), 577–638.

Jaeger, R. C. *Introduction to Microelectronic Fabrication*, Addison-Wesley, Reading, MA, **1993**, pp. 111–112.

Jasinski, J.; Liliental-Weber, Z.; Maruska, H.-P.; Chai, B. H.; Hill, D. W.; Chou, M. M. C.; Gallagher, J. J.; Brown, S. Paper LBNL-53098, Lawrence Berkeley National Laboratory, Berkeley, CA, 27, **2005**. http://repositories.cdlib.org/lbnl/LBNL-53098.

Johansson, S.; Schweitz, J.-A.; Westberg, H.; Boman, H.; Boman M. *J. Appl. Phys*. **1992**, *72*(12), 5956–5963.

Joyce, B. A.; Joyce T. B. *J. Cryst. Growth*. **2004**, *264*(4), 605–619.

Jurczyk, M. *Curr. Top. Electrochem*. **2003**, *9*, 105–116.

Kimura, T. *J. Ceram. Soc. Jpn*. **2006**, *114*(2), 161–166.

Lade, R. J.; Claeyssens, R.; Rosser, K. N.; Ashfold, M. N. R. *Appl. Phys. A*. **1999**, *69*(Suppl.), S935–S939.

Langmuir, I. *J. Am. Chem. Soc*. **1918**, *40*, 1361–1402.

Lee, K.; Auh, K. *MRS Internet J. Nitride Semicond. Res*. **2001** 6, 9.

Leskelä, M.; Ritala, M. *Angew. Chem. Int. Ed*. **2003**, *42*, 5548–5554.

Levine, I. N. *Physical Chemistry*, 5th ed., McGraw-Hill, New York, **2002**, pp. 397–402.

Lim, B. S.; Rahtu, A.; Gordon, R. G. *Nat. Mater*. **2003**, *2*, 749–754.

Luca, V.; Thomson, S. *J. Mater. Chem*. **2000**, *10*, 2121–2126.

Manca, P.; Sanna, S.; Calestani, G.; Migliori, A.; De Renzi, R.; Allodi, G. Los Alamos National Laboratory, Preprint Archive, Condensed Matter, **2000**.

May, P. W. *Philos. Trans. R. Soc. London A*. **2000**, *358*, 473–495.

McKelvy, M. J.; Glaunsinger, W. S. *Annu. Rev. Phys. Chem*. **1990**, *41*, 497–523.

McQuarrie, D. A; Simon, J. D. *Physical Chemistry: A Molecular Approach*, University Science Books, Sausalito, CA, **1997**, pp. 1297–1303.

Mulenko, S. A.; Mygashko, V. P. *Appl. Surf. Sci*. **2006**, *252*, 4449–4452.

Nakamura, S. *IEEE J. Sel. Top. Quantum Electron*. **1997**, *3*(2), 435–442.

Park, S.M.; Moon, J. Y. *J. Chem. Phys*. **1998**, *109*, 8124.

Ratsch, C.; Venables, J. *J. Vac. Sci. Technol. A* **2003**, *21*(5), S96–S109.

Ritala, M.; Leskelä, M. In *Handbook of Thin Film Materials*, Vol. 1, H. S. Nalwa, Ed., Academic Press, San Diego, CA, **2001**, pp. 103–156.

Robbie, K.; Sit, J. C.; Brett, M. J. *J. Vac. Sci. Technol. B*. **1998**, *16*(3), 1115–1122.

Siadati, M. H.; Ward, T. L.; Martus, J.; Atanasova, P.; Xia, C.; Schwartz, R. W. *Chem. Vap. Depos*. **2004**, *3*(6), 311–317.

Sing, K. S. W.; Everett, D. H.; Haul, R. A. W.; Moscou, L.; Pierotti, R. A.; Rouquerol, J.; Siemieniewska, T. *Pure Appl. Chem*. **1985**, *57*(4), 603–619.

Singh, J.; Wolfe, D. E. *J. Mater. Sci*. **2005**, *40*, 1–26.

Suntola, T.; Antson, J. U.S. patent 4,058,430, **1977**.

Thomas, R. L.; Cleary, D. A. *Sensors Actuators B: Chem*. **1996**, *32*(1), 19–22.

Whittingham, M. S.; Jacobson, A. J., Eds. *Intercalation Chemistry*, Academic Press, New York, **1982**.

Xu, X.; Vaudo, R. P.; Brandes, G. R. *Opt. Mater*. **2003**, *23*(1–2), 1–5.

Zhang, J. X.; Cheng, H.; Chen, Y. Z.; Uddin, A.; Yuan, S.; Geng, S. J.; Zhang, S. *Surf. Coat. Technol*. **2005**, *198*(1–3), 68–73.

4 Solid–Liquid Reactions

In this chapter we discuss preparative routes for inorganic materials in three basic types of systems involving the presence of a distinct solid–liquid interface: those in which the liquid and solid phases are of the same chemical identity (solidification and vitrification processes), those in which the liquid and solid phases are not of the same chemical identity (crystallization, precipitation), and the special case in which the liquid phase is a pure ionic liquid or molten salt. Ionic liquids can serve as the solvent as well as a templating agent, and the liquid components may or may not become incorporated into the final solid product. We also discuss two areas where the distinct solid–liquid interface becomes somewhat blurred: namely, sol–gel and solvothermal processes.

In many respects, crystallization from the molten state is analogous to that from a solvent phase. For example, in both processes slow cooling tends to result in larger crystals, while faster cooling typically gives smaller crystals. Another important similarity involves the behavior of impurities. Impurity atoms usually do not "fit" into the crystal lattice of the solute crystallizing from a solvent or melt. In a polycrystalline metal, impurities tend to segregate at grain boundaries during solidification because of their mismatch with the lattice of the metal atoms. This behavior allows crystallization from a solvent to be used as a purification technique, and it is the basis of zone-melting and zone-refining techniques used in the production of very high purity monocrystalline materials. However, there are important differences between the solid–melt and solid–solvent interface as well.

4.1. SOLID LIQUID INTERFACE

The cohesive forces between the particles of a liquid are intermediate in strength compared to those in a solid or gas. The particles of a liquid have more freedom of movement, and hence more kinetic energy and internal energy than those of a solid, but less than those of a gas. As a direct consequence of this greater energy, there are two major structural differences between a solid and a liquid. First, liquids exhibit variation in short-range order. For example, a liquid network may be described as being comprised of various types of polyhedral "holes," with the constituent atoms or molecules at the vertices. Examples of these polyhedra include tetrahedra, octahedra, trigonal prisms, and Archimedean

Inorganic Materials Synthesis and Fabrication, By John N. Lalena, David. A. Cleary, Everett E. Carpenter, and Nancy F. Dean
Copyright © 2008 John Wiley & Sons, Inc.

antiprisms. The second key structural feature differentiating liquids from solids is the absence of long-range order or, more precisely, translational symmetry. In both conventional (i.e., non-glass-forming) solidification and crystallization processes, liquid-phase solute atoms or molecules become rearranged into a more favorable, lower-energy, crystalline phase with long-range order. In fact, that portion of the liquid phase itself in the vicinity of the solid–liquid interface exhibits crystal-induced ordering in the first three to four atomic layers away from the interface. Much less is known about the precise structures of ionic liquids and molten salts. However, some experimental evidence exists suggesting that there are strong similarities between the crystalline and liquid arrangements.

4.2. CRYSTALLIZATION, PRECIPITATION, AND SOLIDIFICATION

Although there are analogies between solidification and crystallization, there are also some important structural differences between the solid–melt interface (also called the *solidification front*) and the solid–solvent interface due to the differing concentration, viscosity, and temperature of the respective liquid phase. These may be summarized as follows:

- A conventional melt is a nominally pure phase for which the thermodynamic activity is constant (equal to unity), whereas a solute of low concentration in a solvent has an activity less than the corresponding Gibbs free energy of the pure phase.
- The melt viscosity determines the likelihood that a glass will form with rapid cooling. In highly viscous melts, the atomic mobility is reduced substantially, suppressing the homogeneous nucleation rate and crystallizability, which are dependent on the ease with which atomic rearrangement can occur.
- The effect of temperature on liquid structure was pioneered by John Desmond Bernal (1901–1971), who suggested that the effect of heating a liquid could be described as a volume expansion of the interstices (holes) existing between the constituent atoms while maintaining the same minimal nearest-neighbor distance between coordinated atoms (Bernal, 1964). For example, a heated liquid contains a greater percentage of larger polyhedral holes (e.g., trigonal prisms, Archimedean antiprisms) than smaller polyhedral holes (e.g., tetrahedra, octahedra).

Let's first focus our attention on crystallization.

4.2.1. Ostwald Rule of Stages and Ostwald Ripening

In 1897, Friedrich Wilhelm Ostwald (1853–1932) published his now famous study of crystallization processes, which led to the *Ostwald rule of stages* or *Ostwald step rule* (Ostwald, 1897). Ostwald noticed that the course of transformation of unstable (or metastable) states into stable states normally occurs in stages, where the first products formed are usually not the most thermodynamically stable products but rather, intermediate products having free energies closest to the initial states (Madras and McCoy, 2001; Stoica et al.,2005). Take, for example, the crystallization of citric acid. Crystallization above $34°C$ leads to a thermodynamically stable anhydrous crystal, whereas crystallization at room temperature leads to a monohydrate polymorph. In detailed crystallization experiments, a probing of the metastable border between the two polymorphs followed the solubility curve of anhydrous citric acid (Nyvlt, 1995).

In subsequent experiments, using other crystal systems, such as ferrous sulfate and sodium hydrogen phosphate, it was similarly observed that the first crystallization product to form was the one most closely resembling the structure of the solvent (Nyvlt, 1995). For the case of citric acid, this is the monohydrate, which more closely resembles the aqueous structure. As the temperature of the solution is increased, the structure of the solvent, as well as the solubility of the crystal, changes, resulting in a more thermodynamically stable anhydrous product. This conversion between the kinetic and thermodynamic product occurs at a critical transition temperature, below which the structure of the solution favors the formation of the hydrated product. As the transition temperature is surpassed, the anhydrous product becomes favored.

Another important issue relating to metastability is the contribution of the interfacial free-energy to the total free energy of a system. The interfacial free-energy contribution is negligible for crystals larger than 1 μm (Apps and Sonnenthal, 2004). By a process occurring near equilibrium conditions, known as *Ostwald ripening*, larger clusters grow at the expense of dissolving smaller clusters. Hence, although many small crystals may form in a system initially, most of them slowly disappear, except for a few that grow larger by consuming the others. The phenomenon was first described for small molecules by Ostwald in 1896. However, it has a wide range of applicability, from the growth of single crystals in solution under ambient conditions to the coarsening of polycrystalline geochemical samples in a subcritical aqueous phase (water temperature $>374°C$, pressure >221 atm) and of polycrystalline solidification products. In polycrystals, the larger crystallites, which usually possess a lower surface free energy, destabilize smaller coexisting crystallites of the same phase. This results in the growth of progressively fewer, larger crystallites, as well as a smaller crystallite size distribution. A fully satisfactory physical basis for Ostwald ripening also remains elusive (Baldan, 2002). In addition, although only a few violations have been cited, there is still no undisputed theoretical basis for the Ostwald step rule.

If we examine the competitive growth of multiple-size crystallites in solutions, we find the growth occurring along concentration gradients. The concentration gradients around the particles are explained with the *Gibbs–Thomson equation:*

$$C_r = C_\infty \exp\left(\frac{2\gamma\Omega}{rRT}\right) \qquad (4.1)$$

where C_r is the solubility at the surface of a spherical particle with radius r, C_∞ the solute concentration (mole fraction) at a plane interface at equilibrium with a particle of infinite radius, γ the specific interfacial energy of the solution–precipitate particle boundary, Ω the mean atomic volume of a particle, R the gas constant [8.314×10^3 J/(K· kmol)], and T the absolute temperature. Equation 4.1 is actually only one of several possible forms of the Gibbs–Thomson equation [named after Josiah Gibbs and three Thomsons: James (1822–1892), his brother Williams (1824–1907), and Joseph John (1856–1940)]; it is also known as the *Ostwald–Freundlich equation* after Ostwald and Herbert Freundlich (1880–1941), who had corrected Ostwald's original expression. At equilibrium, $2/r \sim dA/dV$ (A is the particle surface area and V the particle volume), which implies that small particles are more soluble and hence inherently less stable than larger particles. This observation had actually been made as early as 1813 by William Hyde Wollaston (1766–1828) and again in 1885 by Pierre Curie.

4.2.2. LaMer Crystallization Model

For several decades, Ostwald's work was the foundation of crystallization theory. In 1950, an American chemist, Victor Kuhn LaMer (1895–1966), advanced our understanding of crystallization of colloidal sols. LaMer assumed that crystallization was a diffusion-driven process (LaMer, 1952; Boistelle and Astier, 1988), as described in previous chapters. LaMer's model is described in three steps, presented in Figure 4.1. In the initial steps of the reaction, the concentration of the molecular materials builds until a critical concentration is reached. This critical concentration, called *supersaturation*, can be written as

$$\Delta\mu = k_{\mathrm{B}} T \ln \frac{C}{C_s} \tag{4.2}$$

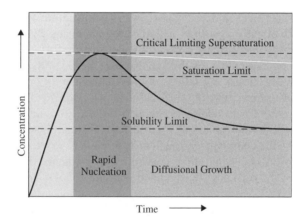

Figure 4.1 Nucleation process. As time proceeds, the concentration increases until eventually reaching the supersaturation limit. At this point, nucleation occurs, relieving the high concentration stress.

Victor Kuhn LaMer (1895–1966) obtained his Ph.D. in food and nutritional chemistry under Henry Sherman in 1921 from Columbia University. His interests soon shifted to physical chemistry, which he taught at Columbia until his retirement in 1961. LaMer was awarded a traveling fellowship to Cambridge University in 1922 and to the University of Copenhagen in 1923, where he worked with Brønsted. LaMer contributed to our current understanding of activity coefficients of multiply charged ions and their reaction rates. He also studied the properties of solutions in heavy water, as well as acid–base equilibria in heavy water and benzene. During World War II he studied smokes and other fine dispersions and, with Sinclair, discovered the higher-order Tyndall spectra that enable rapid particle-size measurement. After the war, LaMer studied liquid dispersions, sedimentation, filtration, and flocculation. He was the founding editor of the *Journal of Colloid and Interface Science* from 1946 to 1965. LaMer was elected to the U.S. National Academy of Sciences in 1948. He was also elected to the Royal Belgian Academy of Arts, Letters, and Sciences and to the Royal Danish Academy of Sciences.

Source: Louis P. Hammett, *Biographical Memoirs of the U.S. National Academy of Sciences*, Vol. 45, 1974, pp. 203–226.

where k_B is the Boltzmann constant and T is temperature. Supersaturation is defined via this relationship: namely, as the relationship where

$$\frac{\Delta\mu}{k_B T} = \ln \frac{C}{C_s} \tag{4.3}$$

At this point, the molecular materials begin to collide and form molecular clusters that are in equilibrium with isolated molecular material. It is the second step that the clusters begin to have enough stability for nucleation to occur, thereby creating the first stable nuclei. At this point, the concentration of free molecular material decreases partially, relieving the supersaturation. The nucleation can occur in two fashions: homogeneous or heterogeneous nucleation. Once the concentration drops below the supersaturation concentration, nucleation can no longer occur and the stable nuclei grow following diffusion kinetics and Ostwald ripening. As the crystal continues to grow, the concentration of free molecular material continues to decrease until eventually, the solubility limit will be reached and the crystal will reach equilibrium with dissolution of the crystal.

More recent modeling work has cast doubt on LaMer's simple model for growth of crystals. The primary reason for the LaMer model failure is the principal assumption that the growth is a diffusion-controlled process relying only on unimolar reactions. In more recent work, reactions are found to consist of multiple steps where the concentration of molecular material is being generated, continually resulting in a significantly higher concentration around the supersaturation limit. This leads to multiple nucleation events and a wide size of crystals being formed.

4.2.3. Nucleation Process

The reversible work, W, to homogeneously nucleate a cluster of a new phase in (unstable) equilibrium with the parent phase is the change in the Gibbs free energy, ΔG:

$$W = \Delta G = -n\,\Delta\mu + S\gamma \tag{4.4}$$

In this equation, n is the number of molecules in the cluster, $\Delta\mu$ the difference of the bulk chemical potentials of the two phases, S the area of the cluster, and γ the interfacial free energy. Equation 4.4 applies to both homogeneous nucleation in solidification and crystallization from a saturated solution. However, if we are talking about homogeneous nucleation of a solute from a saturated solution, the difference between the chemical potential of a solute in the supersaturated and saturated solutions is given by Eq. 4.2. Supersaturation is defined via a rearrangement of Eq. 4.2 (i.e., Eq. 4.4), which is given again here:

$$\frac{\Delta\mu}{k_B T} = \ln\frac{C}{C_s} \tag{4.5}$$

When sufficient supersaturation is reached, nuclei begin to form. The formation of a nucleus must be accompanied by reduction in free energy; otherwise, the nucleus just returns to solution. Using Eqs. 4.2 and 4.5 and substituting β for C/C_s, it is seen that in order to nucleate a cluster of volume V and area S, the change in Gibbs free energy is given by

$$\Delta G = -\frac{V}{\Omega}k_B T \ln\beta + S\gamma \tag{4.6}$$

The coefficient V/Ω represents the number of molecules inside the nucleus. If we initially consider a spherical nucleus with a volume $4\pi r^3/3$ and a surface area $4\pi r^2$, Eq. 4.6 becomes

$$\Delta G = -\frac{4\pi r^3}{3\Omega}k_B T \ln \beta + 4\pi r^2 \gamma \qquad (4.7)$$

The cost associated with a change in energy becomes smaller as the particle surface/volume ratio decreases. Since the volume and surface terms are in competition, there is a certain radius beyond which growth is favored, given by

$$r^* = \frac{2\Omega\gamma}{k_B T \ln \beta} \qquad (4.8)$$

This radius is called the *critical radius*, for which the critical activation energy for nucleation, ΔG^*, is

$$\Delta G^* = \frac{16\pi\Omega^2\gamma^3}{3(k_B T \ln \beta)^2} \qquad (4.9)$$

By utilizing Eq. 4.8, Eq. 4.9 is also seen to be

$$\Delta G^* = \tfrac{1}{3}(4\pi r^{*2}\gamma) \qquad (4.10)$$

which shows that the activation energy required to nucleate a cluster, stable at supersaturation β, is one-third the energy required to create its surface. If one solute particle is withdrawn from the nucleus, the cluster dissolves. If one is added, the cluster continues to grow spontaneously. As the temperature is increased, or as the concentration is increased past supersaturation, the activation energy required decreases. Thus, we would expect a faster nucleation rate at higher concentrations or higher temperatures. It is worth noting that the $16\pi/3$ in Eq. 4.9 is the shape factor specific to a sphere. For cubic nuclei, $V = 8r^3$ and $S = 24r^2$ in Eq. 4.7.

We can thus see that it is easiest to think of nucleation in terms of two competing processes, the coalescing and re-dissolution of clusters. Solute particles coalesce spontaneously to form small clusters (e.g., dimers, trimers), but as long as the cluster size is below the critical size, it is not thermodynamically favored, so it redissolves. However, stable numbers of solute particles tend to stay associated. These stable numbers are often called the *magic numbers* (Echt et al., 1981). Moreover, they represent the most stable geometric configurations. For spherical crystallites, the magic numbers represent the number of atoms necessary to fill in symmetric geometric forms, leading to isocohedral clusters, which then start to adapt the final crystal structure.

When the concentration is closer to the supersaturation limit, heterogeneous nucleation occurs most often. The nucleus develops onto the substrate, with which it makes a contact angle α. Solution of the equations for the nucleus size and activation energy imply that the critical radius is the same as for homonuclear nucleation. On the other hand, the activation energy for heteronuclear nucleation is the product of the activation energy for homogeneous nucleation and a term depending on the value of α, since

$$\Delta G^*_{\text{het}} = \Delta G^* \left(\frac{1}{2} - \frac{3}{4} \cos \alpha + \frac{1}{4} \cos^3 \alpha \right) \tag{4.11}$$

The influence of the contact angle, α, is worth demonstrating. For example, if α is $180°$, the term in parentheses is 1 and $\Delta G_{\text{het}} = \Delta G^*$. If α is $90*$, $\Delta G_{\text{het}} = \Delta G^*/2$, and so on. The smaller the angle, the greater the contact the substrate has with the supersaturated solution, and the lower the energy required for nucleation.

While thermodynamics dictates whether or not nucleus formation occurs, the nucleation rate is often the more important factor. The nucleation rate, J, can be defined as the product of a kinetic factor and the activation energy:

$$J = K_0 \exp \left(\frac{-\Delta G^*}{k_{\text{B}} T} \right) \tag{4.12}$$

In this case, the kinetic factor is dependent on the solubility of the materials and the frequency at which critical nuclei become supersaturated and transform into crystals. Hence,

$$K_0 = N_0 v_0 \tag{4.13}$$

From these last two equations, it becomes clear that all other factors being equal, homogeneous nucleation is more rapid in solutions where the solubility is high. If the solubility is high, the interfacial energy (the cost associated with nucleus formation) between the crystal and the solution is low. In the opposite situation, when the material is sparingly soluble, nucleation is more difficult and occurs only with a high supersaturation concentration. In this situation, nucleation most often is a catastrophic situation, leading to smaller, poor-quality crystallites. Often in small systems it is very important to tailor the concentration carefully around the supersaturation limit. Although the nucleation rate is strongly dependent on the concentration, it also depends on the way in which the high concentration was achieved and the rate at which it was approached.

4.2.4. Crystal Growth

Eventually, upon continued growth, stable clusters begin to adapt the final crystal form. Gibbs demonstrated that in the case of growing crystals there is a condition

$$\sum \sigma_i F_i \tag{4.14}$$

which is a minimum. In this condition, the subscript i represents the ith face of a crystal and F is the surface area of the crystal, while σ is the specific free energy. Von Laue's work (Laue, 1943) allows us to rephrase this condition and say that,

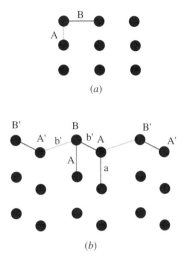

Figure 4.2 (*a*) Two-dimensional crystal representation in which bond *a* is stronger than bond *b*; (*b*) Two-dimensional crystal representation in which bonds *a* and *b″* are the same strength while *b′* is much weaker.

in general, as surface energy increases, the area of the crystal face decreases. This condition has undergone several adaptations over the years, first by Born in his lattice theory. In modern crystal theory, what is more important is not the direct relationship between the surface free energy, but instead, the energy released when a building block is attached to the surface of a growing crystal. The principal consequence of this adaptation is that, as attachment energy increases, the energy required for a bond to form decreases. As a result, the displacement velocity, or the rate of filling of the face, also increases. To understand this further, let's examine a 2D crystal. In our first crystal (Figure 4.2*a*), there are two different types of bonds: *a* in the [10] direction and *b* in the [01]. Now assume that the attachment energy resulting from formation of *a* bonds is greater than that of *b*. This would result in the [10] face having a greater displacement velocity (i.e., growing faster than the [01] face). In our second example (Figure 4.2*b*), there are two types of bonds, a strong A–B bond, and a weaker A–B″ bond along the [10] face, and then the *a* bonds on the [01] face are equal to the weaker A–B bonds.

Now the growth is governed by the formation of two bonds in the [01] direction, one stronger and one weaker, compared to two weaker bonds in the [10] direction. The displacement velocity in the [01] direction is, however, governed by the weaker bond length. Consequently, the crystal grows nearly equally in both the [01] and [10] directions, which leads to two observations. First, for a crystal to grow in the direction of a strong bond, these bonds must be uniform and form an uninterrupted chain. If the bond chain contains bonds of different types, the influence on the shape is determined by the weakest bond present in the chain.

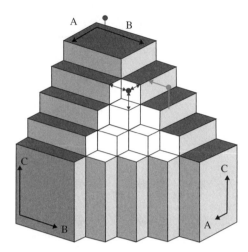

Figure 4.3   Three-dimensional crystal with three periodic bond chain vectors. The figure shows F-faces (100), (010), and (001); S-faces (110), (101), and (011); and the K-face (111).

These observations were made and explained by P. Hartman and W.G. Perdok in their *periodic bond chain vector* (PBC) *theory* (see the Chapter 1 references), which we introduced in Section 1.2.1. This theory takes into account the fact that 3D crystals are far more complicated structures, which are full of partial bonds and preferred directions. When the PBC vector is parallel with a crystal face, there is maximum growth along that crystal face.

Using PBC theory, the faces of a growing crystal can be divided into three categories: flat faces (class F), stepped faces (class S), and kinked faces (class K). In Figure 4.3, there are three PBC vectors, which define six class F faces. Any face that is parallel to one of the A, B, C vectors is a class S face. Faces that are not parallel to any of the vectors become class K faces. As a crystal grows along the F face, a relatively small amount of attachment energy is released as the building block is attached to the crystal face. For class S, two stronger bonds are formed as the building block is attached, whereas in class F, three bonds are formed. This results in, acceleration during the growth; the displacement velocity of class K is faster than that of class S, which is faster than that of class F. As the class K faces get filled in more quickly, they become class S faces, which, in turn, become F faces in the final crystal. Class K faces are rarely seen in the final crystal structure. It is important to note here that we are talking about the final crystallization thermodynamics. In the initial stages, the growth is governed nucleation. Specifically, the crystallite germ grows initially in all directions, resulting in an isometric crystallite. As the crystallite grows into a bigger crystal, the faces start to emerge and the final product is governed by the attachment energy.

In the case of a perfect crystal (i.e., a crystal where there are no defects in a crystal face), the growth takes place by two-dimensional nucleation. To create the

2D nucleus, building blocks just adsorb on the faces and then diffuse together and cluster. Once the nucleus has reached critical size, it becomes stable and exhibits some class K faces, where growth units can be incorporated that convert the class K face to class S. The process continues until all the K faces are filled; the S sites then begin to fill, resulting in a perfect crystal that has all F faces. The kinetics of the growth is typically diffusion-controlled at the rate R:

$$R = J_2 \, dS \tag{4.15}$$

is equal to the 2D nucleation rate (J) and the height of the layer (d) and area (S) of the face.

4.2.5. Precipitation Reactions

One of the simplest and oldest solution-phase synthetic methods is the precipitation reaction. Precipitation is similar to crystallization except that the driving force in the latter is solubility, whereas in the former it is a chemical reaction. In either case, the solid produced is stable and insoluble in a given liquid-phase solvent if the solvation energy (the solute–solvent interaction) is lower in strength than the cohesive or binding energy between the particles of the solid. There are three types of solution-phase chemical reactions taught in introductory chemistry classes: metathesis (double exchange), redox, and acid–base reactions. When the product of any of these solution-phase chemical reactions is an insoluble product, it is called a *precipitation reaction*. In a metathesis reaction, parts of two reactants switch places (e.g., AB + CD → AD + CB). In an oxidation–reduction reaction, electrons are transferred from one species to another with concomitant changes in oxidation states, while in the similar acid–base reaction (there are many definitions of acid–base reactions!) no species undergoes a change in oxidation state. Because of the ability to control kinetics with variations in temperature and solvent type, precipitation reactions are among the most versatile synthetic methods.

As a metathesis reaction proceeds, the concentration of the products increase until nucleation and, eventually, crystallization can occur. The rate of the reaction, and hence nucleation and growth, can be controlled through temperature and concentration. Working with differential solubilities is the simplest method for the synthesis of materials. For example, the semiconductor CdS can be synthesized by the reaction:

$$Cd(NO_3)_2(aq) + Na_2S(aq) \rightarrow CdS(s) + 2NaNO_3(aq)$$

The reversibility of a precipitation reaction is dependent on the solubility of the product. In the case of CdS, $K_{sp} = 3.6 \times 10^{-29}$, so the equilibrium lies far to the right, making the reverse reaction negligible.

Oxidation–reduction reactions are also commonly utilized in synthesis at the solid–liquid interface. Again, there is tremendous variability. Alkali metals (e.g., lithium, sodium) are often used as reducing agents, due to their high reduction potentials:

$$Co(C_2H_3O_2)_{2(toluene)} + 2Na(s) \rightarrow Co(s) + 2Na(C_2H_3O_2)_{2(toluene)}$$

Of course, in working with alkali metals, one typically works in hydrocarbons or ammonia, taking great care to avoid protonic solvents such as water or methanol. The reactions proceed slowly since they are controlled by the surface area of the metal. As the reaction proceeds, the surface becomes oxidized and no longer able to reduce. Typically, for these reactions, the solution is heated to the melting point of the sodium, which allows for a continuous fresh surface for reactions.

For reductions under aqueous conditions, there are several very good reducing agents, such as sodium borohydride ($NaBH_4$):

$$CoCl_{2(aq)} + 2NaBH_{4(aq)} \rightarrow Co_{(s)} + 2NaCl_{(aq)} + 2BH_{3(aq)} + H_{2(g)}$$

In aqueous reactions, it is also very important to follow the side reactions. In borohydride reductions, there is a side reaction with water:

$$BH_{4(aq)}^- + 2H_2O \xrightarrow{\text{catalyst}} BO_{2(aq)}^- + 4H_{2(g)}$$

This, in turn, makes the following reaction the dominate reaction producing the metal boride:

$$2CoCl_2 + 4NaBH_4 + 9H_2O \rightarrow Co_2B + 4NaCl + 12.5H_2 + 3B(OH)_3$$

What is interesting to note here is that the catalyst that facilitates the conversion of the borohydride to borate can be an acid, metal, or metal boride. So, as this reaction proceeds, it is producing the very catalyst that degrades the starting reactant. It is possible to control the boride formation with the use of borohydride esters such as lithium triethoxyborohydride, commonly referred to as superhydride.

Decomposition reactions are another reaction class often employed in nonaqueous solvents. In these reactions, the starting materials are decomposed to create the final product. The preferred starting materials have ligands that are very good leaving groups, such as carbonates, carbonyls, and acetates. The decomposition is facilitated by several different techniques, such as heat in thermolysis, light in photolysis, and sound in sonolysis. The reaction is the same in almost every case:

$$Co_2(CO)_{9(DPE)} \rightarrow Co_{(s)} + CO_{(g)}$$

This reaction produces activated cobalt metal powder following decomposition of the carbonyl at $200°C$ in diphenyl ether (DPE). The metal carbonyls are typically very stable at room temperature; at elevated temperatures they dimerize and decompose, leaving the zero-valent metal. Photochemical decomposition, or photolysis, is common with coinage metals such as silver and gold. While light will reduce these metals, typically they are complexed with photoabsorbing ligand in order to accelerate the decomposition:

$$AgNO_3 + PAMAM \xrightarrow{254\,nm} Ag + NO_3(g)$$

Sonochemical reactions are characterized by extreme conditions. As a cavitation bubble collapses, there is adiabatic conservation of energy, which can often translate to local temperatures in excess of 5200 K. Since the process is adiabatic, heating is not transferred to the solvent system, and there is a subsequent extreme cooling rate of $10^5 - 10^8$ K/s. The reactions using sonolysis are the same as any other decomposition reaction, typically carbonyls. It is important in sonolysis reactions that the leaving ligand not be present in the system after the cavitation bubble collapses (i.e., as the next bubble grows). Its presence will result in a reverse of the decomposition reaction and just re-form the starting reagent. As a result, the most common ligands are those that form gases on decomposition, such as carbonyls.

In many synthetic techniques, reactions are carried out in combination. A starting material can be precipitated as a precursor, which is then thermally decomposed to form the final material. Sonochemical activation of a starting material can lead to a precipitation reaction that improves the homogeneity of the final product. Most often, the precursor formation is used to form a more uniform starting material for a solid–solid synthesis reaction. For example, the synthesis of the superconductor $YBa_2Cu_3O_{9-x}$ is often achieved through a high-temperature ceramic technique. The starting oxides are ground together, fired in air at 950°C, pressed into pellets, sintered under flowing oxygen for 16 hours, then cooled in oxygen to 200°C, followed by additional overnight firing at 700°C to get the optimal properties. The elaborate firing conditions are to aid migration of the cations into the perovskite structure. In contrast, if yttrium, barium, and copper are precipitated together to form a mixed-metal carbonate/oxide (the yttrium will form oxides in aqueous media), during firing at 750°C perovskite is formed. This is dramatically different from the elaborate firing protocol necessary for the ceramic method. This reduced-temperature preparation was possible due to the porous nature of the barium carbonate, caused by the escaping carbon dioxide and the more uniform mixture of the starting cations. Although carbonates are the most common ligand for the formation of precursors; oxalates, acetates, and other organics that are easily converted into carbon dioxide are also used.

Carbonates are a favorite in combination precipitation–decomposition reactions, but hydroxides are almost as common. In these precursors, the hydroxide is converted into water, which is driven off at elevated temperatures. For example, the synthesis of Fe_3O_4, or magnetite can be accomplished by precipitation the $Fe^{2+/3+}$ hydroxide, which with gentle heating is converted into the oxide:

$$3OH^- + Fe_2(OH)_5 \xrightarrow{200°C} Fe_3O_4 + 4H_2O$$

The use of hydroxides as precursors is often very tricky, due to the presence of side reactions. Metal hydroxides such as ferrous hydroxide often will oxide to ferric oxide if oxygen is present, which would yield

$$3OH^- + Fe(OH)_3 \xrightarrow{200°C} \alpha\text{-}Fe_2O_3 + 3H_2O$$

These reactions will proceed without heating under nitrogen, where the driving force is the more favorable iron oxide precipitate. Following Ostwald's rule of stages, the hydroxide would be the kinetic product and the oxide would be the thermodynamic product.

It is often helpful to review Pourbaix diagrams in order to predict some of the side reactions. In his 1945 doctoral thesis, Marcel Pourbaix (1904–1998) examined the relationships between electrochemical potential and pH, correlating them in a two-dimensional plot called a *predominance diagram* (later also called a *Pourbaix diagram*). In 1966 the Centre Belge d'Étude de la Corrison (CEBELCOR), with the help of many other contributors, compiled thermodynamic and structural data for virtually the entire periodic table in the *Atlas of Electrochemical Equilibria in Aqueous Solutions*.

Using a Pourbaix diagram of iron (Figure 4.4), for example, it is seen that to form and stabilize the ferrous hydroxide, one must stay at or below 0 V (i.e., in a reducing environment). Otherwise, at elevated pH and oxidizing conditions, it is impossible to stabilize the Fe^{2+} cation. Pourbaix diagrams have been used over and over again in the formation of metal oxides. In solutions, the standard free energy of the reaction is related to the standard cell potential via

$$\Delta G^\circ = -nFE^\circ$$

The standard free energy is also related to the equilibrium constant

$$\Delta G^\circ = -RT \ln K$$

When these two equations are combined, we get the *Nernst equation:*

$$E^\circ = \frac{RT}{nF} \ln K$$

In a given reaction, for example, iron and hydroxide,

$$Fe^{3+} + 3OH^- \rightarrow Fe(OH)_3$$

the concentration of OH^- is related to the pH via the dissociation constant of water, and the reaction equilibrium, in this case a solubility product, is known. Therefore, it is possible to plot the cell potential versus the pH at a constant concentration of Fe^{3+}, yielding the boundary between the precipitated hydroxide and the soluble cation.

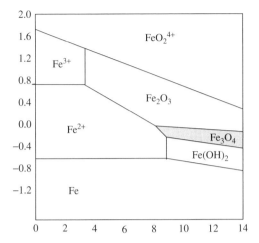

Figure 4.4   Representative Pourbaix diagram of iron. The small gray section represents the region of cell potential and pH where Fe_3O_4 can exist.

Often, with precipitation reactions the starting materials are limited to whatever salts are soluble in the solvent of choice. For water systems this is often limited to metal salts of halides, nitrates, and some sulfates and phosphates. Halides, in particular chlorides, have a pronounced effect on precipitation reactions. Chlorine is able to form bridged complexes much like the hydroxides or oxides of the desired compounds. In addition, acidic environments make possible the oxidation of chloride to chlorine gas, which can further complicate the synthesis. Sulfates and phosphates are typically easier to work with since they do not have the complicated redox behavior of the halides, but they typically have reduced solubilities. Nitrates, although they do not have the solubility concerns of sulfates and phosphates, do have redox complications, which typically result in oxidation of cations. So, the anion, which is expected to act solely as a spectator, in many cases is actually acting as a catalyst.

Organometallic salts in hydrocarbons often have similar complicated problems. Many salts, such as acetates or acetonates, are typically used. Precipitation reactions typically are slower in hydrocarbon media with these salts if the reaction happens at all, due to the chelation effect. Ligands that can multiply-bond the metal tend to form more stable complexes than do singly bound salts. For these reason, most of the precipitation reactions in hydrocarbons are carried out at elevated temperatures.

4.2.6. Conventional Solidification

Let us now switch to a discussion of the solidification process, which occurs at a solid–melt interface. In principle, solidification processes are equally applicable to *congruently melting* metals and nonmetals: in other words, to any substance that melts uniformly and forms a liquid with the same composition as the solid.

Most ceramics melt *incongruently*, meaning that they decompose into another substance. However, even most congruently melting ceramics are very seldom prepared by solidification because of their high melting points. In solidification processes, the molten material is typically poured into a thermally conducting mold to solidify. This is termed *casting*, and it is the most economical and hence most common method for fabricating metal pieces with a predefined size and shape. The quantitative mathematical relations governing this phenomenon are complicated moving interface diffusion equations. Fortunately, we need not present such mathematical expressions in any detail here. Casting will be revisited from a materials science perspective in Section 7.2. Here, we provide some information pertinent to our present discussion.

Conventional industrial casting processes usually involve directional heterogeneous solidification occurring in three stages: nucleus formation, crystal growth, and grain boundary formation. When enough heat is extracted, stable nuclei form in the liquid either on solid-phase impurities near the walls of the mold or on the mold itself, since this is the first region to cool sufficiently for crystals to form. Heterogeneous nucleation can also occur at the surface of the melt on solid-phase metal oxide particles. Oxides typically have much higher melting points than those of their parent metals. Other possible nucleation sites are inclusions and intentionally added grain refiners. At any rate, the solidification begins near the exterior edges and the solid–liquid interface subsequently moves inward towards the casting's center as heat is conducted through the freshly grown solid out through the mold. The nuclei consist of tiny aggregates of atoms arranged in the most favorable lattice under the process conditions. Crystals grow in all directions near the liquid–container interface. Hence, this region, shown in Figure 4.5, is called the *equiaxed zone*.

As solidification continues, an increasing number of atoms lose their kinetic energy, making the process exothermic. For a pure metal, the temperature of the melt will remain constant while the latent heat is given off (until freezing is complete). As the atoms coalesce, they may attach themselves to existing nuclei or form new nuclei. The process continues, with each crystal acquiring a random orientation, and as the gaps between crystals fill in, each grain acquires an irregular shape. The growth morphology is probably under kinetic control; that is, the grain morphology that appears is the one with the maximum growth rate. Eventually, those grains that have a preferred growth direction will eliminate the others, resulting in the formation of a *columnar zone* where the crystals are elongated, or columnlike. The growth direction is typically in the direction of heat flow. For alloys, an inner equiaxed zone can sometimes form in the casting's center, resulting from the growth of detached pieces of the columnar grains. This will depend on the degree of heat convection (extraction) in that region.

The rate of heat extraction is, in turn, dependent on the properties of the cooling medium (e.g., water, air), specimen size, and geometry. Heat energy must be transported to the surface to be dissipated to the surroundings. The surface itself, which is in direct contact with the quenching medium, experiences the fastest cooling rate. The cooling rate throughout the interior of a sample varies

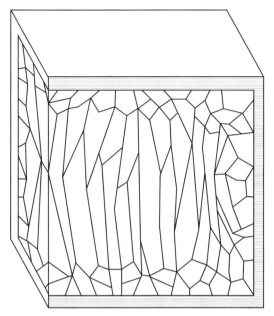

Figure 4.5 Section through a solidification ingot showing the various solidification zones. The equiaxed zone forms near the mold walls. In the interior is the columnar zone, where crystal growth is in the direction of heat flow. (After Lalena and Cleary, 2005. Copyright © John Wiley & Sons, Inc. Reproduced with permission.)

with position, depending on the size and geometry (Callister, 2005). In making microstructure predictions, therefore, it is important to realize that different spatial locations in the melt experience different cooling rates, and the cooling rates will, similarly, change with time and surface temperature. As a consequence, microstructural variations will exist in different regions of a solidification product.

Conventional casting procedures for kilogram-scale quantities typically produce average cooling rates up to about 10^{-3} to 10^0 K/s, resulting in relatively coarse grains, with an average size in the range of several millimeters to several hundred micrometers. As the cooling rate increases, the length scale of the microstructure (e.g., dendrite arm spacing) decreases. The interfaces between grains, formed by the last liquid to solidify, are the grain boundaries. A grain boundary is comprised of atoms that are not exactly aligned with the crystalline grains on either side of it. Hence, the grain boundaries have some degree of disorder and tend to contain a higher concentration of impurity atoms, which do not "fit" into the crystal lattices on either side of them (a melt is never *entirely* pure). The grain boundary has a slightly higher free energy, due to the presence of defects. Next, we investigate how grain morphology and composition, collectively termed *constitution*, are affected by the solidification rate, which is, in turn, determined by the cooling rate. Of the many parameters affecting the

TABLE 4.1 Various Cooling-Rate Regimes with Some Typical Products Obtained

Cooling Rate (K/s)	Techniques Used	Typical Products Obtained
	Rapid Solidification	
10^9-10^{11}	Pulsed laser melting of a solid surface	Amorphous and nanocrystalline thin films
10^5-10^7	Melt spinning; splat quenching	Glassy metallic alloy ribbon
10^2-10^3	Water quenching	Fine-grained polycrystalline bulk solids
	Conventional (Nonequilibrium, or Scheil) Solidification	
$10^{-3}-10^0$	Conventional casting (air-cooled)	Coarse-grained polycrystalline bulk solids
	Equilibrium Solidification	
10^{-6}	Flux growth (e.g., top-seeded solution growth)	Large single crystals

development of the microstructure, the cooling rate is among the most important. Table 4.1 lists some typical products obtained by various cooling-rate regimes.

4.2.6.1. Grain Homogeneity　There are two limiting cases to consider. The first is equilibrium solidification, when the cooling rate is slow enough that solid-state diffusion can act to redistribute atoms and result in homogeneous crystals. In this case, complete diffusion occurs in both the liquid and the solid. Under these conditions, the solid absorbs solute atoms from the liquid and solute atoms within the solid diffuse from the previously frozen material into subsequently deposited layers. The chemical compositions of the solid and liquid at any given temperature then follow the solidus and liquidus lines, respectively, of the equilibrium phase diagram. Hence, it is termed *equilibrium solidification*.

Use of tie lines and the lever rule enable one to determine those compositions, as illustrated in Figure 4.6a for a binary system. The composition of the solid (C_s) as a function of the fraction of solid transformed (f_s), assuming linear solidus and liquidus lines, is given by

$$C_s = \frac{kC_0}{f_s(k-1)+1} \tag{4.16}$$

where k is the partition coefficient (the ratio of the solute concentration in the solid to that in the liquid) and C_0 is the composition of the original liquid alloy. The first crystals to freeze out have composition α_1. As the temperature is reduced to T_2, the liquid composition shifts to L_2. The compositions of the freezing solid and remaining liquid shift continuously to higher B contents and leaner A contents. The average solid composition follows the solidus line to T_4, where it equals the bulk composition of the alloy.

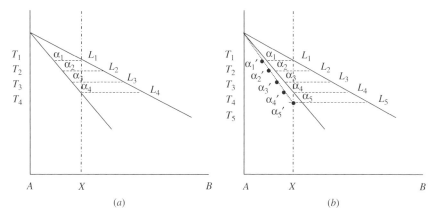

Figure 4.6   (*a*) Portion of a binary-phase diagram illustrating equilibrium solidification; (*b*) nonequilibrium (rapid) solidification, which results in a chemical composition gradient in the crystals, a condition known as coring. (After Lalena and Cleary, 2005. Copyright © John Wiley & Sons, Inc. Reproduced with permission.)

To qualify for equilibrium solidification, the solidification rate must be slower than the solute diffusivity in the solid:

$$D_s \gg L_x v \tag{4.17}$$

where D_s is the solute diffusivity in the solid, L_x the system length scale in one dimension, and v the solidification speed (Phanikumar and Chattopadhyay, 2001). The self-diffusivities of most pure metals at their freezing points (T_f) are in the range 10^{-9} to 10^{-6} cm^2/s. For a system length scale of 1 cm, the solidification rate (cm/s) must be lower than these numerical values for the diffusivities, which are very slow rates indeed. In other words, equilibrium solidification occurs only when the melt is cooled extremely slowly!

The second limiting case approximates conventional metallurgical casting processes in which the cooling rate is on the order of 10^{-3} to 10^0 K/s. As a result, the solidification rate is several orders of magnitude too fast to maintain equilibrium. The most widely used classical treatment of nonequilibrium solidification is by Erich Scheil (Scheil, 1942), who was at the Max-Planck-Institute for Metals Research in Stuttgart. The model assumes negligible solute diffusion in the solid phase, complete diffusion in the liquid phase, and equilibrium at the solid–liquid interface. In this case, Eq. 4.16 can be rewritten as

$$C_s = kC_0(1 - f_s)^{k-1} \tag{4.18}$$

When the solid–liquid interface moves too fast to maintain equilibrium, it results in a chemical composition gradient within each grain, a condition known as *coring* (Figure 4.6*b*). Without solid-state diffusion of the solute atoms in the material solidified at T_1 into the layers subsequently freezing out at T_2, the average composition of the crystals does not follow the solidus line from α_1 to α_4, but rather, follows the line α_1 to α_5, which is shifted to the left of the equilibrium solidus line. The faster the cooling rate, the greater the magnitude of the shift.

Note also that final freezing does not occur until a lower temperature, T_5, in Figure 4.6*b*, so that non–equilibrium solidification happens over a greater temperature range than does equilibrium solidification. Because the time scale is too short for solid-state diffusion to homogenize the grains, their centers are enriched in the higher freezing component while the lowest freezing material gets segregated to the edges (recall how grain boundaries are formed from the last liquid to solidify). Grain boundary melting, *liquation*, can occur when subsequently heating such an alloy to temperatures below the equilibrium solidus line, which can have devastating consequences for metals used in structural applications.

4.2.6.2. Grain Morphology   In addition to controlling the compositional profile of the grains, the solidification velocity also determines the shape of the solidification front (solid–liquid interface), which, in turn, determines grain morphology. The resulting structure arises from the competition between two effects. Undercooling of the liquid adjacent to the interface favors protrusions of the growing solid, which gives rise to *dendrites* with a characteristic treelike shape, while surface tension tends to restore the minimum surface configuration: a planar interface.

Consider the case of a molten pure metal cooling to its freezing point. When the temperature gradient across the interface is positive (the solid is below the freezing temperature, the interface is at the freezing temperature, and the liquid is above the freezing temperature), a planar solidification front is most stable. However, with only a very small number of impurities present in a pure melt on which nuclei can form, the bulk liquid becomes kinetically undercooled. Diffusion of the latent heat away from the solid–liquid interface via the liquid phase favors the formation of protrusions of the growing solid (dendrites) into the undercooled liquid; the undercooled liquid is a very effective medium for heat conduction. Ivantsov, who first developed the diffusive transport theory of dendritic growth, modeled it mathematically for branchless needle dendrites (paraboloids of revolution) over half a century ago (Ivantsov, 1947). It is now known that this is true so long as the solidification velocity is not *too* fast. At the high velocities observed in some rapid quenching processes (e.g., $10 \ ms^{-1}$) dendritic growth becomes unstable, as the perturbation wavelengths become small enough that surface tension can act to restore planarity (Mullins and Sekerka, 1963; Hoglund et al., 1998).

Because of the small number of impurities in a pure metal, the undercooling can be quite large. There aren't many nuclei on which dendrites can form. For dendrites that do form, in Ivantsov's diffusion-limited dendritic growth model, growth is a function of the rate of latent heat removal from the interface. Hence, for a pure metal, one would expect a small number of large dendrites. The dendritic shape maximizes the surface area for dissipating the latent heat to the undercooled liquid. Dendrites are therefore the most common crystal form in solidification products because this shape is best suited for efficient heat and mass transfer at small scales.

In his original model, Ivantsov described steady-state growth, that is, in which the dendrite grows at a constant speed in a shape-preserving manner. Furthermore, he assumed that the solid–liquid interface temperature is the equilibrium melting point. This precluded the role of interfacial physics in his formulation. Subsequent work was performed by researchers who suggested that dendrites grow at some critical value of the dendrite tip radius or wavelength (Langer, 1978). Our modern understanding of steady-state dendritic growth—the microscopic solvability theory—builds on both this and Ivantsov's earlier work. An important parameter in this model is the crystalline anisotropy or, rather, the accompanying anisotropy to the interfacial energy (Langer, 1989). Directional solidification of anisotropic materials thus exhibit two competing directions that determine the dendrite growth direction—the heat flow direction and the preferred crystalline orientation. When they differ, the growth direction is found to rotate from the thermal gradient direction to the preferred crystalline orientation as the growth velocity is increased. This is accompanied by dramatic changes to dendritic morphology, particularly asymmetry and secondary branching (Pocheau et al., 2007). Dendrites have a tendency to develop along directions corresponding to convexities (i.e., maximum curvature regions) of the equilibrium shape of the crystal, which typically correspond to the maxima in the interfacial energy. For face-centered-cubic and body-centered-cubic metals, dendrite growth directions are <100> (Gonzales and Rappaz, 2006).

In Ivantsov's model for a pure metal melt, dendritic growth is a function of the rate of latent heat removal from the interface. However, when we turn our attention to alloys, we see a slightly different situation. Here, in addition to heat flow (undercooling), one must consider mass transport (tip supersaturation). In fact, the planar interface-destabilizing event primarily responsible for dendritic morphology in conventional alloy casting is termed *constitutional undercooling*, to distinguish it from kinetic undercooling. The kinetic undercooling contribution can still be significant in some cases. In most models for two component melts, it is assumed that the solid–liquid interface is in local equilibrium even under nonequilibrium solidification conditions, based on the concept that interfaces will equilibrate much more rapidly than bulk phases. Solute atoms thus partition into a liquid boundary layer a few micrometers thick adjacent to the interface, slightly depressing the freezing point in that region. As in the case for pure meals, the positive temperature gradient criterion for planar interface stability still holds. However, although the bulk liquid is above the freezing point, once the boundary layer becomes undercooled, there is a large driving force for solidification *ahead* of the interface into the thin boundary layer.

The critical growth velocity, v, above which the planar interface in a two-component melt becomes unstable is related to the undercooling, ΔT_c, by an equation given by Tiller:

$$\frac{G_L}{v} \geq \frac{\Delta T_c}{D_L} \tag{4.19}$$

where $\Delta T_c = m_L C_0 (1 - k)/k$, G_L is the thermal gradient in the liquid ahead of the interface, v the solidification speed, m_L the liquidus slope, C_0 the initial liquid composition, k the partition coefficient (defined previously), and D_L the solute diffusivity in the liquid (Tiller et al., 1953).

Constitutional undercooling is difficult to avoid except with very slow growth rates. With moderate undercooling, a cellular structure, resembling arrays of parallel prisms, results. As the undercooling grows stronger, the interface breaks down completely as anisotropies in the surface tension and crystal structure lead to side branches at the growing tip of the cells along the "easy-growth" directions (<100> for FCC and BCC, <1010> for HCP), marking a transition from cellular to dendritic. In a polycrystalline substance, the dendrites grow until they impinge on one another. The resulting microstructure of the solidification product may not reveal the original dendritic growth upon simple visual examination. Nevertheless, the dendritic pattern strongly influences the material's mechanical, physical, and chemical properties.

Over the last 50 years, a large amount of work has gone into obtaining accurate mathematical descriptions of dendrite morphologies as functions of the solidification and materials parameters. Dendritic growth is well understood at a basic level. However, most solidification models fail to accurately predict *exact* dendrite morphology without taking into account such effects as melt flow. In the presence of gravity, density gradients due to solute partitioning produce convective stirring in the lower undercooling range corresponding to typical conditions encountered in the solidification of industrial alloys (Huang and Glicksman, 1981). Melt flow is a very effective heat transport mechanism during dendritic growth, which may result in variations in the dendrite morphology as well as spatially varying composition (macrosegregation).

4.2.6.3. Zone Melting Techniques

4.2.6.3. Zone Melting Techniques   Once an ingot is produced, it can be purified by any of a group of techniques known as *zone melting*. The basis of these methods is founded on the tendency of impurities to concentrate in the liquid phase. Hence, if a zone of the ingot is melted and this liquefied zone is then made to slowly traverse the length of the ingot by moving either the heating system or the ingot itself, the impurities are carried in the liquid phase to one end of the ingot. In the semiconductor industry, this technique is known as the *float zone process*. A rod of polycrystalline material (e.g., silicon) is held vertically inside a furnace by clamps at each end. On one end of the rod is placed a short single-crystal seed. As the rod is rotated, a narrow region, partially in the seed and partially in the rod, is melted initially. To avoid contact with any container material, the molten zone is freely suspended and touched only by the ambient gas. It is then moved slowly over the length of the rod by moving the induction coils: thus the name *floating zone*. As the floating zone is moved to the opposite end of the rod, a high-purity single crystal is obtained in which the impurities are concentrated in one end. The impurities travel with, or against, the direction of the floating zone depending on whether they lower or raise the melting point of the rod, respectively. Typically, growth rates are on the order of millimeters per hour. The float zone technique has been used to grow crystals of several oxides as well. To control the escape of volatile constituents, encapsulants with a slightly lower melting point than that of the crystal are inserted between the feed rod and the container wall so that the float zone is surrounded concentrically by an immiscible liquid.

In the related *zone refining technique*, a solid is refined by multiple floating zones, or multiple passes, as opposed to a single pass, in a single direction. Each zone carries a fraction of the impurities to the end of the solid charge, thereby purifying the remainder. Zone refining was first described and used by the American metallurgist William Gardner Pfann (1917–1982) at Bell Labs in 1952 to purify germanium to the parts per billion impurity level for producing transistors (Pfann, 1952).

4.2.6.4. Rapid Solidification: Vitrification and Quasicrystal Formation

Although cooling rates as high as 10^{11} K/s have been obtained on solid surfaces with pulsed laser melting, for bulk phases rapid solidification typically refers to cooling rates in the range 10^2 to 10^7 K/s. On the low end of this range, very fine-grained crystalline substances are produced. On the high end, the formation of either a glassy phase or quasicrystal is favored over a crystalline phase. Glass formation, or *vitrification*, can be compared to crystallization by referring to Figure 4.7, which is applicable to both metallic and nonmetallic systems. Crystallization follows path *abcd*. As the temperature of a non-glass-forming melt is lowered, the molar volume of the alloy decreases continuously until it reaches the melting point, where it changes discontinuously (i.e., where it experiences a first-order phase transition). The enthalpy and entropy behave similarly. In glass formation, rapid cooling forces the melt to follow path *abef* with decreasing temperature. The liquid remains undercooled (it does not solidify) in the region *be*, below the melting point. The molar volume decreases continuously in the undercooled region, and the viscosity increases rapidly. At the point T_g, called the *glass transition temperature*, the atomic arrangement becomes frozen into a rigid mass that is so viscous that it behaves like a solid. Rapid cooling reduces the mobility of the material's atoms before they can pack into a more thermodynamically favorable crystalline state.

The state of our current scientific understanding of glass formation is founded to a large extent on the theoretical work of Harvard researcher David Turnbull (1915–2007). Turnbull's criterion for the ease of glass formation in supercooled melts predicts that a liquid will form a glass, if solidified rapidly, as the ratio of the glass transition temperature, T_g, to the liquidus temperature, T_l, becomes equal to or greater than two-thirds (Turnbull, 1949, 1950). The T_g/T_l ratio is referred to as the *reduced glass transition temperature*, T_{rg}. The rate of homogeneous nucleation is dependent on the ease with which atomic rearrangement can occur (commonly taken as the atomic diffusion coefficient), which scales inversely with fluidity or viscosity. Easy glass-forming substances form highly viscous melts (e.g., $>10^2$ poise), compared to non-glass-forming substances (e.g., water, with $\eta \sim 10^{-2}$ poise). In highly viscous melts, the atomic mobility is

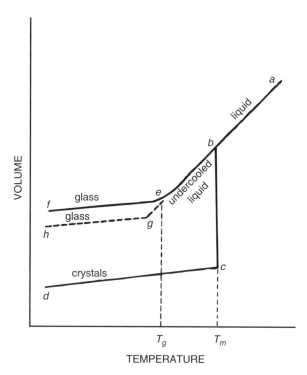

Figure 4.7 Comparison of glass formation (curve *abef*) and crystallization (curve *abcd*). The point T_g is the glass transition temperature and T_m is the melting temperature. (After J. West, *Solid State Chemistry and Its Applications*. Copyright © 1985 John Wiley & Sons, Inc. Reproduced with permission.)

reduced substantially, which suppresses the homogeneous nucleation rate and hence crystallization. In fact, Igor Evgenevich Tammann (1861–1938) pointed out, as early as 1904, that the higher the viscosity of a melt, the lower its crystallizability (Tammann, 1904). The homogeneous nucleation rate is therefore highly dependent on T_{rg}. The $T_{rg} > \frac{2}{3}$ criterion successfully predicts glass formation in metallic and nonmetallic liquids. It must be noted, however, that heterogeneous nucleation (e.g., on "seed" particles present inadvertently) may prevent glass formation. Indeed, crystallization is usually initiated in this manner.

The arguments above are based on kinetics. It may also be shown on thermodynamic grounds that a high value for T_{rg} (and therefore the tendency to form a glass at lower cooling rates) is obtained for deep eutectic systems (i.e., where the melting point of some alloy composition is substantially lowered compared to the melting points of the pure components). These systems tend to be those with very little solid solubility between the components. When atoms do not "fit" together in the lattice (due to mismatches in size, valence, etc.), the tendency for crystallization diminishes. This is due to both a large negative heat of mixing and entropy of mixing for the liquid compared with the competing crystalline phase.

Although in common parlance, the term *glass* has come to refer to silicate glass of one kind or another, in the early 1950s, German scientists succeeded in preparing amorphous tin and lead by cooling vapor at about 10^{12} degrees per second. However, theses metallic glasses were much thinner than aluminum foil and they crystallized well below room temperature. In 1959, Pol E. Duwez (1907–1984), a Belgian-born materials scientist at Cal Tech, was able to prepare a thin gold–silicon metallic glass that remained amorphous at room temperature. Working simultaneously and independently, Turnbull and Morrel H. Cohen (b. 1927) indicated that alloy systems with deep eutectic compositions should exhibit a strong tendency for glass formation with sufficiently fast cooling rates (e.g., in the range 10^5 to 10^6 K/s). Such cooling rates are obtainable in industrial melt spinning and splat quenching techniques. Under these processing conditions, the highly disordered state of the supercooled liquid phase becomes "configurationally frozen" into a rigid amorphous (glassy) state.

Historically, most glass-forming alloys were metal–metalloid and metal–metal binary systems (where the metal is usually a transition element and the metalloid is B, Si, C, or P) with a T_g value well above room temperature, in the range 300 to 700 K. With the exception of the group 12 elements (Zn, Cd, Hg) the transition metals have melting points exceeding 1200 K. Hence, those alloy systems containing very low melting eutectics (e.g., 636 K in the Au–Si system) tend to satisfy the Turnbull criterion. Examples of binary metallic glasses include $Fe_{80}B_{20}$, $Ni_{60}Nb_{40}$, $Ni_{63}Zr27$, and $Ca_{65}Al_{35}$. The compositions of these glasses are near eutectic points (Turnbull, 1981). Turnbull's criterion has thus been validated in systems at cooling rates attainable by "conventional" casting procedures ($\sim 10^6$ K/s). Some alloy systems, such as $Cu_{60}Zr_{40}$, exhibit glass formation over composition ranges extending well beyond a eutectic point. By contrast, the T_{rg} value of pure metals seem to be much smaller than two-thirds. Furthermore, pure metallic liquids ($\sim 10^{-2}$ poise) have much lower viscosities than those of the glass-forming alloys. Therefore, glass formation from pure metal melts by rapid solidification requires extremely high cooling rates, on the order of $\sim 10^{12}$ K/s. In 2004, however, scientists at the Los Alamos National Laboratory succeeded in producing millimeter-sized metallic glass samples of zirconium by placing zirconium crystals under a pressure of 80,000 atm and a temperature of 700°C (Zhao and Zhang, 2004).

In 1990, Japanese researcher Akahisa Inoue and his team at Tohoku University began casting bulk metallic glasses (BMGs) up to 0.6 cm thick. They found they could make these by using three or more elements that differ from one another in atomic size by at least 12%. Today, some alloys can be prepared as bulk metallic glasses in ribbons or rods with thickness of several centimeters and at substantially lower cooling rates. Further advances have since been made with the discovery of new families of multicomponent alloys with significantly improved glass-forming ability. Ternary glass formers include those systems in which the binary subsets exhibit limited mutual solid solubility, such as $Pd_{77.5}Cu_6Si_{6.5}$ and $Pd_{40}Ni_{40}P_{20}$. These systems have been found to form glasses at cooling rates as low as 1000 to 10 K/s, respectively. This is due to both an increased frustration of the homogeneous nucleation process and to the greater suppression of the liquidus temperature as the number of components is increased. Hence, the glass-forming ability appears to be even further enhanced in yet higher-order systems, such as $Pd_{40}Cu_{30}Ni_{10}P_{20}$ (Inoue et al., 1997) and $Zr_{41.2}Ti_{13.8}Cu_{12.5}Ni_{10}Be_{22.5}$ (Vitreloy 1) (Perker and Johnson, 1993). These alloys have T_g values of about 582 and 639 K, respectively, and critical cooling rates of just 1 K/s!

In the supercooled liquid state, BMGs have very high yield strength and a high elastic-strain limit (often exceeding 2%, compared with crystalline materials, which are almost always less than 1%), which makes them very "springy." However, under tensile loads, bulk metallic glasses normally lack any significant global plasticity, which limits applications as structural materials (Johnson, 1999). A recently discovered exception appears to be ZrCuNiAl (Lin etal., 2007). Current efforts have focused on the development of engineering applications for metallic glass–containing composite materials. Such composites have been found to exhibit greatly enhanced ductility and impact resistance compared to monolithic glasses.

Like nonmetallic ones, metallic glasses are thermodynamically metastable states. However, metallic glasses appear to be more susceptible than nonmetallic glasses to *devitrification*, or crystallization at temperatures above T_g, transforming to more stable crystalline phases, typically around 300 to $450°C$. Nanocrystalline grains (grain size < 100 nm) can sometimes be obtained from a metallic glass when it is annealed at temperatures at which primary crystallization can occur. Nanocrystalline phases have been under increased study in recent years because they often have improved properties over their coarse-grained counterparts. Nanocrystalline alloys themselves, however, are also metastable phases, with a tendency toward grain growth.

The devitrification (annealing) of some metallic glasses can lead to the formation of quasicrystals. Examples include icosahedral quasicrystals from the metallic glasses $Pd_{60}U_{20}Si_{20}$, $Al_{75}Cu_{15}V_{10}$, $Ti_{45}Zr_{38}Ni_{17}$, and $Zr_{65}Ni_{10}Ag_{10}Al_{7.5}Ni_{7.5}$. Mechanical alloying has also been used. A more direct route to quasicrystals, introduced in Section 1.6, involves the solidification of certain metallic alloys. Both conventional cooling rates and rapid cooling rates have been used. Only a few studies have looked at the general effect of cooling rates (Xing et al., 1998, 1999). The first reported quasicrystals were metastable phases at room temperature produced by rapid solidification and were, consequently, of poor quality. Stable quasicrystals have since been discovered that revealed very high structural perfection, comparable to that of single crystals. This discovery made it possible to apply conventional solidification techniques. The preferred method appears to be system-specific, as it depends on the temperature stability of the quasicrystalline phase. If the quasicrystal is stable only at elevated temperatures, for example, it can decompose into a crystalline phase if the melt is solidified slowly. If the phase is thermodynamically stable down to room temperature, as is the case for Al–Pd–Mn, quasicrystals can be grown with conventional cooling rates (e.g., 10 degrees per hour).

Most quasicrystals have been obtained by solidifying phases with equilibrium crystal structures containing icosahedrally packed groups of atoms (i.e., phases containing icosahedral point group symmetry). The quasicrystalline phases form at compositions close to the related crystalline phases. Quasicrystals then form by a peritectic reaction of a melt with the crystalline phase. A *peritectic*, or *incongruent, reaction* is defined as an isothermal reversible reaction between a liquid and a solid phase, resulting in a new solid phase. Both the quasicrystalline and crystalline phases span a narrow range in composition and are thermodynamically stable. If the liquid has the correct composition and is cooled below the liquidus line, the quasicrystalline phase may be the only one to grow. Cooling must be performed very slowly to minimize coring (a chemical composition gradient within each grain) and not to induce constitutional undercooling at the liquid–solid interface that will initiate dendritic growth. Centimeter-size single grains have been grown using this approach.

Sir Frederick Charles Frank of the University of Bristol and John Kasper of the General Electric Research Laboratories showed in 1958 that icosahedral coordination ($Z = 12$), as well as other coordination polytetrahedra with coordination numbers $Z = 14$, 15, and 16, are a major structural components of some melts. Such tetrahedrally close-packed structures, in which atoms are located at the vertices and centers of various space-filling arrangements of polytetrahedra, are now called *Frank–Kasper phases*. These structures are, in fact, more stable than close-packed solid-phase crystal structures. When these liquid-phase structures are solidified, they resist construction into a crystalline unit with three-dimensional translational periodicity. Instead, the resulting solid has icosahedra threaded by a network of wedge disclinations. The first such quasicrystalline phases were obtained by rapidly solidifying Al–Mn alloys, which resulted in a phase with fivefold rotational symmetry. Thermodynamically stable ternary icosahedral quasicrystals of many intermetallic systems, including but not limited to, Al–Li–Cu, Al–Pd–Mn, and Zn–Mg–Ln, have since been prepared. Some stable ternary intermetallic phases have been found that are quasiperiodic in two dimensions and periodic in the third. These are from the systems Al–Ni–Co, Al–Cu–Co, and Al–Pd–Mn, These structures contain decagonally packed groups of atoms (tenfold rotational symmetry). Metastable quasicrystals with octagonal (eightfold) and dodecagonal (12-fold) rotational symmetry are also known.

4.3. SOL–GEL PROCESSING

A very interesting solution-based synthetic technique for the low-temperature preparation of both noncrystalline solids (glass and glasslike materials) and crystalline ceramics is the *sol-gel process*. The process gets its name from the two major stages involved: formation of a *sol* (a colloidal suspension of solid particles in a liquid phase) through hydrolysis and polymerization reactions of appropriate precursors; and subsequent attainment of a semirigid porous *gel*. The gelation point is generally accepted as that point at which the sol's viscosity reaches a plateau, where it is able to support stress elasticity. The crystallinity of the final product, which is obtained after removal of solvent and residuals from the pores by aging, drying, and annealing, is highly dependent on the experimental conditions used. Moreover, sol–gels have been used as templates for directing micro- and nano-scale materials morphology, which will be explained further shortly.

Conventional glass preparation requires melting of the precursors at high temperatures, followed by rapid cooling and vitrification of the monolithic glassy material. This technique is restricted to inorganic substances that can survive these harsh conditions without thermally decomposing (e.g., metals, metal oxides). Sol–gel processing got its start in the mid-nineteenth century with work by the French chemist Jacques Joseph Ebelman (1814–1852) and the Scottish chemist Thomas Graham (1805–1869), each of whom observed the formation of a glasslike monolithic substance upon hydrolysis and condensation of tetraethylorthosilicate [also known as tetraethoxysilane (TEOS)] over several months (Ebelman, 1846; Graham, 1864). However, the extended drying time limited this technique to a scientific curiosity. Graham was also the first to use the terms *colloid, sol*, and *gel*. He deduced that colloids were about 1 to 100 nm in size, based on their slow diffusion and lack of sedimentation. Throughout the remainder of the nineteenth into the early twentieth century, further exploration revealed the fundamental nature of the sol–gel process. Finally, from the 1950s to the 1970s, other inorganic substances were discovered that had gelling properties like those of silica. It was during this time that the sol–gel technique started to come of age and find applications.

The sol–gel technique can be used to form powders, films, or monolithic castings. Although the synthesis can lead to several different forms of the same starting materials, they are all made in the same way as the monolith. Two approaches are used in making sol–gel monoliths: gelation of colloidal powders, or hydrolysis and polymerization of alkoxide precursors. When the solvent and water are removed as a gas with elevated temperature, the resulting structure is referred to as an *aerogel*. The resulting solid can have up to 98% of the structure as hollow pore volume and have densities as low as 0.08 g/cm^3. If the solvent and water are removed at or very near room temperature by simple evaporation, the solid is called a *xerogel*. At any rate, the pores themselves within such monoliths may be used as templates for directing the synthesis of other micro- and nanoscaled materials.

Sol–gel processing is driven by the hydrolysis and condensation reactions. *Olation* is the condensation in which a hydroxyl bridge is formed between two metal centers. For coordinately saturated metals, olation occurs through the nucelophilic substitution of hydroxyl groups for water groups on the metal. Olation is believed to occur via a H_3O_2 intermediate. The rate of olation is related to the size, electronegativity and electronic configuration of the metal: In general, the lower the charge density, the greater the rate of olation.

Oxolation is a condensation reaction in which an oxo bridge is formed. Oxolation is favored when the metal center is coordinatively unsaturated. Oxolation is a nucleophilic addition that results in edge- or face-shared polyhedra. Oxolation is also possible in coordinately saturated metals. However, it occurs via a hydroxyl intermediate followed by water elimination. Oxolation can be carried out over a wider range of pH, where olation is more limited to acidic conditions. Since oxolation is a two-step process, the kinetic rate is slower and never diffusion limited. Oxolation is minimized around the isoelectric point.

Although the metal has the most pronounced effect on the rate of condensation, the counterion present can also have a major impact on the morphology and stability of sols that are forming. Counterions such as halides can often compete with water for coordination to the metal center and in addition, they can often act as bridges themselves. The ability of the anion to form complexes with the metal center is driven in part by the ability of the anion to donate charge to the electropositive metal center. In general, stability is conferred when the anion is less electronegative than water. This situation results in a M–X bond that is more stable against dissociation. In the opposite case, where the anion is more electronegative than water, the M–X bond remains polarized, resulting in positively charged species still being attracted to the complex.

The results of these observations were the basis for the partial charge model. The ability of the anion to form a complex is dependent on the amount of partial charge that is transferred to the negatively charged anion. This in turn reduces the ability to pull electrons from the water or hydroxyl species, and thus the ability to form bridges is greatly reduced. Since water is both a high dielectric solvent and a strong nucleophile, any change to the partial charge of the metal can affect the overall morphology and composition of the gel. The dielectric character of the solvent directly affects the possibiliity of nucleophile substitution reactions. During a reaction, the water substitutes for the anion, resulting in hydrolysis and condensation reactions that form the stable sol. When the solvent system is modified through the introduction of low-dielectric solvents, the hydrolysis reactions are typically retarded.

It is possible to form sol–gel reactions using metal alkoxides, solvated metal cations, or organometallic precursors. In the case of mixed-metal systems, where there are different ligands on the metal, the less electronegative ligand is removed first during hydrolysis. Alkoxides are by far the most common starting material for sol–gel reactions. This preference for the alkoxides stems from the ability to control the hydrolysis rate with the addition of alcohols. The rate of reaction is controlled by the ratio of alcohol to water. The reaction is catalyzed by the addition of either an acid or a base. Alkoxide precursors are used as the starting materials, due to the stable nature and ease of hydrolysis.

$$\underset{\overset{|}{OCH_3}}{\overset{\overset{OCH_3}{|}}{H_3CO-Si-OCH_3}} + 4H_2O \longrightarrow \underset{\overset{|}{OH}}{\overset{\overset{OH}{|}}{HO-Si-OH}} + 4CH_3OH$$

There are seven steps to making a sol–gel material: mixing, casting, gelating, aging, drying, dehydration or stabilization, and finally, densification. *Mixing* is the key to the size and composition uniformity of the grains in the colloidal sol.

$$\underset{\substack{| \\ OH}}{\overset{\substack{OH \\ |}}{HO-Si-OH}} + \underset{\substack{| \\ OH}}{\overset{\substack{OH \\ |}}{HO-Si-OH}} \longrightarrow \underset{\substack{| \\ OH}}{\overset{\substack{OH \\ |}}{HO-Si-O-}} \underset{\substack{| \\ OH}}{\overset{\substack{OH \\ |}}{Si-OH}} + H_2O$$

Either colloidal powders, stabilized to prevent precipitation, or alkoxide precursors can be used in this initial step. If alkoxides are used during the mixing, they are hydrolyzed, yielding hydrated silica. The hydrated silica tetrahedra are then interconnected by condensation, creating a silica sol. The sol grows according to the LaMer theory discussed earlier. Rapid nucleation followed by slow growth yields a very uniform-sized colloid. During hydrolysis many different species are present, resulting from the polymerization of the alkoxide. The kinetics are a very complicated picture, which has largely been overlooked in the literature. However, the variables of temperature, nature and concentration of the acid–base that is used in the hydrolysis, solvent, and the type of precursor used all have been shown to affect the rate of hydrolysis. At this stage, the solution is a low-viscosity liquid which is easily *cast* to form monoliths. Care must be taken during casting to prevent the gel from adhering to the mold.

As the sol is allowed to react, the sol becomes more and more connected, resulting in *gelation*. At the point at which the sol supports stress it is not longer a sol but becomes a gel. Because there is no clear point at which a sol is converted to a gel, but rather, a slow gradual conversion, there is no clear thermodynamic picture of the process. The time for gelation is dependent on the concentration of the precursors (or the size of the sol) and the structure of the alcohol used. Longer-chain or branched alcohols result in longer gelation times, as it becomes more difficult for the alcohol to escape from the pores. The greater the concentration of the precursors or the larger the sol, the more the initial sol is cross-linked and the more rapid the gelation. During gelation there is a significant increase in the viscosity as the sol is converted into a gel. The solution begins to retain the shape of the mold. During this stage the gels can be drawn into fibers. In the absence of experimental data on the thermodynamics of the gelation, several theoretical models have been suggested. These models suggest that during the hydrolysis process pentavalent and trivalent silicon transition states are formed. These states then coalesce rapidly to form a silica chain. After three or four chains are formed, a ring structure becomes more stable, resulting in a spherical colloid. It is these transition states that help the condensation reactions precede, resulting in the expulsion of water from the gel.

Aging of the gel is called *syneresis*, which is the process of separating the liquid from the gel to allow further solidification. As the sols become interconnected, the solvent and water from the condensation are pushed outside the pores. The condensation reactions continue as the colloidal sols are brought closer together. Aging determines the average pore size and the resulting density of the monolith. The longer the gel is allowed to age, the stronger the monolith as a result of the greater extent of cross-linking from the condensation. It is during syneresis that the most shrinkage occurs. The condensation reaction helps to pull adjacent sols together, pushing out water from the pore size. The contraction of the pore volume is driven by the tendency to reduce the solid–liquid interface and reduce the surface potential. The strain experienced during syneresis is greater at lower temperatures when the gel is more flexible. At elevated temperatures the condensation reactions are faster, resulting in stiffer gel structure.

Drying of the gel is the final removal of the solvent and water from the structure. It is during the point that the monolith experiences the greatest stresses and cracks often form. Several techniques have been developed to help prevent cracking of the structure, including supercritical drying. In supercritical drying, the pressure on the system is increased in order to reach the supercritical point of the solvent. This allows the solvent to escape as a gas rather than as a liquid, dramatically reducing the stresses on the monolith. The dry monolith is still very reactive, due to the surface silanol bonds in the pores. These surface silanol bonds are stabilized through thermal treatment, which converts them to the silica. *Densification* is the last step in creating a sol–gel structure where the density needs to be higher. In this case the monolith structure is heated to 1000 to 1700°C to collapse the pore structure and create a solid glassy structure.

4.4. SOLVOTHERMAL AND HYDROTHERMAL TECHNIQUES

A *solvothermal process* is one in which a material is either recrystallized or chemically synthesized from solution in a sealed container above ambient temperature and pressure. The recrystallization process was discussed in Section 1.5.1. In the present chapter we consider synthesis. The first solvothermal syntheses were carried out by Robert Wilhelm Bunsen (1811–1899) in 1839 at the University of Marburg. Bunsen grew barium carbonate and strontium carbonate at temperatures above 200°C and pressures above 100 bar (Laudise, 1987). In 1845, C. E. Shafhäutl observed tiny quartz crystals upon transformation of freshly precipitated silicic acid in a Papin's digester or pressure cooker (Rabenau, 1985). Often, the name *solvothermal* is replaced with a term to more closely refer to the solvent used. For example, solvothermal becomes *hydrothermal* if an aqueous solution is used as the solvent, or *ammothermal* if ammonia is used. In extreme cases, solvothermal synthesis takes place at or over the supercritical point of the solvent. But in most cases, the pressures and temperatures are in the subcritical realm, where the physical properties of the solvent (e.g., density, viscosity, dielectric constant) can be controlled as a function of temperature and pressure. By far, most syntheses have taken place in the subcritical realm of water. Therefore, we focus our discussion of the materials synthesis on the hydrothermal process.

The physical properties of water at or near the critical point are dramatically different from those of the water typically used in aqueous reactions. At the critical point, water becomes far less dense (0.3 g/cm^3), as the hydrogen bonding is disrupted dramatically by the high pressure. This disruption is partly responsible for the water dissociation constant (pK_W) changing from 14 to 21, indicating an increase in the number of protons and hydroxyl anions available in the system. The lack of hydrogen bonding also results in a decrease in the dielectric constant of water, reflecting the reduced ability to shield ions.

Although the lack of hydrogen bonding in water at the critical point may seem like a small effect, it is very apparent with solubilities. For example, in liquid water at ambient conditions, the solubility of gases such as oxygen is low, $\sim 9 \times 10^{-6}$ g/cm^3. As one approaches the critical point, however, the solubility increases until it reaches a value near infinity. With many cations (e.g., Ni^{2+}), water at ambient conditions forms a first hydration shell, in most cases, of six water molecules. With Raman studies conducted at the critical point, the waters in the hydration shell were found replaced by the counterions, leaving, on average, only 2.5 water molecules coordinated to the metal center. Raman spectroscopy revealed similar behavior in sodium chloride, strontium nitrate, and zinc nitrate. These observations lead to the conclusion that in supercritical water, ion pairs are more tightly bound, resulting form enhanced cation–anion interactions. The decrease in the number of hydration water molecules has a very practical effect on synthesis: namely, the reduced solubility of cations. Supercritical water behaves more like a hydrocarbon than like a polar solvent!

Hydrothermal synthesis is often applied to the preparation of oxides. The synthesis of metal oxides in hydrothermal conditions is believed to occur in a two-step process. In the first step, there is a fast hydrolysis of a metal salt solution to give the metal hydroxides. During the second step, the hydroxide is dehydrated, yielding the metal oxide desired. The overall rate is a function of the temperature, the ion product of water, and the dielectric constant of the solvent. The two steps are in balance during the reaction. The hydroxide of the metal salt is favored by a high dielectric constant, while the dehydration of the metal hydroxide is favored by a low dielectric constant. Since the fast reaction is the first step, it is expected that as one approaches supercritical conditions, the rate of reaction increases.

There are several precursors that can be used as starting materials. The resulting properties of the oxide are determined, in part, by the choice of precursor. For example, starting with ferric nitrate, ferric sulfate, or ferrous chloride under hydrothermal conditions results in the formation of α-Fe_2O_3. Starting with ferrous ammonium citrate, under the same hydrothermal conditions, yields Fe_3O_4. Under hydrothermal conditions, organic precursors such as acetates, formats, or oxalates tend to undergo further conversions to carbon monoxide, which in turn reduces ferric to ferrous cations. This can often be manipulated to create mixed oxidation states in the final oxide, as in the magnetite example above. In mixed-metal systems where different metal oxides are introduced to a hydrothermal reaction, it is necessary to start with coprecipitated gels. Different metals undergo hydrolysis at different rates. This leads to segregation of the metals into different metal oxides. To prevent this, hydrolysis is carried out first to form a mixed-metal oxide gel.

This gel is then hydrothermally treated to convert it to the oxide. The gel helps to trap the cations and prevent phase separation.

Hydrothermal synthesis offers a degree of morphology control not easily available with other techniques, due to the ability to tune the dielectric constant. Zeolites are another example of a material commonly synthesized under hydrothermal conditions. The structure of zeolites is controlled, in part, by the reaction of silica and alumina in the starting precursor. This leads to a very open framework structure during the dehydration steps. Metal fluoride structures and metal organic frameworks are other common examples of the open structure, produced in supercritical water. It is also possible to create more condensed structure, such as diamond or GaN. At $800°C$ and 1.4 kbar, the diamond structure is the more thermodynamically stable structure. The greater pressure further reduces the dielectric constant of the water and nearly eliminates the hydrogen bonding. This allows the formation of more complicated structures such as diamond.

4.5. MOLTEN SALTS AND ROOM-TEMPERATURE IONIC LIQUIDS

Salts, both those with high melting points and those with low melting points, have several different uses in the preparation of inorganic materials. The term *molten salt* is generally taken to mean an inorganic salt with a high melting point relative to room temperature. The term *ionic liquid* is generally understood to imply an organic salt that is a nonvolatile liquid at room temperature.

4.5.1. Molten Salts

Molten salts may be used in topochemical and nontopochemical reactions of existing solids. *Topochemical reactions* (sometimes called *topotactic reactions*) follow the pathway of minimum atomic or molecular movement (Elizabé et al., 1997). Consequently, the products are isostructural, or nearly so, with the starting solid phase. In a topochemical synthesis, a molten salt is *reactive* (i.e., it serves as a reactant) because a component of it is incorporated into the final product. An example of a topochemical route utilizing a reactive molten salt is ion exchange in pure alkali metal halides and nitrates. In this method, a solid phase, which is usually in powder form because of the large surface area/volume ratio per particle, with easily exchangeable cations (e.g., those with layered and tunnel structures) is immersed in the molten salt. The salt functions as a source of mobile cations that interchange with those of the solid, to an *extent* determined by the equilibrium constant for the reaction. The *rate* of ion exchange is determined largely by the ion mobilities, which is a function of temperature and ion charge (higher charged ions are less mobile), while thermodynamics usually favor exchange of ions of similar size and/or higher charge. It is possible to carry out *aliovalent ion exchange*, whereby cations of different charges are exchanged (e.g., an alkali metal cation and an alkaline-earth metal cation) with the concurrent creation of vacancies (v) to maintain overall electroneutrality in the final product. For example:

$$Na_2La_2Ti_3O_{10}(s) + Ca(NO_3)_2(l) \rightleftharpoons (Ca, v)La_2Ti_3O_{10}(s) + 2NaNO_3(l)$$

$$K = \frac{a_{Na^+}^2 a_{Ca^{2+}}}{a_{Na^+} a_{Ca^{2+}}} \tag{4.20}$$

In this equilibrium constant expression, the activity coefficient, a, is specified for the exchangeable cations (Na^+, Ca^{2+}) in each phase. The equilibrium may lie to the left or right, depending on the value of K. Accordingly, the product is more accurately represented as $Na_{2-x}Ca_{x/2}v_{x/2}La_2Ti_3O_{10}$, to signify possible incomplete, or partial, ion exchange. The creation of vacancies in a host by aliovalent ion exchange opens the material to the possibility of subsequent intercalation reactions, where another species can be inserted into the location once occupied by the original cation. In some cases the intercalation can introduce mixed valency on transition metal cations that are present in the structure. This will depend on the inserted species' reduction potential relative to those of the transition metal in its various oxidation states. It is noteworthy that many of these products are kinetically stable, thermodynamically metastable phases that cannot be prepared by the direct high-temperature ceramic route (Lalena et al., 1998).

A momentary digression is warranted here concerning the phase purity of powder products. What one hopes to obtain in a process such as the ion-exchange reaction discussed above is a homogeneous single-phase solid solution, as opposed to a multiphase mixture of product(s) and parent phase. The definition of *phase*, unfortunately, is somewhat ambiguous, as it gives no criterion for the length scale or degree of homogeneity. Does homogeneous require that every unit cell in every crystallite be identical? Generally speaking, we accept a "phase" as a state of matter in which the physical structure is, on average, *microscopically* homogeneous and, as such, one that has uniform thermodynamic properties. Phase purity is typically ascertained by conventional x-ray powder diffractometry. A powder diffraction pattern contains peaks whose positions depend on the size and shape of the unit cell, while their intensities are dependent on the identities of the unit cell contents. For coarse-grained powders (crystallite size >1 μm), small differences in peak positions, intensities, and asymmetry are easily discernible, which makes the technique well suited for checking phase purity. However, the possible presence of isostructural phases with only slight differences in unit cell dimensions and/or with isoelectronic species (and thus very similar atomic form factors or scattering factors) does present challenges with regard to verifying phase purity.

Both reactive and *nonreactive* molten salts can be used in nontopochemical routes. An example of a nontopochemical route to inorganic materials utilizing reactive molten salts is when a metallic element is reduced in a low-melting alkali metal polychalcogenide (A_2Q_n, where $Q = O$, S, Se, Te) to form a ternary metal chalcogenide. Potassium bismuth sulfide (KBi_3S_5) has been prepared in this way from the reaction of bismuth metal in potassium sulfide (Kanatzidis, 1990). Superconductive magnesium boride films (MgB_2) have been prepared by electrolysis from molten mixtures of $MgCl_2$, KCl, and MgB_2O_4, as has CaB_6 from $CaCl_2$ and CaB_4O_7 (Abe et al., 2002). $BaBiO_3$-derived superconducting oxides have been synthesized from anodic oxidation of molten salt mixtures containing $Ba(OH)_2$, KOH, and Bi_2O_3 (Therese and Kamath, 2000, and references therein).

In nonreactive molten salts, on the other hand, flux components are not incorporated into the product phase. Here, the molten salt acts more in the classical sense as a reagent to promote the reaction at a lower temperature than would be required by the ceramic, or direct, route (Section 5.2). This is accomplished by two attributes of molten salts: an acid–base equilibrium that enables the general dissolution–recrystallization of metal oxides; and a highly electropositive (oxidizing) environment that stabilizes the highest oxidation state of many transition metals (Gopalakrishnan, 1995), which can lead to mixed valency. A plethora of complex transition metal oxides have been synthesized in nonreactive molten alkali metal hydroxides, carbonates, and hypochlorites. Examples of such molten salt routes to mixed transition metal oxides include (Rao and Raveau, 1998):

- La_2CuO_4: from a mixture of La_2O_3 and CuO in molten NaOH
- $YBa_2Cu_4O_8$: from Y_2O_3, BaO, and CuO in Na_2CO_3/K_2CO_3 flux under flowing oxygen
- $La_4Ni_3O_{10}$: created by bubbling chlorine gas through an NaOH solution of lanthanum and nickel nitrates

4.5.2. Ionic Liquids

A more recent development, which has garnered some attention within the materials community, is the use of nonvolatile room-temperature ionic liquids (ILs) for the "bottom-up" synthesis of nanostructured inorganic materials. As mentioned previously, ILs are highly polar organic solvents, often possessing highly organized hydrogen-bonded networks. Although this field is still in its very early infancy, review articles (Antonietti et al., 2004; Taubert and Li, 2007) have highlighted some of the more recently uncovered aspects of room-temperature IL reaction media, including:

- Highly ordered ILs can serve as morphology-modifying templates for obtaining morphologies inaccessible by other synthetic routes.
- Shorter reaction times and milder conditions are generally required.
- Lower nucleation energies are observed.
- The IL reaction media itself can serve simultaneously as a reactant.
- There is polymorph selectivity.

We can see from this list that the main advantage of ILs is their ability to serve as an "all-in-one" solvent–reactant–template system. The function(s) played by the IL are, of course, determined by the anion–cation combination, which has virtually unlimited flexibility (Taubert and Li, 2007). The most studied ILs are the 1-alkly-3-methylimidiazolium salts. The liquid-phase tetrafluoroborate salt, with a trace amount of water, has been used to hydrolyze titanium tetrachloride, yielding crystalline anatase powder consisting of 2- to 3-nm particles that were assembled into spongelike superstructures (Zhou and Antonietti, 2003). Although anatase can also be prepared at low temperatures with the sol–gel technique, an amorphous phase with an average particle size of about 20 nm is produced. The nucleation rate of anatase in the IL must therefore be increased 1000-fold. Moreover, the amorphous sol–gel product carries the disadvantage of requiring calcination above 350°C in order to obtain the crystalline phase (Antonietti et al., 2004). Hollow titania microspheres have been produced by hydrolysis in immiscible IL–toluene mixtures (Nakashima and Kimizuku, 2003).

It is also possible to carry out reactions in ionic liquids at temperatures much higher than the melting point of the IL. Called *ionothermal reactions*, these differ from hydrothermal and solvothermal reactions in that they may be carried out at atmospheric pressure, due to the low vapor pressures of ionic liquids. Ionothermal reactions have led to novel materials containing cations with unusual coordination numbers (e.g., four-, five-, and six-coordinate aluminum) and morphologies.

4.6. ELECTROCHEMICAL SYNTHESIS

The process of forcing a nonspontaneous oxidation–reduction reaction to occur by means of electrical energy is termed *electrolysis*. The oldest and most common form of electrolysis is *electroplating* (galvanoplasty), which is the electrochemical deposition of a thin layer of metal or alloy (e.g., brass, solder) on a conductive substrate or a nonconductive substrate that has had a conductive coating applied. The advantage of electroplating over the vapor deposition of thin films is that the solid–liquid interface facilitates the growth of very conformal coatings (i.e., films with good step coverage) on substrates of any shape.

Electrolysis requires an *electrolytic cell*, generally consisting of two electrodes (usually made of metal or carbon), connected to an external source of direct current, immersed in an electrolyte bath (a liquid that can conduct electricity by the movement of ions). The electrodes may be of the same material in an electrolytic cell, whereas in a *galvanic (voltaic) cell*, which produces electricity, they are necessarily different. The metal to be deposited is initially dissolved in the plating bath, and the cell may be divided or undivided (Figure 4.8). If divided, a tube of an electrolyte in a gel, called a *salt bridge*, is used to allow the flow of ions (while preventing mixing) between the various solutions. Electroplating was invented in 1805 by the Italian chemist Luigi Brugnatelli (1761–1818), who electrodeposited gold onto silver using a voltaic pile. However, Humphry Davy (1778–1829), who electrolyzed molten sodium hydroxide to obtain sodium metal in 1807, is often credited as the father of electroplating because Napoleon rejected Brugnatelli's work and denied him permission to publish it in the French Academy (Witt, 2006).

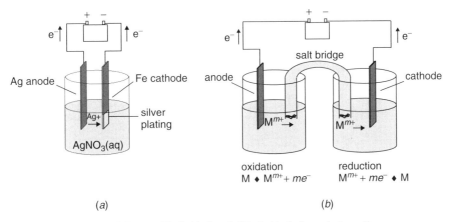

Figure 4.8 (*a*) Undivided and (*b*) divided electrolytic cells.

In a galvanic cell, which produces an electromotive force (emf), or potential difference across its terminals, the direction of electron flow is from the anode (the negative terminal) through the external circuit toward the cathode (the positive terminal). Opposite processes, each one in and of itself called a *half-reaction*, occur at the two electrode–solution interfaces within the cell. One half-reaction, reduction, occurs at the cathode, while the opposite, oxidation, occurs at the anode. By convention, the emf of galvanic cells is positive; voltages for spontaneous reactions have + signs. The galvanic cell voltage is the difference between the reduction potentials of the two half-cells:

$$E^{\circ}_{cell} = E^{\circ}_{cathode} - E^{\circ}_{anode} \tag{4.21}$$

By convention, the potentials of all half-reactions, E°, are found tabulated for the reduction process under standard conditions of temperature (298.15 K), pressure (1 atm), and solute concentrations (1 molar). For nonstandard conditions, the reduction potentials, and hence the cell voltage, will differ. The concentration dependence on the cell voltage is given by the *Nernst equation*:

$$E_{cell} = E^{\circ}_{cell} - \frac{RT}{nF} \ln K \tag{4.22}$$

A few standard reduction potentials are shown in Table 4.2. In the table, species on the left-hand side of the reduction half-reaction are strong oxidizing agents if their reduction potential is highly positive (e.g., F_2). Alternatively, species on the right-hand side of the half-reaction are strong reducing agents if the reduction potential of their oxidized state is highly negative (e.g., Na metal). More extensive tabulations are readily available in a number of sources: for example, the *CRC Handbook of Chemistry and Physics*.

In an electrolytic cell, electrical energy is used to force a current through the cell to produce a chemical change for which the cell potential is negative;

TABLE 4.2   Some Standard Electrode (Reduction) Potentials in Aqueous Solution at 25°C

	Electrode Potential (V)
$F_2 + 2H^+ + 2e = 2HF$	3.053
$Au^+ + e = Au$	1.692
$Au^{3+} + 3e = Au$	1.498
$ClO_3^- + 6H^+ + 6e = Cl^- + 3H_2O$	1.451
$Cl_2(g) + 2e = 2Cl^-$	1.35827
$NO_3^- + 4H^+ + 3e = NO + 2H_2O$	0.957
$Ag^+ + e = Ag$	0.7996
$ClO_3^- + 3H_2O + 6e = Cl^- + 6OH^-$	0.62
$Cu^+ + e = Cu$	0.521
$ClO_4^- + H_2O + 2e = ClO_3^- + 2OH^-$	0.36
$Cu^{2+} + 2e = Cu$	0.3419
$Bi^{3+} + 3e = Bi$	0.308
$W^{3+} + 3e = W$	0.1
$NO_3^- + H_2O + 2e = NO_2^- + 2OH^-$	0.01
$Fe^{3+} + 3e = Fe$	−0.037
$Ni^{2+} + 2e = Ni$	−0.257
$Fe^{2+} + 2e = Fe$	−0.447
$2H_2O + 2e = H_2 + 2OH^-$	−0.83
$2NO_3^- + 2H_2O + 2e = N_2O_4 + 4OH^-$	−0.85
$SO_4^{2-} + H_2O + 2e = SO_3^{2-} + 2OH^-$	−0.93
$Mg^{2+} + 2e = Mg$	−2.372
$Na^+ + e = Na$	−2.71

that is, electrical work causes an otherwise nonspontaneous chemical reaction to occur. As shown in Figure 4.8, in the external circuit the direction of electron current flow is from the negative terminal of the source to the cathode and away from the anode to the positive terminal of the source. The minimum voltage necessary to bring about electrolysis under standard conditions is the standard cell potential calculated by Eq. 4.21. However, to overcome the internal cell resistance, the voltage actually needed to force electrolysis is always somewhat higher (typically, several tenths of a volt) than that calculated by Eq. 4.21. The extra voltage is termed the *overpotential*.

Many plating baths contain cyanide, phosphate, or carbonate anions to facilitate the half-reactions and to increase conductivity. In making predictions about overall cell reactions, all possible half-reactions should be considered, including oxidation or reduction of the anions present in solution or even of the solvent species itself. In general, the most easily oxidized species (the one with the most negative reduction potential) will be oxidized, and the most easily reduced species (the one with the most positive reduction potential) will be reduced. Of the possible half-reactions, the ones with the most positive (least negative) potential will usually take place. Hence, we can see from the reduction potentials listed in Table 4.2 that like electrolysis of molten sodium chloride, electrolysis of an aqueous sodium chloride solution does not yield sodium metal, but rather, water is reduced to hydrogen gas and hydroxide anion.

In electroplating, the metal film is formed from metal ions dissolved in the electrolyte when they are reduced to the insoluble metallic state by the reduction half-reaction occurring at the cathode. As just described, various other half-reaction processes may be in competition with the metal ion reduction, such as electrolysis of the solvent itself (usually, water), reactions that consume H^+ ions or generate OH^- ions (termed *electrogeneration of base* since they raise the pH), or anion reduction. The best known of these, perhaps, are various possible reactions of the chlorate, perchlorate, and nitrate ions, which may all serve as strong oxidizing agents (i.e., species that are easily reduced), depending on the pH. Hydroxide ion is generated directly or indirectly as a consequence of H^+ consumption by the reduction of these anions in aqueous solutions. The standard reduction potentials for these half-reactions are more positive than those for the reduction of most metal ions to the metallic state, the exceptions being Ag^+, Au^+, Au^{3+}, Bi^+, Bi^{3+}, Cu^{2+}, Ir^{3+}, Pt^{2+}, Tc^{2+}, Te^{4+}, Tl^{3+}, Rh^+, Rh^{2+}, Rh^{3+}, Ru^{2+}, and W^{3+}. Other than in these cases, baths containing nitrate, chlorate, and perchlorate salts would not be expected to deposit metal but rather, insoluble metal hydroxide by cathodic reduction. Phosphates and several binary and ternary oxides have also been synthesized by cathodic reduction via electrogeneration of base.

Cathodic reduction is also used to purify some metals, such as copper. Slabs of impure copper serve as the anode, while a pure copper sheet serves as the cathode in an undivided electrolytic cell. The electrolytic bath is copper(II) sulfate. During electrolysis, Cu^{2+} ions leave the anode and plate on the cathode. Impurity metals more reactive than copper are oxidized and stay in solution. Less reactive metals collect at the bottom of the cell. After about a month, the enlarged copper cathodes are removed (Ebbing and Gammon, 2005). Metals can also be oxidized electrolytically at the anode (anodized). It is even possible to further oxidize some metals in a low oxidation state to a higher oxidation state.

There are number of experimental parameters in electrochemical synthesis, which often must be selected empirically through trial and error, including deposition current, deposition time, deposition temperature, bath composition, choice of cell (divided or undivided), and choice of electrode (bulk inert, bulk reactive, or electrodes with preadsorbed reactive films). The morphology of the final product obtained (e.g., crystallinity, adherent film versus polycrystalline powder) is highly dependent on all of these factors (Therese and Kamath, 2000).

REFERENCES

Abe, H.; Yoshii, K.; Nishida, K. *Jpn. J. Appl. Phys.* **2002**, *41*, L685.

Antonietti, M.; Smarsly, B.; Zhou, Y. *Angew. Chem. Int. Ed.* **2004**, *43*, 4988–4995.

Apps, J. A.; Sonnenthral, E. L. Paper LBNL-56070, Lawrence Berkeley National Laboratory, Berkeley, CA, Apr. 1, **2004**. http://repositories.cdlib.org/lbnl/LBNL-56070.

Baldan, A. *J. Materi. Sci*. **2002**, *37*, 2171–2202.

Bernal, J. D. *Proc. R. Soc. London*. **1964**, *280A*, 299.

Boistelle, R.; Astier, J. P. *J. Cryst. Growth*. **1988**, *90*, 14–30.

Calheter, J. W., Jr. *Fundamentals of Materials Science and Engineering: An Integrated Approach*, Wiley, Hoboken, NJ, **2005**, p. 593.

Ebbing, D. D.; Gammon, S. D. *General Chemistry*, 8th ed., Houghton Mifflin, New York, **2005**.

Ebelman, M. *Ann. Chem. Phys*. **1846**, *57*, 319.

Echt, O.; Sattler, K.; Recknagel, E. *Phys. Rev. Lett*. **1981**, *47*, 1121–1125.

Elizabé, L.; Kariuke, B. M.; Harris, K. D. M.; Tremayne, M.; Epple, M.; Thomas, J. M. *J. Phys. Chem. B*. **1997**, *101*, 8827.

Gopalakrishnan, *J. Chem. Mater*. **1995**, *7*, 1265.

Graham, T. *J. Chem. Soc*. **1864**, *17*, 318.

Hoglund, D. E.; Thompson, M. O.; Aziz, M. *J. Phys. Rev. B*. **1998**, *58*, 189.

Huang, S. C; Glicksman, M. E. *Acta Metall*. **1981**, *29*, 71.

Inoue, A.; Nishiyama, N.; Kimura, H. *Mater. Trans. JIM*. **1997**, *38*, 179.

Ivantsov, G. P. *Dokl. Akad. Nauk*. **1947**, *58*, 56.

Johnson, W. L. *Mater. Res. Soc. Bull*. **1999**, *24*, 42.

Kanatzidis, M. G. *Chem. Mater*. **1990**, *2*, 353.

Lalena, J. N.; Cushing, B. L.; Falster, A. U.; Simmons, W. B., Jr.; Seip, C. T.; Carpenter, E. E.; O'Connor, C. J.; Wiley, J. B. *Inorg. Chem*. **1998**, *37*, 4484.

LaMer, V. K. *Ind. Eng. Chem. Res*. **1952**, *44*, 1270–1278.

Laudise, R. A. *Chem. Eng. News*. **1987**, *56*, 30.

Laue, M. von Z. *Kristallogr*. **1943**, *104*, 124.

Lin, Y. H.; Wang, G.; Wang, R. J.; Zhao, D. Q.; Pan, M. X.; Wang, W. H. *Sience*. **2007**, *315*, 1385.

Madras, G.; McCoy, B. J. *J. Chem. Phys*. **2001**, *115*, 6699–6706.

Mullins, W. W.; Sekerka, R. F. *J. Appl. Phys*. **1963**, *34*, 323.

Nakashima, T.; Kimizuku, N. *J. Am. Chem. Soc*. **2003**, *125*, 6386.

Nyvlt, J. *Cryst. Res. Technol*. **1995**, *30*, 443–449.

Ostwald, W. *Lehrbuck der Allgemeinen Chemie*, Vol. 2, Part 1, Leipzig, Germany, **1896**.

Ostwald, W. *Z. Phys. Chem*. **1897**, *22*, 289–330.

Perker, A.; Johnson, W. L. *Appl. Phys. Lett*. **1993**, *63*, 2342.

Pfann, W. G. *Trans. AIME*. **1952**, *194*, 747.

Phanikumar, G.; Chattopadhyay, P. P. *Sādhanā*, **2001**, *26*, 25.

Rabenau, A. *Angew. Chem*. **1985**, 97, 1017; *Angew. Chem. Int. Ed*. **1985**, *24*, 1026.

Rao, C. N. R.; Raveau, B. *Transition Metal Oxides: Structure, Properties, and Synthesis of Ceramic Oxides*, 2nd ed., Wiley-VCH, New York, **1998**.

Scheil, E. Z. *Metallkd*. **1942**, *34*, 70.

Stoica, C.; Tinnemans, P.; Meekes, H.; Vlieg, E.; van Hoof, P. J. C. M.; Kaspersen, F. M. *Cryst. Growth Des*. **2005**, *5*, 975–981.

Tammann, G. Z. *Elektrochem*. **1904**, *10*, 532.

Taubert, A.; Li, Z. *Dalton Trans*. **2007**, DOI: 10.1039/b616593a.

Therese, G. H. A.; Kamath, P. V. *Chem. Mater*. **2000**, *12*, 1195.

Tiller, W. A.; Jackson, K. A.; Rutter, R. W.; Chalmers, B. *Acta Metall*. **1953**, *1*, 50.

Turnfull, D.; Fisher, J. *J. Chem. Phys*. **1949**, *17*, 17.

Turnbull, D. *J. Chem. Phys*. **1950**, *18*, 198.

Turnbull, D. *Met. Trans. A* **1981**, *12*, 695.

Witt, K. https://www.wesrch.com/User_images/Pdf/10_1168903056.pdf, **2006**.

Xing, L. Q.; Eckert, J.; Loser, W.; Schultz, L. *Appl. Phys. Lett*. **1998**, *73*, 2110.

Xing, L. Q.; Eckert, J.; Loser, W.; Schultz, L.; Herlach, D. M. *Philos. Mag. B*. **1999**, *79*, 1095.

Zhao, Y.; Zhang, J. *Nature.* **2004**, *430*, 332–335.

Zhou, Y.; Antonietti, M. *J. Am. Chem. Soc*. **2003**, *125*, 14960.

5 Solid–Solid Reactions

5.1. SOLID–SOLID INTERFACE

In Chapters 3 and 4 we considered the solid–vapor and solid–liquid interfaces. In both of these cases, one of the components, the vapor or the liquid, is a mobile phase where the molecules move rapidly with respect to the atoms in the solid phase. In Chapter 3 we examined cases where the mobile phase penetrated the solid as well as cases where it adhered to the solid. In either case we viewed the solid as remaining intact, a recognizable and distinct component of the system. In this chapter, where we examine the solid–solid interface and review some solid–solid synthetic techniques, neither solid will remain chemically identifiable after the reaction. Moreover, the slow time scale on which atoms in the solid phase move will require high temperatures and small reaction distances for reasonable reaction times.

Before discussing a few specific synthetic techniques, we examine the generic solid–solid interface. For simplicity, consider a solid–solid interface composed of two crystalline materials, as shown in Figure 5.1. The most obvious feature of the solid–solid interface is that both phases are tightly packed, and in order for one phase to move into the other, vacancies or interstices are necessary. Moving atoms in one phase out of the way to accommodate atoms from the other phase will take a great deal of energy. The diffusion coefficient for atoms in this scenario is low. Diffusion coefficients, D, for atoms in a solid are many orders of magnitude less than what is observed in liquids and gases. Typical values are of D are: gases, 10^{-1} cm^2/s; liquids, 10^{-5} cm^2/s; and solids, 10^{-20} cm^2/s. The units on the diffusion coefficient, cm^2/s, are such that Dt, where t is time in seconds, gives the square of the average distance traveled by a particular atom. Hence, in 1 minute, a molecule in a liquid with $D = 10^{-5}$ cm^2/s will diffuse, on average, 0.02 cm. In a solid, however, on the same time scale, an atom will diffuse only 7.7×10^{-10} cm, or about 0.1 Å. In Figure 5.1, the distance between the atoms is typically 2 Å. Because there is an energy barrier to diffusion, the diffusion coefficient has a temperature dependence (Fisler and Cygan, 1999) similar to the Arrhenius behavior observed in chemical reaction rates: namely,

$$D(T) = D(0)e^{-E_a/RT} \tag{5.1}$$

Inorganic Materials Synthesis and Fabrication, By John N. Lalena, David. A. Cleary, Everett E. Carpenter, and Nancy F. Dean
Copyright © 2008 John Wiley & Sons, Inc.

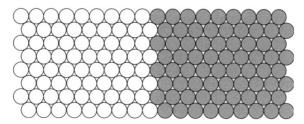

Figure 5.1   Interface of two different crystals prior to a chemical reaction occurring.

where E_a is the activation energy for diffusion. This equation, and the one relating displacement to D,

$$\langle \Delta x^2 \rangle = 2Dt \tag{5.2}$$

form the basis for all solid–solid synthetic strategies.

Because diffusion has an activation energy barrier, higher temperatures result in higher diffusion coefficients. Furthermore, smaller particle sizes mean smaller Δx required for a complete reaction. Therefore, the traditional approach to solid–solid reactions is to use fine powders and high reaction temperatures. This direct combination of two or more powdered solids at high temperature, referred to as the *ceramic method* of synthesis, is discussed in Section 5.2.

As discussed in Chapter 3, the deposition in molecular beam epitaxy is kept to small thicknesses, on the order of a monolayer, so that a very small diffusion coefficient, D, does not prevent a reaction from occurring on a reasonable time scale. In the ceramic method, we are again presented with the interdependence of the diffusion coefficient, reaction temperature, and displacement distance, Eqs. 5.1 and 5.2.

5.2. CERAMIC METHOD

The ceramic method is the most common method used in the synthesis of solids from solids. In this section we review the basic components of this method. This will provide a point of reference for our subsequent discussion of other methods, such as mechanical alloying and combustion synthesis. A ceramic is a compound comprising a metal and a nonmetal; an example is magnesium oxide, MgO. Because of the large difference in electronegativity between metals and nonmetals, most ceramics are also ionic compounds. As such, they have the typical properties of ionic compounds that are discussed in general chemistry courses: they have high melting points, are hard, and do not conduct electricity.

A ceramic can be formed simply by burning a metal in air:

$$2Mg(s) + O_2(g) \rightarrow 2MgO(s)$$

When ceramic compounds are used as reactants in solid-state synthesis, the reaction conditions are often extreme because of the refractory nature of ceramics.

For example, the spinel $ZnCr_2O_4$ can be prepared from ZnO and Cr_2O_3 (Mancić et al., 2004):

$$ZnO(s) + Cr_2O_3(s) \rightarrow ZnCr_2O_4(s)$$

In this case, both starting materials are oxides, and the reaction can be carried out open to the atmosphere. However, regardless of how fine the powdered starting materials may be, at some point a crystallite of ZnO and Cr_2O_3 will be in contact, and it is at this point that the problem illustrated in Figure 5.1 becomes important.

Zinc(II) and chromium(IV) cations are smaller (0.88 and 0.69 Å respectively) than oxygen(II) anions (1.24 Å), so one way to view this reaction is through cation migration. The zinc cations need to extract themselves from their oxide octahedral hole and move to the chromium compound to find an available tetrahedral hole. Simultaneously, chromium cations are moving in the direction of the zinc compound. After a few exchanges of cations, the two crystallites look as shown in Figure 5.2.

At this point, the zinc and chromium cations must not only extract themselves from their own coordination sites, they must travel a significant distance through the product materials that are forming at their interface. Figures 5.1 and 5.2 present an oversimplified version of what happens at the atomic level. For example, no defects are shown in Figure 5.1 or 5.2. Smigelskas and Kirkendall (1947) showed that defects are important for understanding atomic migration in solids. In this classic experiment, Kirkendall showed that when copper and zinc interdiffused, they did so at different rates, leaving the faster-diffusing material with additional vacancies. This effect is now known as the *Kirkendall effect*. It has been observed in a dramatic fashion in the synthesis of nanoparticles. Yin et al. (2004) showed that when nanoparticles of cobalt sulfide (Co_3S_4 and Co_9S_8) are formed by adding sulfur (outer circle in Figure 5.3*a*) to cobalt nanoparticles (gray), the higher diffusion coefficient of the cobalt results in a hollow nanoparticle of cobalt sulfide (black) because the vacancies created in the cobalt coalesce to form a large single void in the center of the nanoparticle.

The traditional ceramic method of solid-state synthesis is designed to overcome the problems of slow ionic migration over long distances. We discuss the classic

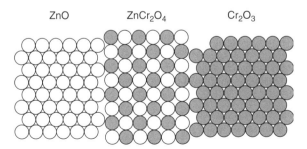

Figure 5.2   Interface of two different crystals where a reaction has taken place, resulting in product material separating the unreacted crystals.

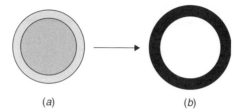

(a)   (b)

Figure 5.3 *(a)* Sulfur surrounding a cobalt nanocrystal; *(b)* hollow nanocrystal of cobalt sulfide.

method first and then examine some modifications that have been developed to facilitate faster or more complete reactions. The first step in the ceramic method is mixing of the starting materials. Even at this early stage, the unique characteristics of an all-solid reaction must be recognized. The purity of the starting materials will affect the purity of the product. There is not a method to purify the product after the synthesis. This is in stark contrast to a typical organic synthesis, where the desired product can often be purified to a level that is determined only by the experimentalist's patience and need for product. In solid–solid reactions, the impurities in the starting material will probably appear in the product.

The starting materials are mixed in the molar ratio needed in the final product. In the case of $ZnCr_2O_4$, 1 mol of ZnO and 1 mol of Cr_2O_3 would be needed. Fine powders of each would be used. Powders are normally sold with purity and mesh designations. The purity may read 3 N7, which means three nines followed by a seven: 99.97%. Sometimes all the digits are nines: 5 N means 99.999%. The mesh designation follows one of the standard mesh sizes. Sample U.S. standard sieve sizes include:

Mesh No.	Size (μm)
20	840
100	149
400	37

As the mesh number increases, the openings in the mesh decrease. In Figure 5.4a the mesh number is small, and in Figure 5.4b the mesh number is large. A powder listed as −100 mesh means that all the particles passed through a 100 mesh; therefore, the diameter of all is less than 149 μm. No lower limit on their size is implied. Sometimes a more specific designation is provided and might read −100 +400, which means that the particles have diameters between 37 and 149 μm. One might assume that in all solid–solid reactions, one always uses the highest mesh number possible to keep particle sizes small and therefore keep diffusion lengths minimized. This is often true, but unfortunately, some reactions carried out with too fine particles convert to an explosive regime. In such a case, larger particle sizes (wire, foil, or chunks) are used to restrain the reaction rate.

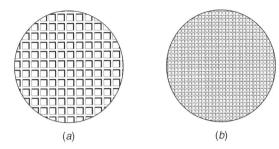

Figure 5.4   (a) Small and (b) large mesh numbers.

With the correct purity, mesh size, and molar ratios, the starting materials are mixed. Because they are solids, this mixing process is not spontaneous on any reasonable time scale, and therefore the mixing must be accomplished be mechanical means. Several minutes of stirring the powder mixture with a lab spoon is not likely to achieve a homogeneous mixture, especially if the mesh sizes of the reactants are significantly different. Instead, ball mills are often used and mixing may take several hours. After mixing, the powder has a lot of trapped air. Given that the crystallites must be in contact with each other in order to react, this air should be removed and the crystallites brought into as close contact as possible. Therefore, the powder mixture is often pressed into a pellet to get the crystallites in close contact and eliminate the air. Finally, the pellet is ready to be heated to begin the reaction.

After the reaction has proceeded for some time, the situation shown in Figure 5.2 develops. Therefore, it is often necessary to stop the reaction, break up the pellet, regrind the material, which now consists of product and unreacted starting material, remake the pellet, and continue heating. Heating times can go for weeks at temperatures in excess of 1000°C. The standard method for determining the extent of reaction, and hence the need for additional reaction time, is powder x-ray diffraction (XRD), discussed in Section 1.6. In much the same way that the organic chemist uses nuclear magnetic resonance (NMR) to establish the absence of starting material in a mixture, the solid-state chemist uses powder XRD to determine that all the starting material has been consumed.

Powder XRD is useful for establishing the presence and absence of phases. In addition, it can be used to determine particle size. A common method for particle size measurement uses the *Scherrer formula*,

$$d = \frac{K\lambda}{\text{FWHM}\cos\theta} \tag{5.3}$$

where K is a shape factor and is usually set equal to 0.9, λ is the x-ray wavelength, and FWHM is the full width at half maximum (in radians) for a diffraction peak occurring at 2θ (in degrees). As a crystallite gets larger, the width of the diffraction peak decreases. The linewidth also contains contributions from instrumental broadening. These can be accounted for by using a standard material such as those available from the National Institute of Standards and Technology. Materials such as LaB_6 and ZnO/CeO_2, calibrated with respect to particle size, are useful for this purpose. A more exact determination of particle size requires a technique such as electron microscopy. In addition to Scherrer linewidth analysis, powders used in ceramic reactions are routinely characterized with respect to surface area. The BET analysis discussed in Section 3.2 is the method of choice for determining the surface area, and hence average particle size, of a powdered ceramic material. This attention to particle size highlights the importance of this parameter in the ceramic method.

An alternative to pressing the reactants into a pellet is to add the powders to a nonreactive flux agent, as discussed in Section 4.5. A nonreactive flux agent is usually a salt or a mixture of salts. The reactants are mixed with the salt, and when the mixture is heated, the salt melts and essentially acts as a solvent for the reactants. Ionic liquid solvents have two attributes that facilitate synthesis: (1) an acid–base character, which can be described by chemical equilibria, that enables the general dissolution and recrystallization of metal oxides; and (2) a highly oxidizing environment, which stabilizes the higher oxidation states for many metals. As with any solvent, it is usually important that the components of the [nonreactive] flux not become part of the product. As an example, consider the synthesis of barium titanate. Barium titanate, $BaTiO_3$, one of the most studied ceramic materials because of its ferroelectric properties, can be prepared by reacting $BaCO_3$ and TiO_2 in a KF flux (Galasso, 1973):

$$BaCO_3(s) + TiO_2(s) \xrightarrow{\text{KF}} BaTiO_3(s) + CO_2(g)$$

The mixture of barium carbonate, titanium oxide, and potassium fluoride is heated slowly at first, allowing the carbon dioxide to escape. Potassium fluoride has a melting point of $860°C$. After the sample has finished outgassing and the potassium fluoride has melted, the reaction continues at $1160°C$ for 12 hours, followed by cooling at $25°C/hour$ to $900°C$, where the liquid flux can be poured off. The product crystals are cooled slowly to room temperature, and the remaining flux is washed off with hot water. Molten salts continue to be investigated as flux agents to provide a method for preparing ceramics, especially single-crystalline samples of ceramics (Xing et al., 2006). With a flux agent, the synthetic chemist is avoiding the need to reduce particle size down to the unit cell scale in order to observe a reaction. Unfortunately, discovering the optimum flux agents in a ceramic synthesis, if such flux agents exist, is a highly empirical task (Geselbracht et al., 2004).

For the high temperatures used in the ceramic method, special attention must be paid to the reaction vessel. Materials that are normally considered inert, such as platinum, will react with certain elements (e.g., phosphorus) at high-temperatures. Much of this interesting high-temperature chemistry is learned the hard way after an unexpected reaction of the crucible. For open crucible reactions, the common crucible compositions are alumina, zirconia, quartz, and platinum, all available commercially in sizes ranging from under 1 inch in diameter up to

2 feet in diameter for quartz crucibles used in commercial silicon crystal pulling operations. Quartz and platinum are also used in closed (sealed) reaction vessels, which are discussed below. Whether closed or sealed, the chemical compatibility of the crucible with the reactants and products is an important issue, particularly in the sealed case, where failure of the reaction vessel can cause serious injuries. In addition to mechanical strength, another issue concerning the selection of a crucible is the concern for contamination. A Pyrex® vessel will contain boron, sodium, and aluminum, and these elements can leach from the glass at high temperature given the right contents of the vessel. Quartz is chemically cleaner but still contains trace amounts of other elements, especially sodium. High-purity quartz has been developed for the semiconductor industry. A major problem with quartz is that if it is exposed to alkali metal ions at high temperatures ($>1200°C$), these ions can diffuse into the quartz and cause devitrification and ultimately, vessel failure.

Materials prepared at high temperature in an open crucible are typically quite stable with respect to room temperature and ambient conditions. The ceramic method is also used to prepare materials that are not stable with respect to ambient conditions, particularly with respect to oxygen and water reactivity. In addition to products having such sensitivities, reactants also have the potential for such reactivity. In these cases, inert atmospheres are required for the mixing and firing of the reactants. Inert atmosphere manipulations require special training (Shriver and Drezdzon, 1986). They can range from a disposable plastic glove bag (Figure 5.5) which is purged with an inert gas, to a fully controlled automatic atmosphere system (Figure 5.6) commercially produced. The emphasis so far has been on the small particle size necessary to produce a reasonable reaction rate. These same particles also have a high surface area/mass ratio, and therefore if the material is water or oxygen sensitive, the fine-powdered form of the material will accelerate this decomposition reaction. Nonetheless, the ceramic method can

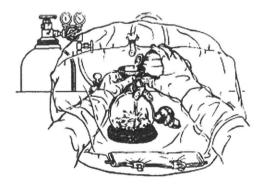

Figure 5.5   Glove bag used for inert-atmosphere manipulations. (Courtesy of Glas-Col, LLC, Terre Haute, IN.)

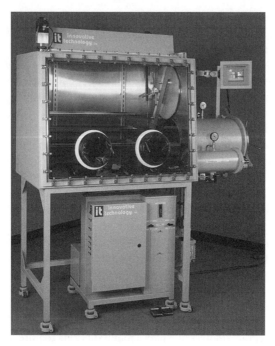

Figure 5.6 Commercial inert-atmosphere chamber also know as a glove box. (Courtesy of Innovative Technologies, Inc., Newburyport, MA.)

be used with air-sensitive powders when appropriate inert atmosphere techniques are used.

Normally, the air-sensitive powders are weighed and loaded into a reaction vessel that can be sealed. Often, a quartz glass ampoule is used because sealing a glass tube is easier than sealing (welding) a metal tube. If the glass is likely to react with the materials being added, several options are available. One is to switch to a metal reaction tube such as gold or platinum. This, however, will require welding to seal the tube. Another option is to coat the inside of the glass tube with carbon or metal. A carbon coating can be achieved by first coating the inside of the glass with an organic liquid rich in carbon, such as benzene, C_6H_6. The excess hydrocarbon is poured out and the coating is pyrolzed to produce the carbon film.

A metal film can be produced chemically, such as the reaction of silver nitrate and formaldehyde in an aqueous ammonia solution:

$$2Ag(NH_3)_2^+(aq) + H_2CO(aq) + H_2O \rightarrow 2Ag(s) + H_2CO_2(aq) \\ + 2H^+(aq) + 4NH_3(aq)$$

Grinding powders in an inert atmosphere can be difficult unless a ball mill has been introduced into a dry box. However, if the one of the reactants melts, mixing is not as critical. For example, the reaction of sulfur and tantalum to form TaS_2 involves mixing elemental sulfur and tantalum. Both reactants are stable in air, but at the temperature necessary for the reaction to occur, both reactants will react with oxygen before reacting with each other. This type of ceramic reaction is done in a sealed glass tube.

The common allotrope of sulfur, α-S_8, melts at 120°C (Greenwood and Earnshaw, 1984). For a reaction in a sealed tube at 1000°C, the vapors generated by this liquid can present a serious overpressure situation. As a rough estimate, consider a reaction tube of internal diameter (25 mm) and length 200 mm. If 1 g of α-S_8 is brought to 1000°C is converted to $S_2(g)$, and the ideal gas law (Eq. 3.1) is used, the internal pressure would be 16.6 atm. Assuming that the bursting pressure of glass tubing is 1000 psi (68 atm), the tube should hold. However, the strength of glass tubing is unpredictable and will be reduced by scratches, nicks, poor annealing, and most important, the seal of the tube is typically much weaker.

Ideally, the sulfur vapor will react with the tantalum metal and therefore the gas pressure will be reduced. Also, not all of the sulfur vapor will be S_2. Formation of gas-phase species S_n, where $n \geq 3$, will also significantly reduce the pressure. However, the gas pressure inside the tube can be increased because of the presence of a chemical transporting agent. Chemical transporting agents are often used to promote single-crystal growth in closed reaction tubes (Schäfer, 1964). We discuss the mechanism of these agents below, but for now we confine our attention to their contribution to the pressure inside the reaction tube. One approach to chemical vapor transport is adding a slight excess of one of the reactants. Another approach is to add a separate transporting agent. If the transporting agent iodine is added to a tube at concentration of 5 mg/cm^3 of tube volume, then at 800°C, the pressure due to the iodine will be 1.73 atm. This assumes that the iodine remains as the diatomic gas, $I_2(g)$. At 800°C, the equilibrium constant for the dissociation of iodine molecules to atoms,

$$I_2(g) \rightleftharpoons 2I(g)$$

is 0.0114 (Brown, 1931). An initial concentration of 5 mg $I_2(s)$/cm^3 results in an almost equal concentration of $I_2(g)$ and $I(g)$ at 800°C. Hence, the pressure due to $I_2(g)$ + $I(g)$ is $\frac{4}{3} \times 1.73$ atm = 2.3 atm.

At the reaction temperature, the crystallographic phase of the TaS_2 formed is trigonal and is designated 1T-TaS_2 (Revelli, 1979). If the reaction tube is removed from the furnace, this phase persists at room temperature. It is a golden metallic lustrous material that crystallizes in thin-layered opaque crystals. It has been used as a host for intercalation reactions as well as studied for its own intrinsic properties, which include charge density waves. If the reaction tube, however, is cooled slowly, over a period of days from 850 to 550°C, including several one-day annealing periods at 750 and 650°C, a different phase of TaS_2 is obtained. This phase is hexagonal and is designated 2H-TaS_2. This material is blue-black. The mirrorlike faces of the 1T-TaS2 are converted to a rough and rumpled coallike surface in 2H-TaS$_2$. This is but one example of the importance of the cooling procedure used in solid-state synthesis. A programmable furnace, whether of the muffle or tube type, simplifies a slow cooling rate and any annealing operations. Not all compounds undergo phase transitions with temperature, but as in the case of TaS$_2$, when they do, the results can be quite dramatic.

Reactions in sealed glass tubes must be recovered from the glass tube. This presents another difficultly in that the long time exposure to high temperature and reactive elements can weaken the glass significantly. Normally, one would score the tube with a glass cutting knife, wrap the tube in weighing paper, and either break the tube by hand or with a mallet. However, the scoring can unexpectantly rupture the tube explosively. Needless to say, eye, hand, and arm protection are required when working with sealed glass tubes after the reaction is completed.

We have considered ceramic synthesis in an open crucible and closed glass tube. A third possibility is a flow system. Chromium chloride, CrCl$_3$, can be prepared by passing carbon tetrachloride over a sample of chromium oxide maintained at 900°C (Angelici, 1986):

$$Cr_2O_3(s) + 3CCl_4(g) \rightarrow 2CrCl_3(s) + 3COCl_2(g)$$

This requires the use of a tube furnace. Tube furnaces have an advantage over muffle or box furnaces in that it is relatively easy to control the atmosphere in a tube furnace. As shown in Figure 5.7, an insert tube, typically quartz, can be inserted over the entire length of the furnace and extended enough beyond so that the ends are cool. Using standard gas-handling techniques, the atmosphere in the furnace can be controlled or the tube can be evacuated. The flow system approach has the advantage of being able to conduct air-sensitive reactions without a sealed reaction tube with its overpressure dangers and need for breakage to recover the product. A flow system does require a significantly greater amount of the volatile reactants since the pressure in a flow tube is much less than the pressure in a sealed tube and much of the volatile reactant will pass through the furnace unreacted and potentially, be wasted. A method for delivering the volatile reactants must be constructed. The flow approach is much like the chemical vapor deposition described in Section 3.6, although the emphasis here is on

Figure 5.7   Tube furnace with insert tube.

bulk synthesis and not on thin-film production, and hence the hardware is not as sophisticated.

Another advantage of a tube furnace is that a sample can be subjected to a temperature gradient in a tube furnace. Commercially available furnaces have shunts mounted on the exterior of the furnace so that current can be diverted from the heating element. With this arrangement, and a lot of patience, the user can shape a temperature profile to optimize the chemical processing. Alternatively, multiple heating elements, each with a separate controller, can be packaged as a single-tube furnace with independent heating zones. The mechanism of chemical vapor transport, mentioned above, provides an example of using a variable-temperature profile to optimize a chemical process. This process relies on the temperature dependence of a gas-phase equilibrium. Suppose that a metal, M, and a halogen such as iodine have the following temperature-dependent equilibrium:

$$M(s) + I_2(g) \rightleftharpoons MI_2(g)$$

Let's further suppose that the reaction as written is endothermic and therefore that as the temperature is increased, the equilibrium shifts to the right. For a reaction tube containing M and I_2 in a temperature gradient, the following processes will occur. At the high-temperature end of the reaction tube, the concentration of the halogen (solid dots in Figure 5.8) will be low because of the shift to the right in the equilibrium. The metal halide vapor (open dots) concentration in the high-temperature end will increase and hence diffuse to the low-temperature end of the reaction tube. Here it encounters a smaller equilibrium constant, and therefore the equilibrium shifts to the left, depositing M(s) (rhombi in Figure 5.8). When the steady state is achieved, halogen vapor is diffusing from the low-temperature end of the tube to the high-temperature end, and metal halide vapor is diffusing in the opposite direction. To facilitate these gas flows, the tube furnace is often inclined at a slight angle. What makes this a particularly interesting method for preparing compounds and growing single crystals is that the temperatures used are less than the sublimation temperatures of the reactants and products, and more important, they are less than the decomposition temperatures.

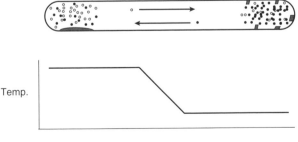

Distance along furnace

Figure 5.8   Chemical vapor transport. In this example, material in the hot region, left side of tube, is transported to the cool region, right side of tube.

The tube furnace shown in Figure 5.7 without shunts and a single heating element will still have a temperature gradient because the ends are open to the atmosphere. By keeping the reaction confined to a small distance relative to the length of the tube, essentially isothermal conditions can be maintained. If a gradient is needed, the sample can be repositioned to a portion of the tube furnace with a temperature gradient.

In the previous experimental techniques, high temperatures were required to overcome the small diffusion coefficients found in solid–solid reactions. An alternative approach is to reduce the particle size of the reactants such that only very small diffusion distances are needed to achieve complete chemical reaction. Clearly, particle size and reaction temperature are related and could be regarded as the two most important experimental parameters in the ceramic method. We examine both of these in some detail, starting with particle size. The most obvious method for reducing particle size is to grind a material. Yet what appears to the naked eye as a fine powder resulting from mechanical grinding actually contains individual crystallites with many unit cells. A cube 1 μm on edge (a mesh size of 2500 passes particles smaller than 5 μm) contains 1 billion cubic unit cells 10 Å on edge. Hence, mechanical grinding, despite its general applicability, has a limit of usefulness. Several other techniques, each with limited applicability, have been developed to reduce particle size below 1 μm. The two methods that we discuss here are sol–gel processing and coprecipitation.

Sol–gel as a synthetic method was discussed in Section 4.3. Here we revisit the topic, with the emphasis shifted to particle size. Sol–gel techniques have been reviewed extensively (Wright and Gommerdijk, 2001). The basic principle underlying the sol–gel method involves the condensation of metal alkoxides to form metal oxide networks with entrapped alcohol and water:

$$M(OR)_n(soln) + nH_2O(l) \rightarrow M(OH)_n(soln) + nROH(soln) \rightarrow MO_m(s) + nH_2O(soln)$$

As the reaction proceeds, it passes through a number of stages, shown in Figure 5.9. The first is referred to as the *sol*. At this point small colloidal particles of $MO_m(s)$ are suspended in the solution, but they are not connected to each other. As this point in the reaction, the particles can be precipitated (flocculated) out of solution or allowed to form a three-dimensional net. It is this net with the entrapped water and alcohol that is referred to as a *gel*. Again, the synthetic chemist has two options: to allow the liquid component of the gel to evaporate, producing a *xerogel*, or to remove the liquid component via supercritical extraction, producing an *aerogel*.

In ceramic processing, it is the alkoxide solution and xerogel that are most useful for producing small particle sizes. The solution can be coated on a substrate, dried to form the xerogel, and then heated to form a dense ceramic film. Alternatively, the dried xerogel itself, in bulk form, can be used as a starting material that produces a dense ceramic when heated. In either case, the particle size is much less than what can be achieved through mechanical grinding. The

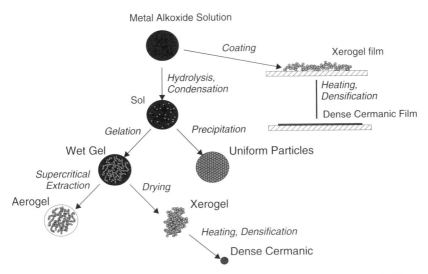

Figure 5.9   Sol–gel processing, leading to dense ceramics or dense ceramic films.

advantage of sol–gel processing goes beyond small particle size. The gel can be produced with a variety of metal alkoxides to produce a specific stoichiometry in the final product (Goosey et al., 1988). If a particular metal component is not easily prepared or handled as an alkoxide, the sol–gel approach can still be used with a mix of metal alkoxides and metal salts (Johnson, 1985).

Sol–gel processing requires the existence of a suitable metal-containing species that will undergo the proper chemistry to produce the desired ceramic material. Hence, considerable research is directed at preparing the precursor molecules for the hydrolysis reaction, in much the same way that research into volatile metal-containing species is required to advance the field of chemical vapor deposition discussed in Section 3.6. An alternative to sol–gel processing for the purpose of producing small particle sizes is coprecipitation. In this approach, the metal-containing species are dissolved in a solvent, precipitated together, and the resultant precipitate is fired. The recent example published by Babooram and Ye (2006) illustrates the technique. The authors are interested in preparing $SrBi_2Ta_2O_9$, a ferroelectric material, sometimes abbreviated SBT for strontium bismuth tantalate. This material has gained attention as an alternative to lead zirconium titanate (PZT) because of its potential applications in nonvolatile ferroelectric memory applications.

$SrBi_2Ta_2O_9$ can be prepared using the traditional ceramic method (Shimakawa et al., 1999) from a mixture of carbonate and oxides:

$$SrCO_3(s) + Bi_2O_3(s) + Ta_2O_5(s) \rightarrow SrBi_2Ta_2O_9(s) + CO_2(g)$$

In this example, the mixture was heated to 1000 to 1100°C for 30 hours in air. The heating process was stopped several times to allow for a regrinding of the reactant–product mixture. After 30 hours it was determined by x-ray powder diffraction that only $SrBi_2Ta_2O_9$ was present.

In the *coprecipitation synthesis* of $SrBi_2Ta_2O_9$, reported by Babooram and Ye, a solution of strontium acetate, $Sr(CH_3CO_2)_2 \cdot \frac{1}{2}H_2O(s)$, bismuth nitrate, $Bi(NO_3)_3 \cdot 5H_2O(s)$, and tantalum(V) chloride, $TaCl_5(s)$, was prepared using poly(ethylene glycol) and methanol as the solvent. The precipitate was formed by adding an oxalic acid/aqueous ammonia solution to the metals solution. The precipitate was washed, heated to $750°C$, and then sintered up to 8 hours at $1200°C$. The coprecipitation method requires that all the components have roughly the same solubilities. Otherwise, when the precipitation agent is added, oxalate in the case above, the less soluble components will precipitate out as separate phases, and no advantage from atomic mixing is realized.

When widely different solubilities prevent use of the coprecipitation technique, there is still another solution-based method that can be applied. This one is known as the *freeze drying* or *spray drying method* and is often used in connection with the sol–gel approach (Sugita, 1992). These drying methods have been exploited in the sol–gel method because one of the problems with removing the solvent from the gel is that the gel undergoes an enormous amount of shrinking which results in cracking. This shrinking and cracking are detrimental to the formation of continuous films produced via the route outlined in Figure 5.9.

5.3. MECHANICAL ALLOYING

The emphasis in the ceramic method is on high reaction temperatures to increase the value of the diffusion coefficient, D, or small particle size to decrease the diffusion distances. An alternative to high temperature is high pressure. In *mechanical alloying*, two components are ground together in order to get them to react with each other. The term *mechanical attrition* is also used. The method has been reviewed by several authors (Koch, 1989; Murty and Ranganathan, 1998; Suryanarayana et al., 2001). Mills for this type of synthetic work are commercially available. Figure 5.10 shows a benchtop unit by Fritsch GmbH. This technique has been used for many years to process elements and single compounds. More recently, it has gained attention for its usefulness in causing chemical reactions between two or more reactants. Consider a specific example reported by Alcalá et al. 2004. They reacted iron and hematite using a mill similar to that shown in Figure 5.10 to produce magnetite:

$$4Fe_2O_3(s) + Fe(s) \rightarrow 3Fe_3O_4(s)$$

The reactant powders were placed in a 45-mL steel container along with seven steel balls each with a diameter of 15 mm. Most impressively, the temperature of the sample was kept below $50°C$. Because of the oxygen sensitivity of iron powder, the milling was done in a nitrogen atmosphere for just over 3 hours for complete conversion of a 5-gram sample. The use of steel milling balls in this

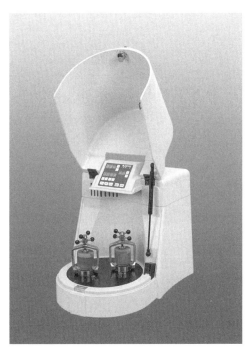

Figure 5.10 Benchtop pulverizing mill used for mechanical alloying. (Courtesy of Fritsch GmbH, Idar-Oberstein, Germany.)

case circumvents a potential problem with this technique: Given the vigorous conditions inside the milling cell, material from the milling apparatus itself can become an impurity in the final product. In this case, where the final product is magnetite and the balls are steel, contamination is less of a concern. Balls and containers are made from a wide variety of materials (Figure 5.11) to prevent unacceptable contamination. These materials include agate, zirconium oxide, silicon nitride, aluminum oxide, and tungsten carbide. The process described above not only produced magnetite but also resulted in a nanocrystalline magnetite. This is one of the reasons that mechanical alloying is enjoying renewed interest: It is a method for producing nanocrystalline materials (Lumley et al., 2006).

Another attractive feature of mechanical alloying is that it has the potential for producing homogeneous phases comprised of immiscible components. This feature has been exploited for the production of *oxide dispersion–strengthened* (ODS) alloys (Bhadeshia, 1997). An ODS alloy is prepared by mechanical alloying of an oxide, such as yttria, Y_2O_3, or titanium(III) oxide, Ti_2O_3, with an alloy. Normally, an oxide such as yttria and metals such as iron and aluminum are immiscible. With this technique, the oxide becomes dispersed along the grain boundaries of the alloy, resulting in an alloy with increased creep strength compared to the same alloys without the oxide dispersed in it. This ability to create homogeneous phases from immiscible components goes beyond the addition of

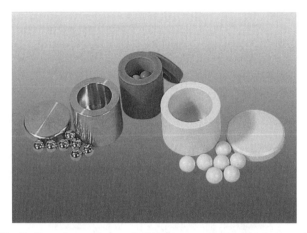

Figure 5.11   Balls and containers of different materials used for benchtop pulverizing mill. (Courtesy of Fritsch GmbH, Idar-Oberstein, Germany.)

small amounts of oxide to an alloy. The iron–copper phase diagram in the International Critical Tables shows limited solubility in either the liquid ($>1500°C$) or solid phase ($<1450°C$). Yet a $50:50$ iron/copper alloy can be produced using mechanical alloying (Benjamin, 1976).

Mechanical alloying is also used to produce amorphous metals (Nandi et al., 2001). Amorphous metals are normally produced by rapid cooling, which prevents crystallization, but in Nandi's report, a mixture of aluminum, copper, and niobium was milled over a period of 52 hours to produce an amorphous alloy, $Al_{65}Cu_{20}Nb_{15}$. In addition, nanocrystalline Cu_9Al_4 was also detected. Amorphous metals are in demand because of their increased strength and easier thermal processing characteristics. Mechanical alloying works because as the small metal particles are flattened between the milling balls, fresh metal surfaces become exposed. The adsorbed material on the metals prior to their being flattened could be adsorbed gases or a metal oxide film. In either case, the mechanical deformation of the particles stretches the adsorbed film, breaking it apart, and allowing fresh metal surfaces to come into contact. This produces what is termed a *cold weld*, and is illustrated in Figure 5.12.

Figure 5.12   Cold weld between two different materials as a result of freshly exposed surfaces and high pressure.

As we saw in Chapter 3 when we considered solid–vapor interactions, regardless of the form of the adatoms, vapor or solid, basic chemical bonding concepts are adhered to when two atoms combine to form a bond. In the mechanical milling process, the high pressure provides the energy necessary to overcome any thermal barrier to bond formation. Larger particles in the mechanical alloying process fracture rather than flatten. These smaller pieces can then undergo the process illustrated in Figure 5.12.

5.4. COMBUSTION SYNTHESIS

Combustion synthesis in solid-state chemistry is also known as metathesis reactions or self-propagating high-temperature synthesis (SHS). Our emphasis in the ceramic method was the need for high reaction temperatures in order to achieve usable diffusion coefficients for the atoms in the reacting solids. Mechanical alloying offers an alternative to high reaction temperatures (i.e., high-pressure pulses), which themselves could be a source for local rapid heating. In the combustion synthesis, high temperatures again take center stage, but this time the high temperature is achieved not by placing the sample in a furnace, but rather, by using the heat generated by the exothermic chemical reaction itself.

In a trivial application of this method, one could consider the synthesis of carbon dioxide from methane as a combustion synthesis. This trivial example provides some framework for additional insights into the solid-state case. The balanced reaction for methane reacting with oxygen is

$$CH_4(g) + 2O_2(g) \rightarrow CO_2(g) + 2H_2O(g)$$

This reaction could be viewed in terms of a *double displacement reaction*: One of the oxygen atoms and one of the hydrogen atoms have displaced each other. One of the most famous examples of this type of exchange reaction, a single exchange, occurs between iron oxide and aluminum:

$$Fe_2O_3(s) + 2Al(s) \rightarrow 2Fe(s) + Al_2O_3(s)$$

This reaction, known as the *thermite reaction*, requires an initial source of heat to begin the reaction, but after ignition, this highly exothermic reaction generates enough heat to proceed in a self-sustained manner. The iron is produced as a liquid because of the intense heat. In Figure 5.13*a*, the reaction is shown occurring. After enough liquid iron has been produced, the bottom is opened and liquid iron flows out (Figure 5.13*b*). In this case, the iron is being used to weld two railroad rails together.

The topic combustion synthesis was first studied by Alexander Merzhavov (b. 1931) in the 1970s has been reviewed by Gillan and Kaner 1996. A typical reaction occurs between a metal halide and an alkali or alkaline earth main group

(a)   (b)

Figure 5.13 Thermite reaction being used to produce liquid iron for welding railroad rails: *(a)* reaction proceeds, producing liquid iron; *(b)* liquid iron is released from the bottom of the reaction pot. (Courtesy of Christoph Schmitz. Copyright © 2006 Christoph Schmitz.)

compound, such as an oxide, nitride, or sulfide. The reaction of $ZrCl_4$ and lithium nitride is such an example:

$$3ZrCl_4(s) + 4Li_3N(s) \rightarrow 3ZrN(s) + 12LiCl(s)$$

Zirconium chloride is a white powder, and lithium nitride is a black powder. When this reaction proceeds, the lithium chloride forms as a liquid just as the iron forms as a liquid in the thermite reaction. The melting point of lithium chloride is $605°C$ (the melting point of iron is $1536°C$!) The temperature of the zirconium chloride/lithium nitride reaction reaches $1370°C$ less than 1 second after the reaction is initiated. This liquid plays the same role as the flux agents already discussed: It provides a medium for rapid mixing of reagents.

Alkali or alkaline earth metals, although common in combustion syntheses, are not required for a combustion synthesis. $TiCl_3$ and $AlCl_3$ can be combined in a combustion reaction to form Ti_3Al, $TiAl_3$ or $TiAl$, depending on the initial molar amounts of the reactants (Blair et al., 2003). Ferrite magnetic materials can be formed from magnesium oxide, zinc oxide, iron, α-Fe_2O_3, and sodium perchlorate (Pankhurst et al., 2000):

$$MgO + ZnO + 2Fe + Fe_2O_3 + NaClO_4 \rightarrow 2Mg_{0.5}Zn_{0.5}Fe_2O_4(s) + NaCl$$
$$+ \tfrac{1}{2}O_2(g)$$

Yttria, Y_2O_3, can be prepared from a reaction of yttrium nitrate and urea (Shea et al., 1996):

$$2Y(NO_3)_3(aq) + CH_4N_2O(aq) \rightarrow Y_2O_3(s) + 2H_2O(g) + CO_2(g)$$
$$+ 6NO_2(g) + N_2(g)$$

The use of metal nitrates provides a particularly convenient method for doping a host oxide with luminescent ions. Along with the yttrium nitrate, a luminescent rare earth ion such as Eu^{3+}, can be added as a nitrate salt. In addition to yttria, alumina can also be prepared from aluminum nitrate and hydrazine (Ozuna et al., 2004):

$$4Al(NO_3)_3(aq) + 15N_2H_4(aq) \rightarrow 2Al_2O_3(s) + 30H_2O(g) + 21N_2(g)$$

In both of these reactions, the metal nitrate and the organic component are prepared as an aqueous solution. The solution is boiled, dried, and the residue is heated, typically to $500°C$. At this point the reaction begins and converts to a self-sustained reaction. Both of these examples are oxidation–reduction reactions, which derive their large exothermic character from the formation of molecular nitrogen as a product.

Given the speed of these combustion reactions, establishing a detailed mechanism presents experimental challenges. Nonetheless, a few basic principles have emerged. In those reactions where a salt is a by-product, the heat released from the reaction melts the salt and provides a liquid medium in which the product and reactants can combine. This has been inferred from the results of syntheses that were deliberately loaded with additional product salt, which increased the heat capacity of the system. This additional heat capacity resulted in a lower peak reaction temperature. If enough salt was added, the peak reaction temperature was insufficient for reaching a self-sustaining reaction.

The temperature that a self-sustaining reaction could reach can be estimated, on the high side, by calculating the *adiabatic flame temperature* (McQuarrie and Simon, 1997). The adiabatic flame temperature is routinely calculated in physical chemistry. An example with the combustion of methane gas will illustrate the salient points. The balanced chemical reaction for the complete combustion of methane is, again

$$CH_4(g) + 2O_2(g) \rightarrow CO_2(g) + 2H_2O(g)$$

The enthalpy change for this reaction at $25°C$ is $\Delta H = -802.2$ kJ. Hence, when 1 mol of methane is converted to 1 mol of carbon dioxide gas and 2 mol of water vapor at $25°C$ and 1 atm, 802.2 kJ of energy is released as heat. If none of this heat escapes to the surroundings (hence the term *adiabatic*), all of it goes into heating the carbon dioxide and water vapor. To calculate the final temperature, the adiabatic flame temperature, one needs to know how the heat capacity of the material being heated, carbon dioxide and water vapor in this case, varies with temperature.

The temperature dependencies of many substances, including water and carbon dioxide, have been empirically fit to regular polynomials to facilitate integration. This interation is necessary because the heat released by the reaction at constant pressure, q_p, will determine the adiabatic temperature, T_f,

$$q_p = \Delta H = \int_{298}^{T_f} C_P(T)dT$$

where $C_p(T)$ is the total heat capacity of the products. Solving this for T_f gives an adiabatic flame temperature of $\approx 2000°C$ for the combustion of methane. This is an upper limit on the true temperature given that a significant fraction of the heat will be transferred to the surroundings instead of being retained to heat only the product gases.

This same calculation can be made for the self-propagating high-temperature reaction if the functional form of the variable-temperature heat capacity is known. From this calculation we see how adding additional product salt reduces the temperature that the reaction can reach by increasing the heat capacity of the system.

5.5. MICROWAVE SYNTHESIS

In Section 5.2 we addressed the issue of particle size as an important parameter in the ceramic method of solid-state synthesis. The other important parameter, temperature, is addressed in a variety of ways beyond conventional resistive heating, including laser heating (Bushma and Krwtsun, 1992; Cauchetier et al., 2003), radio-frequency induction heating (Szepvolgri et al., 2003), and the topic of this section, microwave heating (Mingos and Baghurst, 1997). The concept of using microwaves for the purpose of heating ceramics is a relatively new idea and has gained considerable attention in the recent literature. Several reviews have been published along with more targeted works. Shulman (2003) reviewed the use of microwaves for high-temperature processing on the industrial scale. She emphasized the energy efficiency of this approach to heating. She identified industrial microwave furnace use in the UK, the United States, Japan, Russia, and Germany. Trinh et al. (2003) reviewed microwave heating of ceramics with an emphasis on the *microwave effect*, a nonthermal interaction between the atoms in a sample of the applied microwaves. It leads to lower sintering temperatures, but the effect itself remains controversial. Trinh examined over 100 published papers and concluded that the evidence for the microwave effect was inconclusive. He attributed the effect observed to difficulties with determining the temperature of a sample while it is being exposed to microwave radiation. Sorrell et al. (1999) compared microwave and resistance heating of ceramics, again with a view that the microwave effect may be an experimental artifact. Sorrell et al.'s work emphasized the importance and difficulty of accurate temperature measurements. He examined the bulk density and grain size of zirconia (ZrO_2) and alumina (Al_2O_3) sintered over the temperature range 1000 to 1600°C. The importance of accurate temperature measurements was advanced in the work of Koh et. al (1995), Standard et al. (1997), and Ehsani et al. (1997).

Cozzi et al. (1995) reported on the use of a household microwave oven for the purpose of joining ceramic (Al_2O_3) rods. In this work, silicon carbide was mixed with Al_2O_3 to prepare cement, which was microwave susceptible, for use in joining the pure Al_2O_3 rods. They reported a welding temperature of $1400°C$. Rods became welded after 30 minutes at this temperature with a loading pressure ranging from 9 to 440 psi. Palaith and Silberglitt 1989 also reported on the use of a home microwave oven for ceramic welding. They welded Al_2O_3, mullite, and silicon nitride (Si_3N_4). Palaith and Silberglitt referenced the work of Ewsuk et al. (1989), which reported that the microwave susceptibility of Al_2O_3 could be altered by seeding with metals such as calcium, iron, magnesium, sodium, and potassium. Microwave welding has also been reported by scientists at Toyota in Japan (Fukushima et al., 1990). Oda et al. (1987) reported the use of microwave for the purpose of drying slip-cast ceramics. They employed a combination of microwave pulses, infrared heating, and ambient air to reduce drying times by at least 97%. Rokhvarger (2001) reported the use of microwave for processing ceramics, but more interestingly, he also reported the use of microwaves for the purpose of welding a ceramic lid to a ceramic container, providing complete integrity of a vessel containing nuclear waste. Daneke et al. (2001) discussed the challenge of processing materials with microwaves over a large temperature range. They point out that if a material is rendered too susceptible to microwaves at low temperature, it will potentially arc and reflect microwaves at high temperature. Veronesi et al. (2000) provided additional theoretical consideration with respect to microwave heating of multiphase material, where the behavior of one phase may dominate the bulk material.

If a sample is susceptible to it, Microwave heating is an attractive alternative to resistive heating because of the fundamentally different mechanism of heating that occurs with microwave heating. With resistive heating, the surface of a sample is heated because of the difference in temperature between the surface and the surroundings: basic thermodynamics at work. The entire sample becomes heated because the surface of the sample is hotter than the interior, and therefore heat flows to the cooler interior, as in Figure 5.14a. The heating rate of the sample is determined, in part, by the thermal conductivity of the sample. In microwave heating, by contrast, the entire sample is heated at once (Figure 5.14b).

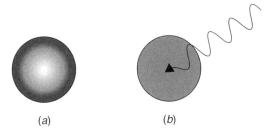

(a) (b)

Figure 5.14 (a) Conductive or surface heating; (b) microwave heating.

The sample absorbs microwave radiation and converts the absorbed electromagnetic radiation to heat that the sample retains. Hence, the sample can be brought to a very high temperature without needing to heat the surroundings. Not all materials absorb microwave radiation. As with any matter–electromagnetic radiation interaction, three possibilities exist. The material can (1) reflect the radiation, (2) transmit the radiation with minimal attenuation, or (3) absorb the radiation.

Several parameters are used to characterize the interaction of microwave radiation and matter: the complex permittivity (ε^*), the dielectric constant (ε'), and the loss tangent (tan δ). The dielectric constant, ε', can be thought of in a straightforward manner, as shown in Figure 5.15. Two parallel plates have a given capacitance, C_0, when there is no material between them: a vacuum. When the vacuum is replaced by a nonconducting medium, a *dielectric*, the new capacitance, C, is greater than C_0. The dielectric constant, ε, is the ratio of these two capacitances:

$$\varepsilon' = \frac{C}{C_0}$$

The dielectric constant is part of a more complicated parameter, the *complex permittivity*, ε^*:

$$\varepsilon^* = \varepsilon' + i\varepsilon'' \tag{5.4}$$

where ε'' is the dielectric loss factor and i is the square root of -1. The two terms in the complex permittivity represent the conductive and capacitive responses of the dielectric material in the presence of an oscillating electromagnetic field. The conductive response is in phase with the applied field and the capacitive response is 90° out of phase with the applied field. With both factors present in a dielectric

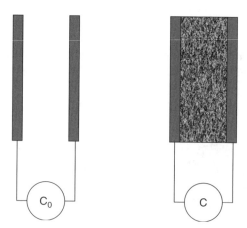

Figure 5.15 Representation of the two capacitances used to determine the dielectric constant.

material, the response of the material, its polarization, will be out of phase with the applied field by an angle δ. The tangent of this angle, the *loss tangent*,

$$\tan \delta = \frac{\varepsilon''}{\varepsilon'} \tag{5.5}$$

is important in microwave heating because it determines how much of the microwave radiation is absorbed by a dielectric material. The loss tangent will be a function of frequency and temperature.

The square of the electric field is a measure of how much energy the sample will aborb. At a given frequency, which will determine ε', the power absorbed is

$$P = 2\pi f \varepsilon_0 \varepsilon'' \left| \vec{E^2} \right| \tag{5.6}$$

where f is the frequency, ε_0 the permittivity of free space, and ε'' is the dielectric loss, as before. Most microwave absorption is measured at 2.45 GHz, the frequency that household microwave ovens use. Although this may not be the optimal frequency for solid-state chemical reactions, varying microwave frequencies is not a trivial task, so synthetic chemists are content to work with what is conveniently available. Takizawa et al. (2002) reported the use of 28 GHz for the microwave processing of ceramics. In this case, a higher frequency was used to subject the sample to a more uniform electric field, and hence a more uniform heating. At a frequency of 2.45 GHz, the wavelength of the radiation is 12.2 cm, significantly greater than that of most small-scale reaction vessels. At 28 GHz, the wavelength is 1.1 cm, smaller than that of most reaction vessels. Whether 2.45 or 28 GHz, the microwave frequency should be absorbed by the sample but not too strongly. If a sample has a microwave absorption spectrum such as that shown schematically in Figure 5.16, irradiation at the lower frequency indicated by the arrow will be better than irradiation at the peak absorbance frequency. If the peak frequency is chosen, the heating pattern will look more like the resistive pattern shown in Figure 5.14 because the outer layer of the sample will absorb

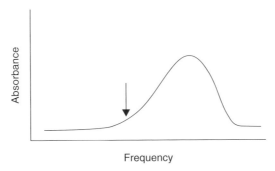

Figure 5.16 Generic microwave absorption spectrum, indicating with an arrow that the ideal irradiation frequency is not necessarily the absorption maximum.

most of the radiation and become hot while the center experiences essentially no microwave radiation.

The temperature dependence of the loss tangent generally follows an exponentially increasing trend. One of the reasons for this is that at higher temperatures, a sample will have a higher concentration of defects, and these defects are the major contribution to the mechanism responsible for microwave absorption in these samples (Meng et al., 1996). The simplest test to determine whether or not a material absorbs microwave radiation is to put it in a microwave oven, turn the oven on, and see what happens. Walkiewicz et al. (1988) tested a wide variety of materials and reported the temperature of the material after a given exposure time using a 1-kW 2.45-GHz commercial oven. After 6.25 minutes, a sample of CuO reached a temperature of $1012°C$. After 6 minutes, a sample of Na_2SO_4 reached $67°C$, and a sample of Al_2O_3 reached $78°C$ after 4.5 minutes. Clearly, the anhydrous sodium sulfate and the alumina do not appreciably absorb 2.45-GHz microwave radiation. However, this does not prevent microwaves from being used to heat these materials. Fang et al. (2004) sintered Al_2O_3 using microwave radiation by using an yttria-stabilized zirconia that does absorb microwave radiation. In their work, samples of Al_2O_3 were heated to $1500°C$ for 10 minutes using a 900-W household microwave oven.

To use microwaves as a source of heating in a solid-state chemical reaction, only one of the reactants needs to be microwave susceptible. This was demonstrated by Baghurst et al. (1988), who prepared $YBa_2Cu_3O_{7-x}$ from CuO, Y_2O_3, and $Ba(NO_3)_2$, taking advantage of the rapid heating of the CuO in the presence of microwave radiation. The advantage of microwave heating in this case is the same as in food preparation: a faster reaction time. The $YBa_2Cu_3O_{7-x}$ synthesis above can be done in about 1 hour compared to 24 hours by conventional resistive heating. Baghurst et al. (1988) also reported the rapid synthesis of $CuFe_2O_4$ from CuO and Fe_2O_3 and the conversion of PbO_2 to Pb_3O_4. The lead(IV) oxide decomposition is interesting because Pb_3O_4 does not absorb microwaves nearly to the degree that PbO_2 does. Hence, as the reaction proceeds, the temperature of the sample drops.

The mechanism by which dielectric materials absorb microwave radiation and convert it to heat is of particular interest to those using this method for heating ceramics. As mentioned above, the microwave effect has been a topic of some controversy in this field. Given that microwave frequencies are significantly less than vibrational frequencies, it is clear that the coupling of microwaves to solids is not through vibrational excitation. Freeman et al. (1998) published a numerical model designed to simulate microwave-induced ionic transport, the *microwave effect*. They identify the *ponderomotive force*, a nonlinear effect that is second order in the strength of the electric field, as responsible for enhanced ionic diffusion in dielectric solids irradiated with microwaves. Under the influence of this effect, as a result of movement in response to an oscillating electric field, ions do not return to their original position. The ponderomotive force is proportional to the gradient of the electric field intensity (Fitzpatrick, 2006). As would be expected at a crystallite–vapor interface, this dependence on the gradient of an inhomogeneous electric field means that the forces on the ion at the two points where the ion changes direction in response to the changing electric field are not equal. The net effect is that the ion will move toward the weaker portion of the field. This effect, well established in the plasma physics community, has been identified as the origin of the microwave effect in the sintering of ceramics.

The microwave effect is a nonthermal effect that results in increased reaction rates as well as properties in the final product which differ from those obtained by thermal processing. For example, ceramics processed with microwave heating display larger grain size, higher densities, and lower sintering temperatures (Shulman, 2002). Establishing that these effects were genuine and a result of microwave interaction (e.g., ponderomotive force) and not the result of improper thermal measurements has been the root of the controversy concerning this effect.

REFERENCES

Alcalá, M. D.; Criado, J. M.; Real, C.; Grygar, T.; Nejezchleba, M.; Subrt, J.; Petrovsky, E. *J. Mater. Res.* **2004**, *39*, 2365–2370.

Angelici, R. J. *Synthesis and Technique in Inorganic Chemistry*, 2nd ed., University Science Books, Mill Valley, CA, **1986**, pp. 33–38.

Babooram, K.; Ye, Z. -G. *Chem. Mater.* **2006**, *18*(2), 532.

Baghurst, D. R.; Chippindale, A. M.; Mingos, D. M. P. *Nature.* **1988**, *332*, 311.

Benjamin, J. S. *Sci. Am*. **1976**, *234*, 40–48.

Bhadeshia, H. K. D. H. *Mater. Sci. Eng. A Struct. Mater. Prop. Microstruct. Process*. **1997**, *223* (1–2), 64–77.

Blair, R. G.; Gillan, E. G.; Nguyen, N. K. B.; Daurio, D.; Kaner, R. B. *Chem. Mater*. **2003**, *15*, 3286–3293.

Brown, W. B. *Phys. Rev*. **1931**, *38*, 709–711.

Bushma, A. I.; Krivtsun, I. V. *Laser Treat. Mater. (Pap. Eur. Conf.).* **1992**, 719–23.

Cauchetier, M.; Musset, E.; Luce, M.; Herlin, N.; Armand, X.; Mayne, M. *Nanostruct. Silicon-Based Powders Compos.* **2003**, 6–24.

Cozzi, A. D.; Clark, D. E.; Ferber, M. K.; Tennery, V. *J. Ceram. Trans*. **1995**, *59*, 389–396.

Daneke, N.; Krause, H.; Walter, G.; Zesch, U.; Wielage, B. *Elektrowaerme Int*. **2001**, *59*(1), 26–30.

Ehsani, N.; Ruys, A. J.; Standard, O. C.; Sorrell, C. C. *Mater. Eng*. **1997**, *8*(2), 89–95.

Ewusk, K. G.; Lanagan, M. T.; Jones, F. Symposium Paper 5-SIX-89, presented at the 91st Annual Meeting of the American Ceramic Society, Indianapolis, IN, Apr. 25, **1989**.

Fang, Y.; Ggrawal, D.; Roy, R. *Matter. Lett.* **2004**, *58*, 498.

Fisler, D. K.; Cygan, R. T. *Am. Mineral.* **1999**, *84*, 1392–1399.

Fitzpatrick, R. *Introduction to Plasma Physics: A Graduate Level Course*, Creative Commons Attribution-NoDerivs 2.0, **2006**, p. 51.

Freeman, S. A.; Brooks, J. H.; Cooper, R. F. *J. Appl. Phys.* **1998**, *83*(11), 5761–5772.

Fukushima, H.; Yamanaka, T.; Matsui, M. J. *Mater. Res*. **1990**, *5*(2), 397–405.

Galasso, F. S. *Inorganic Syntheses*. **1973**, *14*, 142–144.

Geselbracht, M. J.; Noailles, L. D.; Ngo, L. T.; Pikul, J. H.; Walton, R. I.; Cowell, E. S.; Millange, F.; ÓHare, D. *Chem. Mater*. **2004**, *6*(6), 1153–1159.

Gillan, E. G.; Kaner, R. B. *Chem. Mater*. **1996**, *8*, 333–343.

Goosey, M. T.; Patel, A.; Watson, I. M.; Whatmore, R. M. *Chemtronics*. **1988**, *3*, 103–106.

Greenwood, N. N.; Earnshaw, A. *Chemistry of the Elements*, Pergamon Press; Oxford, **1984**, pp. 772–775.

Johnson, D. W. *Am. Ceram. Soc. Bull*. **1985**, *64*(12), 1597–1602.

Koch, C. C. *Annu. Rev. Mater. Sci*. **1989**, *19*, 121–143.

Koh, M. T. K.; Singh, R. K.; Clark, D. E. *Ceram. Trans*. **1995**, *59*, 313–322.

Lumley, R.; Morton, A.; Polmear, I. R. H. J. Hannink and A. J. Hill, Eds., In *Nano-structure Control of Materials*, Woodhead Publishing Ltd.; Cambridge, UK, **2006**, pp. 219–250.

Mancić, L.; Marinkovic, Z.; Vulić, P.; Milosević, O. *Sci. Sinter*. **2004**, *36*, 189–196.

McQuarrie, D. A.; Simon, J. D. *Physical Chemistry: A Molecular Approach*, University Science Books; Sausalito, CA, **1997**, pp. 1297–1303.

Meng, B.; Klein, B. D. B.; Booske, J. H.; Cooper, R. F. *Phys. Rev. B*. **1996**, *53*(19), 12777–12785.

Mingos, D. M. P.; Baghurst, D. R. In *Microwave-Enhanced Chemistry*, H. M. Kingston and S. J. Haswell, Eds.; American Chemical Society, Washington, DC, **1997**, pp. 3–53.

Murty, B. S.; Ranganathan, S. *Int. Mater. Rev*. **1998**, *43*(3), 101–141.

Nandi, P.; Chattopadhyay, P. P.; Pabi, S. K.; Manna, I. *Mater. Phys. Mech*. **2001**, *4*, 116–120.

Oda, S. J.; Woods, B. G.; Foster, J. *J. Can. Ceram. Soc*. **1987**, *56*, 70–73.

Ozuna, O.; Hirata, G. A.; McKittrick, J. *J. Phys. Condens. Matter*. **2004**. *16*, 2585–2591.

Palaith, D.; Silberglitt, R. *Ceram. Bull*. **1989**, *68*(9), 1601–1606.

Pankhurst, Q. A.; Affleck, L.; Aguas, M. D.; Kuznetsov, M. V.; Parkin, I. P.; Barquin, L. F.; Boamfa, M. I.; Perenboom, J. A. A. *Adv. in Sci. Technol*. **2000**, *29*, *B Mass Charge Transport Inorg. Mater*. 925–935.

Revelli, J. F. *Inorg. Synth*. **1979**, *19*, 35–49.

Rokhvarger, A. *Ceram. Trans*. **2001**, *119*, 459–466.

Schäfer, H. *Chemical Transport Reactions, transl*. H. Frankfort, Academic Press; New York, **1964**.

Shea, L. E.; McKittrick, J.; Lopez, O. A. *J. Am. Ceram. Soc*. **1996**, *79*(12), 3257–3265.

Shimakawa, Y.; Kubo, Y.; Nakagawa, Y.; Kamiyama, T.; Asano, H.; Izumi, F. *Appl. Phys. Lett*. **1999**, *74*(13), 1904–1906.

Shriver, D. F.; Drezdzon, M. A. *The Manipulation of Air-Sensitive Compounds*, 2nd ed., Wiley-Interscience; New York, **1986**.

Shulman, H. S. *Microwave Heating Ceramics*, Alfred University Undergraduate Seminar, Dec. 12, **2002**.

Shulman, H. S. *Ind. Heat*. Mar. 10, **2003**.

Smigelskas, A. D.; Kirkendall, E. O. *Trans. AIME* **1947**, *171*, 130–142.

Sorrell, C. C.; Standard, O. C.; Ehsani, N.; Harding, C. M. G.; Ruys, A. J. *Adv. Sci. Technol. B Ceram.*, **1999**, *14*, 665–680.

Standard, O. C.; Ruys, A. J.; Sorrell, C. C. *Mater. Eng.* **1997**, *8*(2), 97–106.

Sugita, K. *Adv. Mater.* **1992**, *4*(9), 582–586.

Suryanarayana, C.; Ivanov, E.; Boldyrev, V. V.; Ansell, G. S. *Mater. Sci. Eng. A Struct. Mater. Prop. Microstruct. Process.* **2001**, 304–306, 151–158.

Szepvolgri, J.; Mohai, I.; Karoly, Z. *In Proc. 48th Internationales Wissenschaftliches Kolloquium*, Technische Universitaet Ilmenau, Ilmenau, Germany, **2003**, pp. 287–288.

Takizawa, H.; Kimura, T.; Iwasaki, M.; Uheda, K.; Endo, T. *Ceram. Trans.* **2002**, *133*, 211–216.

Trinh, D. H.; Standard, O. C.; Sorrell, C. C. *J. Aust. Ceram. Soc.* **2003**, *39*(2), 119–129.

Veronesi, P.; Siligardi, C.; Leonelli, C. *Ceram. Inf.* **2000**, *397*, 397–399.

Walkiewicz, J. W.; Kazonich, G.; McGill, S. L. *Miner. Metall. Process.*, Feb. **1988**, pp. 39–42.

Wright, J. D.; Sommerdijk, N. A. J. M., *Sol–Gel Materials: Chemistry and Applications*, Gordon and Breach, Australia, **2001**.

Xing, X.; Zhang, C.; Qiao, L; Liu, G.; Meng, J. *J. Am. Ceram. Soc.* **2006**, *89*(3), 1150–1152.

Yin, Y.; Rioux, R. M.; Erdonmeg, C. K.; Hughls, S.; Somorjai, G. A.; Glivisatos, A. P. *Science* **2004**, *304*, 311.

6 Nanomaterials Synthesis

Nanoscaled materials were first used centuries ago. An early example was the incorporation of gold nanoparticles in stained glass in A.D. 10. In the nanosize range, gold can exhibit a variety of colors (Hunt, 2004). However, it was not until the 1850s that nanosized particles were first studied systematically, when Michael Faraday prepared a suspension of gold nanoparticles in water by chemical reduction of aqueous gold chloride. Faraday proposed that the various possible colors of the mixture were due to light scattering, which was dependent on the extent of particle coagulation. In 1861, Thomas Graham deduced that colloids were about 1 to 100 nm in size, from their slow diffusion and lack of sedimentation. Today, we know that the size of the gold particles determines the surface plasmon resonance frequency and hence the color of colloidal gold. Synthetic routes have been refined to the point that one can now purchase gold colloids in a variety of sizes (Figure 6.1).

Particles and films of ultrasmall dimensions had actually been obtained systematically throughout the majority of the twentieth century. Monolayer and multilayer organic thin films were first prepared in the 1930s by Irving Langmuir and his colleague Kathleen Blodgett (1898–1979) by repeatedly dipping substrates into water covered with a monolayer organic film. Around the same time, August Herman Pfund (1879–1949) and Carel Hermann Burger (1893–1965) prepared what were described as "extremely fine" and "loosely packed" crystalline deposits of the 15 heavy group elements on chamber walls and mica substrates by direct thermal evaporation of bismuth and antimony in order to improve the radiometric properties of thermopiles and thermocouples. By 1960, several physical methods had been used to prepare submicrometer particles. However the nanotechnology research fervor which has invaded every aspect of modern materials research is attributed to a speech by Richard Feynman (1918–1988) given in 1959 at the annual meeting of the American Physical Society at the California Institute of Technology. The speech, entitled "There's Plenty of Room at the Bottom," was an open invitation for physicists not to look for new physics but rather, to investigate how known properties change at small dimensions.

Inorganic Materials Synthesis and Fabrication, By John N. Lalena, David. A. Cleary, Everett E. Carpenter, and Nancy F. Dean
Copyright © 2008 John Wiley & Sons, Inc.

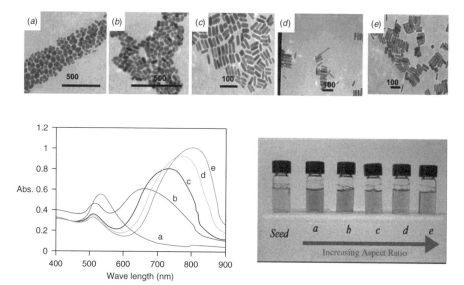

Figure 6.1 Transmission electron micrograph of characteristic orientations self-assembled from different nanoparticles: (a) 13.4-nm g-Fe_2O_3 and 5.0-nm Au; (b) 7.6-nm PbSe and 5.0-nm Au; (c) 6.2-nm PbSe and 3.0-nm Pd; (d) 6.7-nm PbS and 3.0-nm Pd; (e) 6.2-nm PbSe and 3.0-nm Pd. Scale bars: a–c, e: 20 nm. [From *J. Phys. Chem.* **2005**, *109*(29), 13857.]

Our current state of affairs would not be possible without the advancements made in the latter half of the twentieth century of analytical techniques, which have been instrumental in characterizing nanosized particles. The scanning tunneling microscope, invented in 1982, allowed for atomic-scale topographical mapping of metal surfaces and, later, the manipulation of individual atoms. High-resolution mass spectroscopy enabled the 1985 discovery of the carbon allotrope C_{60}, buckminsterfullerene, from the products of laser vaporization of graphite. In 1991, high-resolution transmission electron microscopy enabled the discovery of carbon nanotubes deposited on electrodes during the arc discharge of fullerenes (Lalena and Cleary, 2005).

Nanomaterials can be in the form of fibers (one-dimensional), thin films (two-dimensional), or particles (three-dimensional). A *nanomaterial* is any material that has at least one of its dimensions in the size range 1 to 100 nm (Figure 6.1). Many physical and chemical properties are determined by the very large surface area/volume ratio associated with such ultrasmall particles. There are two major categories into which all nanomaterial preparative techniques can be grouped: the physical, or *top-down*, *approach* and the chemical, or *bottom-up*, *approach*. In this chapter, our primary focus is on chemical synthesis. Nevertheless, we discuss the physical methods briefly, as they have received a great deal more interest in the industrial sector because of their promise to produce large volumes of nanostructured solids.

Richard Phillips Feynman (1918–1988) obtained his bachelor's degree from the Massachusetts Institute of Technology in 1939 and his doctorate in physics from Princeton in 1942. He began his career at the height of World War II, working on the Manhattan project in the Los Alamos National Laboratory. Feynman was a practical joker at the lab, often testing security by picking locks and safes, and he helped relieve tension by playing the drums. After the war, he taught at Cornell before moving to the California Institute of Technology. Feynman was jointly awarded the Nobel Prize in Physics in 1965, along with Julian Schwinger and Sin-Itiro Tomonaga, for expanding the theory of quantum electrodynamics. He was an outstanding educator, taking great care to explain topics to his students. His principle was that if a topic could not be explained in a freshman lecture, it was not yet sufficiently understood. Feynman is also credited for providing the inspiration for nanotechnology with his famous "There's Plenty of Room at the Bottom" speech at the annual meeting of the American Physical Society in 1959.

Source: *Wikipedia*, the free encyclopedia.

6.1. TOP-DOWN METHODS FOR FABRICATION OF NANOCRYSTALLINE MATERIALS

The top-down approach starts with a bulk material and attempts to break it down into nanoscaled materials through physical methods. Hence, *most* of these techniques are really forms of fabrication rather than synthesis. For nanostructured bulk phases, including powders, the common methods are milling, devitrification of metallic glass, and severe plastic deformation. For nanocrystalline thin films (films with nanosized crystallites), methods include thermal vaporization (under high vacuum), laser ablation, and sputtering (thermal plasma), all of which were discussed in Chapter 3. To obtain nanocrystalline films with controlled grain morphologies and sizes from these techniques, the experimental conditions (e.g., the temperature gradient between source and substrate, gas pressure, gas flow rate, sputtering time, source–substrate distance) must be selected carefully for each system under investigation.

6.1.1. Nanostructured Thin Films

For conventional film growth by the sputtering technique, gas pressures of only a few millitorr ($p < 100$ mtorr) are typically used. However, higher pressures are utilized for nanocrystals. The nucleation and growth of sputtered nanoparticles are often very sensitive to gas pressure in a complex way. One of the earliest studies designed to examine the effect of gas pressure on nanoparticle size was on molybdenum (Chow et al. 1990) . Three pressure regimes were found: (1) at pressures below 150 mtorr, conventional film growth was observed; (2) at $200 < p < 400$ mtorr, nanoparticle size decreased with increasing pressure (presumably due to decreased sputtering rate); and (3) at $p > 500$ mtorr, larger particles were able to agglomerate in the vapor phase, which led to a particle size/pressure correlation similar to that observed in thermal evaporation with $p > 1$ torr. In addition, it is usually necessary to use customized *cluster sources* for generating nanoparticles. In these devices, before reaching the deposition chamber where the substrate is located, the ejected target atoms are made to pass through an aggregation or annealing region, where the experimental parameters are carefully controlled, to allow for cluster growth. A very popular cluster source in use today is the magnetron-sputtering source, which utilizes magnets to confine electrons in the vicinity of the target, allowing for higher ion current densities and subsequent deposition rates at lower pressures (Haberland, 1994; Haberland et al., 1996). Once formed, the clusters are jet-propelled through a nozzle to form a cluster beam.

Ablation is a powerful technique that uses high-energy lasers to vaporize or ablate materials from the surface. The wavelength of the laser is tuned for the specific material to achieve maximum absorption of the energy, most often ultraviolet. The target is vaporized, creating a plume of neutral metal atoms. The plume is then cooled with a carrier gas to form clusters. It is possible to couple laser evaporation with laser pyrolysis to form alloys.

6.1.2. Nanocrystalline Bulk Phases

Mechanical attrition has become the most common top-down preparatory route for powders containing nanosized particles. Precursor bulk materials are placed in a mill where their grain size is ground down to between 3 and 25 nm, depending on their crystal structure. Milling beyond this minimum size results in amorphous materials. Mechanical attrition also does not result in uniform crystal sizes or shapes, which is its major disadvantage. There are two ways to synthesize nanoparticles using mechanical attrition: milling of a single precursor into smaller grains, and *mechanochemical synthesis*.

Parameters such as milling time, size and shape of the vessel, number and size of media, and the powder/media ratio are typically taken from empirical research. The larger the milling balls, the more frequent the collisions, up to a maximum, before there is a decrease as the balls become too large for the size of the vessel. But the more balls there are, the lower the impact velocity of the balls. For high-energy milling, impact efficiency is an important consideration, as only those impacts that cause plastic deformation, fracturing or cold welding matter.

For brittle starting materials, the grain size is reduced through a process of fracturing and cold welding of smaller grains. With softer metals and intermetallic compounds, the dominating force reducing the grain size is not fracturing, but instead, creation and aligning of dislocations, which result in shear bands. This process is similar to high-temperature recrystallization but at lower temperatures, restricting grain growth. Low temperature is essential in the formation of nanoparticles, often requiring milling in liquids or in refrigerated mill stations. The minimum grain size obtainable is determined by the balance of the dislocation accumulation and recovery through formation of new grains. In the case of relatively low melting solids, like many with the face-centered-cubic crystalline structure, the minimum grain size scales inversely with the melting point.

Mechanochemical synthesis is an expansion of simple mechanical alloying/attrition. Mechanochemical synthesis is essentially the solid-state reaction of materials facilitated by milling. During the early 1970s, it was observed that the milling of nickel or aluminum alloys in air resulted in the formation of metal oxide nanoparticles. During the milling process, heat was generated which resulted in oxidation of the surface of the initial powders. These surface oxides were incorporated into the initial materials through cold welding or the joining of two interfaces through collision. With ball milling, cold welding was so complete that the oxide layer created on the interface was integrated into the particle completely. In the late 1980s, researchers at the University of West Australia discovered that if calcium is milled with CuO, Cu nanoparticles are formed separated by CaO. The reduction of CuO by calcium typically takes place at 1400 K. This provided a simple method for the synthesis of alloys without first reducing the metal oxides to metals. During collisions there is a $\sim$100-K temperature rise, which results in an order-of-magnitude increase in the reaction enthalpy. Additionally, there is the potential for extreme surface heating, which can lead to increased rate of reaction.

Other top-down methods are used for the production of ultrafine-grained metals and alloys. These include the devitrification of metallic glass (Section 4.2.6.5) and severe plastic deformation, in which a coarse-grained polycrystalline metal or alloy is subjected to large shear strains under pressure, forcing the grains to subdivide into nanosized crystallites. Severe plastic deformation (SPD) typically produces polycrystalline metals and alloys that have grain sizes between 100 and 1000 nm, with subgrain domains. Metals processed by SPD have high dislocation densities and nonequilibrium grain boundaries. The ultrasmall grain size and structural features impart high strength, good ductility, superior superplasticity, low friction coefficients, high wear resistance, enhanced high-cycle fatigue life, and good corrosion resistance, all of which make them attractive in structural applications, medical implants, and the automotive and forming industries (Zhu and Langdon, 2004).

The most promising SPD technique appears to be *equal-channel angular pressing* (ECAP), also known as *equal-channel angular extrusion* (ECAE). We analyze this method in detail in Section 7.2.8. A brief description will suffice for now. In this method, an ingot is pressed repeatedly through a special die with two channels of equal cross section, usually intersecting at an angle of 90°. Elevated temperatures and increased channel intersection angles can be used for hard-to-deform materials. The cumulative strain in the ingot, built up with multiple passes through the die (usually, four to six), leads to a homogeneous ultrafine-grained material. ECAP was invented in 1977 in the former Soviet Union by Vladimir Segal, but its applicability to ultrafine-grained metals was first demonstrated in the early 1990s by Ruslan Z. Valiev (Lowe and Valiev, 2004).

6.2. BOTTOM-UP METHODS FOR SYNTHESIS OF NANOSTRUCTURED SOLIDS

In the chemistry community, most nanomaterials research is centered on the bottom-up approach. In this technique, well-controlled nanoparticles are chemically synthesized in a solution phase and subsequently, compacted or assembled into macroscopic materials. Nanoparticles have very large surface area/volume ratios, which results in a higher chemical reactivity relative to larger particles. Both the ionization energy and the electron affinity of small clusters of atoms (e.g., <100) exhibit particle-size dependency, in contrast to the particle-size independence of the work function in bulk solids. Moreover, since most of the atoms are on the surface, nanoparticles can serve as nearly stoichiometric reagents in chemical reactions. This large surface/interior atom ratio also tends to maximize the surface van der Waals forces, which together with the tendency of the system to minimize the total surface, or interfacial, energy, typically results in agglomerates. Agglomeration can occur at several stages: during the synthesis, drying, handling, or during processing. If the agglomerates are bound to each other physically, as a result of sintering or synthesis, the agglomerate is referred to as *aggregate* and cannot be separated without further chemical treatment to break the bonding between particles.

The tendency for agglomeration can often be reduced through physical separation, such as sonication or mechanical shearing. However, a more effective method for lowering surface chemical reactivity in general utilizes a protective shell of ligand molecules. Chemical methods utilizing surfactants in this manner can produce very homogeneous and well-dispersed nanoparticles. Later the surfactant can be removed for further processing. *Surfactants* (amphiphiles) are a class of chemicals that raise or lower the surface tension of a system. They get their name from the term *surface-active agent* and typically have polar

(thermolysis), light (photolysis), or sound (sonolysis). Advantages of using organometallic compounds are that precursors can be made that have the constituents
in molecular proximity to each other. Then as the precursor is decomposed, the
constituents remain close and the resulting nanoparticles can have a controlled
crystallinity and morphology. Controlling the temperature or exposure to light or
sound provides kinetic control over the growth of nanoparticles. To gain additional thermodynamic controls over the nucleation, it is possible to carry out the
decomposition in the presence of polymers, organic capping agents, or structural hosts. The capping agent provides a means to hinder the growth of the
nanoparticle sterically, preventing coalesce and agglomeration.

Metal carbonyls are the precursor of choice, due in part to ease in decomposition. The carbonyls are decomposed in the presence of stabilizing polymers, and
spherical nanoparticles are formed. Through a combination of surfactants it is
possible to change the morphology. For the decomposition reactions to maintain
uniform size and morphology, it is important to achieve rapid nucleation and
controlled growth. As a result, the metal carbonyl is typically injected into a hot
solution. This provides rapid nucleation. The injection leads to a slight reduction
in the temperature, which coupled with a reduction in temperature provides slow
growth. This rapid-injection style of reaction allows *size distribution focusing*
by separating nucleation from growth. When a surfactant mixture such as oleic
acid and TOPO is used, and quenched shortly after injection, nanorods (4 nm ×
25 nm) of HCP Co are formed. The change from spheres to rods is a result of
the differential absorbance of the surfactant on the facets of the nanoparticles.

It is possible to create more complex compounds and morphologies, such as
alloys or core-shell structures. The nucleation step still occurs with injection of
the metal carbonyl. However, when other organometallic precursors are in the
solution, you can get the alloy to grow on the nuclei through transmetallation. To
prepare core–shell material, the nuclei are allowed to react, then the temperature
is reduced and a second precursor is introduced. The initial injection provided the
nucleation which the additional materials when decomposed grow on the initial
nuclei.

When decomposition is facilitated through the use of the ultrasonic or acoustic
waves it is called *sonolysis* or *sonochemistry*. Ultrasonic irradiation is carried out
with an ultrasound probe, such as a titanium horn operating at 20 kHz. During
sonication, the formation, growth, and collapse of bubbles occur in an adiabatic
fashion and with extreme energies. The acoustic cavitation involved the localized
hot spot of temperatures in excess of 5000 K and a pressure of ~1800 atm. Since
the reaction is adiabatic with little bulk heating, there is a subsequent cooling
rate of 10^9 K/s. The cavitation is dependent on the solvent not coupling with
the precursors. Generally, volatile precursors in low-vapor-pressure solvents are
used to optimize the yield.

Nanostructured particles are easily produced by sonochemically treating
volatile organometallic precursors. The powders formed are usually amorphous,
agglomerated, and porous. To get the crystalline phases, these powders must be
further annealed; however, this annealing temperature is lower than that needed

to do solid-state conversion. These powders had a surface area over 100 times greater than that of powders available commercially.

When the sonolysis is done in high-boiling organics, highly porous amorphous powders are formed. For example, an amorphous iron powder was produced by the sonocation of iron carbonyl in decalin. This powder was comprised of small crystallites (5 Å) and had a surface area of 120 m^2/g. If stabilizers or polymers are added postsonication or during sonication, metal colloids result. These stabilizers could be alkyl thiols, PVP, oleic acid, and SDS. If the sonication is done in the presence of oxygen, oxides are formed. The size of the self-assembled monolayer-coated nanoparticles is determined by the surfactant concentration in the coating solution. If the sonolysis is done in the presence of a support or porous host, colloidal metal particles are formed. These powders had a surface area over 100 times greater than powders available commercially and was amorphous. Such materials are generally considered for catalytic reactions but not for magnetic applications.

6.2.5. Sol–Gel Methods

Sol–gel synthesis was discussed in Chapter 4. In 1967, Maggio P. Pechini (b. 1923) developed a sol–gel method for lead and alkaline earth titanates and niobates, materials that do not have favorable hydrolysis equilibria (Pechini, 1967; Lessing, 1989). This method is also known as the *Pechini*, or *liquid mix, process*. Pechini worked for the Sprague Electric Company in Massachusetts, which was interested in manufacturing ceramic capacitors. In the Pechini method, cations are chelated and then, with the aid of polyalcohols, the chelate is cross-linked to create a gel through esterification. This has the distinct advantage of allowing the use of metals that do not have stable hydroxo species. The chelating agent needs to have multiple carboxylate groups. Initially, Pechini used citric acid. This has often been replaced by EDTA (ethylenediaminetetraacetic acid), which has the advantage of chelating most metals and, with four carboxylate groups, is easily cross-linked to form the gel. It is also possible to use poly(vinyl alcohols), which provide a three-dimensional network during gel formation. The gelled composite is sintered, pyrolyzing the organic and leaving nanoparticles that are reduced by the pyrolyzed gel. Like many techniques, the limitations of the Pechini method, lie in the lack of size and morphological control. With traditional sol–gel methods, the particles are part of the gel structure, in the Pechini method the metal cations are trapped in the polymer gel. This reduces the ability to grow controlled shapes. The size is controlled to an extent by the sintering process and the initial concentration of metals in the gel.

Sol–gel processing can be used to prepare a variety of materials, including glass, powders, films, fibers, and monoliths. Traditionally, the sol–gel process generally involves hydrolysis and condensation of metal alkoxides. Metal alkoxides are good precursors because they readily undergo hydrolysis; that is, the hydrolysis step replaces an alkoxide with a hydroxide group from water and a free alcohol is formed. Once hydrolysis has occurred, the sol can react further and condensation (polymerization) occurs. It is these condensation reactions that lead to gel formation.

Factors that need to be considered in a sol–gel process are solvent, temperature, precursors, catalysts, pH, additives, and mechanical agitation. These factors can influence the kinetics, growth reactions, and hydrolysis and condensation reactions. The solvent influences the kinetics and conformation of the precursors, and the pH affects the hydrolysis and condensation reactions. Acidic conditions favor hydrolysis, which means that fully or nearly fully hydrolyzed species are formed before condensation begins. Under acidic conditions there is a low cross-link density, which yields a denser final product when the gel collapses. Basic conditions favor the condensation reaction. Thus, condensation begins before hydrolysis is complete. The pH also affects the isoelectric point and the stability of the sol. These, in turn, affect the aggregation and particle size. By varying the factors that influence the reaction rates of hydrolysis and condensation, the structure and properties of the gel can be tailored. Because these reactions are carried out at room temperature, further heat treatments need to be done to get to the final crystalline state. Because the as-synthesized particles are amorphous or metastable, annealing/sintering can be done at lower temperatures than those required in conventional solid-state reactions.

Sol–gel routes can be used to prepare pure, stoichiometric, dense, equiaxed, and monodispersed particles of TiO_2 and SiO_2. But this control has not been extended to metal ferrites. Generally, the particles produced are agglomerated. Ultrafine powders of $CoFe_2O_4$ ($\sim$30 nm) and $NiFe_2O_4$ (5 to 30 nm) after being fired at 450 and 400°C, respectively. Most of the ferrite sol–gel synthesis focus has been cobalt ferrite doping studies with Mn, Cr, Bi, Y, La, Gd, Nd, and Zn. Sol–gel routes have been attractive for the preparation of hexagonal ferrites. For example, the M-type hexagonal ferrite $Ba_{1-x}Sr_xFe_{12}O_{19}$ formed 80 to 85-nm hexagonal platelets after a 950°C calcination for 6 hours. Nanospheres of the W-type ferrite $BaZn_{2-x}Co_xFe_{16}O_{27}$ resulted after calcination in air for 4 hours at 650°C. The particle size ranged from 10 to 500 nm (650 to 1250°C) and increased with increasing calcination temperatures. U-type hexagonal ferrite was also prepared, with 10 to 25-nm spherical particles formed at 750°C. The grain size could be changed by increasing the calcination temperature. These calcined powders had an amorphous layer on them. Yttrium iron garnets with particle sizes from 45 to 450 nm have also been prepared. Mathur and Shen (2002) have prepared the manganite perovskite, $La_{67}Ca_{33}MnO_3$ by dissolving the metal precursors in an acidic ethanolic solution. Drying the solution at 120°C and calcining at 300 to 400°C leads to preceramic foam which forms nanocrystalline $La_{67}Ca_{33}MnO_3$ (40 nm) after a 650°C heat treatment.

It should be noted that the sol–gel process is particularly attractive for the synthesis of multicomponent particles with binary or ternary compositions using double alkoxides (two metals in one molecule) or mixed alkoxides (with mixed metaloxane bonds between two metals). Atomic homogeneity is not easily achieved by coprecipitating colloidal hydroxides from a mixture of salt solutions, since it is difficult to construct double metaloxane bonds from metal salt. Hybrid materials such as metal–oxide and organics–oxide can be prepared using the sol–gel approach. For example, controlled nano-heterogeneity can be achieved in metal–ceramic nanocomposites. Reduction of metal oxide particles in hydrogen provided metal–ceramic nanocomposite powders such as Fe in silica, Fe_2O_3, and $NiFe_2O_4$. The metal particles, a few nanometers in size with a very narrow size distribution even for high metal loading, were distributed statistically in the oxide matrix without agglomeration, as a result of anchoring the metal complexes to the oxide matrix. The narrow particle size distribution could not be achieved if the sol–gel processing was performed without complexation of metal ions.

6.2.6. Polyol Method

The term *polyol* is short for *polyalcohol*: for example, ethylene or propylene glycol. In the polyol method, the polyol acts as solvent, reducing agent, and surfactant. The polyol synthesis was first described in 1983 by Michel Figlarz, Fernand Fiévet, and Jean-Pierre Lagier at the University of Paris (Figlarz et al., 1983). The polyol method is an ideal method for the preparation of larger nanoparticles with well-defined shapes and controlled particle sizes. In this method, precursor compounds such as hydroxides, oxides, nitrates, sulfates, and acetates are either dissolved or suspended in a polyol. For example, with CuO, the general metal oxide/polyol ratio is $0.07 : 1$. The reaction mixture is then heated to reflux. As the temperature is increased, the reduction potential of the glycol increases, which leads to nucleation. During the reaction, the metal precursors become solubilized in the diol, forming an intermediate, and are reduced to form metal nuclei, eventually giving metal particles. Submicrometer-sized particles can be synthesized by increasing the reaction temperature or inducing heterogeneous nucleation upon addition of foreign particles or forming foreign nuclei in situ.

Nanoparticle size is controlled by the use of initiators, such as sodium hydroxide. The hydroxide helps to deprotonate the glycol, which increases the reducing power. The reduction using ethylene glycol is a one-electron step in which Cu_2O is the intermediate structure. During the reduction, the CuO is partial reduced, creating Cu_2O with a smaller grain size than the parent precursor. As the reaction approaches completion, the Cu nanoparticle begins to aggregate and sinter. The degree of sintering depends on the temperature, reduction time, and CuO/ethylene glycol ratio. The sintering is due to the high reaction temperature and the Brownian motion of the particles, which leads to increased atomic mobility, which results in an increased probability of particle collision, followed by adhesion and finally, agglomeration.

Increases in the amount of sodium hydroxide cause an increase in the reaction rate. The mean particle size decreases as the hydroxide ratio increases, and the mean particle size becomes less dependent on the initial precursor morphology. While sodium hydroxide helps deprotonates the glycol, it also increases the solubility of the CuO precursor through the creation of hydroxo species. This shifts the rate-determining step from the dissolution process to the reduction process. Adding as little as 0.01 M of NaOH results in the reaction half-life begining to decrease by a factor of 6.

The polyol method has also been shown as a useful preparative technique for the synthesis of nanocrystalline alloys and bimetallic clusters. Nanocrystalline $Fe_{10}Co_{90}$ powder, with a grain size of 20 nm, was prepared by reducing iron chloride and cobalt hydroxide in ethylene glycol without nucleating agents. Nickel clusters were prepared using Pt or Pd as nucleation agents. The nucleating agent was added 10 minutes after the nickel–hydroxide–PVP–ethylene glycol solution began refluxing. The Ni particle size was reduced from about 140 nm to 30 nm when a nucleating agent was used. Reduction in the particle size was also obtained by decreasing the nickel hydroxide concentration and by the use of PVP. Nickel prepared without nucleating agents had an oxidation temperature of $370°C$. Smaller nickel particles synthesized with nucleating aids oxidized at a lower temperature of $260°C$, as expected from the higher surface area of finer particles.

Desorption studies showed that the adsorbed surface species were CO moieties and H_2O, and nitrogen-containing species were not observed. This indicated that ethylene glycol, not the polymer, was adsorbed on the surface of particles. The ethylene glycol had only half-monolayer coverage. When this protective glycol was removed from the surface completely, oxidation occurred. It was suggested that the Ni–Pd and Ni–Pt particles had a 7- to 9-nm Pd nucleus and a 6- to 8-nm Pt nucleus, respectively. Oxidation studies showed that some alloying of Ni with Pt occurred. Cobalt–nickel alloys of 210- to 260-nm particle sizes were also prepared using either silver or iron as nucleating agents.

Polymer-protected bimetallic clusters were also formed using a modified polyol process. The modification included addition of other solvents and sodium hydroxide. In the synthesis of Co–Ni with average diameters between 150 and 500 nm, PVP and ethylene glycol were mixed with either cobalt or nickel acetate with PVP. The glycol and organic solvents were removed from solution by acetone or filtration. The PVP-covered particles were stable in air for extended periods of time (months).

In contrast to aqueous methods, the polyol approach resulted in the synthesis of metallic nanoparticles protected by surface-adsorbed glycol, thus minimizing the oxidation problem. The use of polyol solvent also reduces the hydrolysis problem of ultrafine metal particles, which often occurs in aqueous systems. Oxide nanoparticles can be prepared, however, with the addition of water, which makes the polyol method act more like a sol–gel reaction (forced hydrolysis). For example, 5.5-nm $CoFe_2O_4$ has been prepared by the reaction of ferric chloride and cobalt acetate in 1,2- propanediol with the addition of water and sodium acetate.

6.2.7. High-Temperature Organic Polyol Reactions: IBM Nanoparticle Synthesis

To achieve rapid nucleation in polyol reactions, a general rule of thumb is that the higher the temperature of the glycol, the faster the nucleation and the more uniform the nanoparticles formed. This empirical rule prompted the evaluation of other solvents, such as propylene glycol. In the mid-1990s, researchers at IBM begin using ethylene glycol as the reducing agent, but in higher-boiling-point solvents such as dioctyl ether. In addition, they began using carbonyls and other organometallics in place of precursors such as oxides. In 2000, they reported the synthesis and characterization of monodisperse nanocrystals of metals and nanocrystal hybrid assemblies (Sun et al., 2000) and semiconductors (Murray et al., 2000) using a high-temperature solution-phase approach. Each nanocrystal contains an inorganic crystalline core coordinated by an organic monolayer, which allows for the self-assembly of multiple nanocrystals into a hybrid organic–inorganic superlattice material (Murray et al., 2001).

The IBM nanoparticle synthesis route is a combination of the polyol method and the thermolysis routes. Rapid injection of the organometallic precursor into a hot solution containing polyol stabilizing agents allows for the immediate formation of nuclei. Because capping agents–surfactants are present, the size and shape of the nanoparticles are controlled. By adjusting reaction conditions such as time, temperature, precursor concentration, and surfactant type and concentration, size and morphology can be controlled. In general, increasing the reaction time causes an increase in nanocrystal size, as does increasing the reaction temperature. Ostwald ripening also may force some of the smaller nanocrystals to disappear as they are redeposited onto larger nanocrystals. Higher surfactant/reagent concentration ratios favor the formation of smaller nanocrystals (Sun et al., 1999).

Initially, IBM used this method for the synthesis of FePt nanoparticles and ferromagnetic nanocrystal superlattices. Whereas in a thermolysis reaction, iron carbonyl and platinum organometallics result in the formation of platinum-coated iron nanoparticles, in the presence of a polyol such as 1,2-hexadecanediol, an increase in the rate of platinum decomposition leads to alloy nanoparticles. The chain length of the diol determines the resulting nanoparticle size. Surfactants that bind more tightly to the nanocrystal, or larger molecules, provide greater steric hindrance and slow the deposition rate of new material to the nanocrystal, resulting in smaller average size (Murray et al., 2001). The temperature determines the rate of decomposition and the rate of nucleation. Higher temperatures promote faster nucleation rates, which limit the size distribution of the nanocrystals, resulting in a more monodispersed sample. Superlattices are formed when the nanocrystals self-assemble via interactions between the organic monolayers. For FePt, the result is an HCP structure of nanoparticles that can be over 1 μm in size. Changing the alkyl group on the surfactants can change the interparticle distance. Changing an alkyl group from dodecyl to hexyl results in a particle spacing of 1 nm and a closest-cubic-packed superlattice.

REFERENCES

Brust, M.; Walker, M.; Bethell, D.; schriffrin, D. J.; Whyman, R. *J. Chem. Soc. Commun.* **1994**, 801–802.

Chow, G. M.; Chien, C. L.; Edelstein, A. S. *J. Mater. Res.* **1990**, 6(1), 8–10.

Feng, S. H.; Xu, R. R. *Acc. Chem. Res.* **2001**, 34, 239–247.

Figlarz, M.; Fievet, F.; Lagier, J.-P. U.S. patent 4,539,041, application filed, **1983**; patent granted, 1985.

Haberland, H. *J. Vac. Sci. Technol. A.* **1994**, *12*(5), 2925.

Haberland, H.; Moseler, M.; Qiang, Y.; Rattunde, O.; Reiners, T.; Thurner, Y. *Surf. Rev. Lett.* **1996**, *3*(1), 887.

Hoar, T. P.; Shulman, J. H. *Nature.* **1943**, *152*, 102.

Hunt, W. H. *JOM* **2004**, *56*(10), 13.

Lalena, J. N.; Cleary, D. A. *Principles of Inorganic Materials Design*, John Wiley, New York, **2005**.

Lessing, P. A. *Ceram. Bull.* **1989**, *68*, 1002.

Lowe, T. C.; Valiev, R. Z. *JOM.* **2004**, *56*(10), 64–68.

Massart, R. *IEEE Trans. Magn.* **1981**, *17*, 1247–1248.

Mathur, S.; Shen, H. *J. Sal-Rel Sci. Technal.* **2002**, *25*, 147.

Murray, C. B.; Kagan, C. R.; Bawendi, M. G. *Annu. Rev. Mater. Sci.* **2000**, *30*, 545–610.

Murray, C. B.; Sun, S.; Gaschler, W.; Doyle, H.; Betley, T. A.; Kagan, C. R. *IBM J. Res. Dev.* **2001**, *45*(1), 47–56.

Pechini, M. P. U.S. patent 3,330,697, **1967**.

Shulman, J. H.; Stoeckenius, W.; Prince, L. M. *J. Phys. Chem.* **1959**, *63*, 1677.

Sun, S.; Murray, C. B.; Doyle, H. *Mater. Res. Soc. Symp. Proc.* **1999**, *577*, 385–398.

Sun, S.; Murray, C. B.; Weller, D.; Folks, L.; Moser, A. *Science.* **2000**, *287*, 1989–1992.

Tian, Y.; Newton, T.; Kotov, N.; Guldi, D.; Fendler, J. *J. Phys. Chem.* **1996**, *100*, 8927–8939.

Winsor, P. A. *Trans. Faraday Soc.* **1948**, *44*, 376.

Winsor, P. A. *Trans. Faraday Soc.* **1950**, *46*, 762.

Zhong, Q.; Steinhurst, D.; Carpenter, E.; Owrutsky, J. *Langmuir.* **2002**, *18*, 7401–7408.

Zhu, Y. T.; Langdon, T. G. *JOM.* **2004**, *56*(10), 58–63.

7 Materials Fabrication

With the exception of thin-film processes such as CVD, PVD, or plating, most materials are not synthesized directly into their final form. Additional processing, or fabrication, is required to produce desired shapes, forms, and properties. Just as having the correct ingredients does not guarantee a successful outcome in the kitchen, having the correct chemical composition does not imply that a fabricated material will meet the requirements of an application. The differences between a light fluffy soufflé and a soupy mess, a tender, moist roast or a dry, hard piece of meat, and a fabricated component that meets requirements or one that will fail in an application all often boil down to processing. The broad field of materials science likes to represent the interrelation or interaction between the structure, processing, properties, and performance of a material by a pyramid, as shown in Figure 7.1. Performance of a material in a given application cannot be optimized without building a strong base of understanding of the relationship between the structure (micro and macro), processing (fabrication or synthesis), and properties of each material. Because of these interrelations, a discussion of materials synthesis would not be complete without the additional discussion of how further processing, or fabrication, can be used to produce desired materials properties. In this chapter we provide a broad introduction to several common materials fabrication techniques and how those techniques might affect the usability of a material.

Secondary materials fabrication techniques can be defined broadly by three categories: deformation processing, consolidation processing, and subtractive processing. The classification used to produce a given component depends primarily on the properties of the material to be fabricated and secondarily on the resulting form desired. Deformation processing is used to form ductile materials into useful shapes and comprises several of the oldest material fabrication techniques. Consolidation processing is often used for brittle materials, which do not deform plastically, or to produce ductile materials into complex shapes or forms more easily than can be done with deformation techniques. Consolidation techniques generally involve combining powders or small particles of a material, or constituent materials, then performing a sintering or other diffusional process to join the particles together. Subtractive processing techniques, such as chemical and mechanical machining, can be used in conjunction with the other methods, and typically do not significantly influence the bulk properties of the resulting

Inorganic Materials Synthesis and Fabrication, By John N. Lalena, David. A. Cleary, Everett E. Carpenter, and Nancy F. Dean
Copyright © 2008 John Wiley & Sons, Inc.

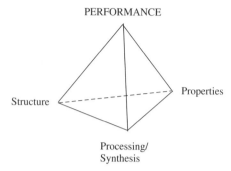

PERFORMANCE

Structure

Properties

Processing/
Synthesis

Figure 7.1 The *materials pyramid*, used by materials scientists to show the interrelation of the four cornerstones of their discipline: processing/synthesis, structure, properties, and ultimately, performance.

component. For this reason, subtractive processes are not reviewed here, but the interested reader is directed to ASM International (1989) and Boothroyd and Knight (1989).

7.1. INFLUENCE OF STRUCTURE ON MATERIALS PROPERTIES

As mentioned above, properties of a material are influenced by that material's structure, which in turn is influenced by the material's processing. To complete the circular discussion, processing methods used for a material are often dictated by the material properties. Processing or fabrication methods and their influence on materials constitute the bulk of this chapter. A brief discussion of the influence of structure on properties will help put these effects into perspective.

It is sometimes useful to distinguish between the intrinsic properties of a material and those properties that depend on microstructure. The intrinsic properties are determined by the structure on the atomic scale and are not susceptible to significant change by modification of the microstructure. These are properties such as melting point, elastic modulus, coefficient of thermal expansion, magnetic properties, and whether the material is conductive or insulative. In contrast, many of the properties critical to engineering applications, such as mechanical strength and dielectric constant, are strongly dependent on structure.

In materials science terms structure can have differing meanings when considering different length scales. Structure can refer to crystallographic orientation, or texture, grain size/shape, or other microstructural features, such as the presence or development of secondary phases, or even intergranular features such as dislocation networks or other defect structures. Each of these categories can have its own influence on a material's properties.

7.1.1. Crystallographic Texture

When discussing the properties of many materials, the term *crystallographic orientation*, or *texture* (introduced in Section 1.2.3), is often used. Crystallographic orientation in either a naturally occurring material or a fabricated component refers to how the atomic planes are oriented relative to a fixed point in space. By definition, each grain in a polycrystalline material has one crystallographic orientation, and surrounding grains will differ by some degree of rotation. This characteristic applies to all solids that are crystalline in nature, whether they are metallic or ceramic, natural crystals or fabricated components. Although we may like to think that most materials have a relatively random arrangement of grains, and hence isotropic properties, that is rarely the case. Most polycrystalline materials exhibit some tendency to display certain orientations, first developed during solidification from the melt and/or subsequently through further processing. This preferred orientation is referred to as a *crystallographic texture*, or simply as a *texture*, by materials scientists.

A component or object in which grain orientations are completely random is said to have no texture. If the material exhibits some preferred orientation, it is said to have a texture. The texture can be weak, moderate, or strong, depending on the percentage of grains that have the preferred orientation. Texture is produced in most materials, either as an unintended by-product of the processing method or with the intention to exploit favorable properties in one orientation. In some cases, processing methods or routes are chosen explicitly to develop a preferred texture.

Whether or not produced intentionally, crystallographic texture can strongly influence several key materials properties. At length scales that are significantly larger than the grain size, a material with a random arrangement of grains will appear isotropic. At the opposite extreme, a single crystal has the strongest possible texture (since there exists only one grain orientation) and the highest possible degree of anisotropy, as the arrangement of atoms differs in different directions. Because many material properties are orientation- or texture- specific, optimization of a component's performance may require that a materials engineer control crystallographic texture between these two extremes. Some properties that depend on texture are elastic (Young's) modulus, yield strength, ductility, toughness, magnetic permeability, electrical conductivity, Poisson's ratio, and for noncubic materials, coefficient of thermal expansion (Randle, 2001).

Unfortunately, few materials or components exhibit either of the two texture extremes (random or single crystal), so a method of quantifying texture or degree of preferred grain orientation must be developed. As described in Chapter 1, diffraction of some source beam (x-ray, electron, or neutron) is utilized to do just this. Because the crystal has planes of atoms arranged at regular known intervals or spacings, we can use Bragg's law to determine the orientation of a given set of planes relative to a fixed reference coordinate system. A texture goniometer that rotates the sample through all orientations while monitoring the intensity of a particular Bragg reflection is used. The results are plotted in two-dimensional stereographic projections known as *pole figures*, described in more detail in the companion volume to this book (Lalena and Cleary, 2005).

A *pole figure* references a particular Bragg reflection; that is, a (100) pole figure in a textured sample would differ from a (111) pole figure (whereas for a random texture it would not). The fixed reference coordinate system is usually chosen to correspond to a physical characteristic of the sample or process under consideration. For example, in material that is rolled, one reference axis would correspond to the rolling direction (RD), one axis to the transverse direction (TD), and the third axis would be in the through-thickness or normal direction (ND). This coordinate system is shown in Figure 7.2. Isotropic materials with random grain orientations may present an almost featureless pole figure, whereas sharply textured materials will show arcs or individual spots representing preferred orientations, as shown schematically in Figure 7.3.

Although very useful in conveying information, a single pole figure does not give complete information about the texture of a sample. Because it is a two-dimensional projection of a three-dimensional object, some information about the material's texture is lost. Further information can be determined by examining multiple pole figures for a given sample. In the last 15 to 20 years, with the use of computers, complex mathematical techniques have been developed to reduce data from a set of pole figures to a three-dimensional representation of material texture, referred to as an *orientation distribution function* (ODF). An ODF analysis takes advantage of the fact that the coordinate axes of any individual grain, or crystal, can be transformed to the sample reference coordinate system through a series of three Euler angle rotations: φ_1, Φ, and φ_2. This gives a true three dimensional representation of intensity data for crystal pole orientations throughout the Euler space. ODF texture data for a sample are usually presented in print as a series of sections of the computed Euler space, to aid in visualization. Beyond this brief mention to familiarize the reader with a concept, ODF analysis and data presentation are beyond the scope of this book. For further

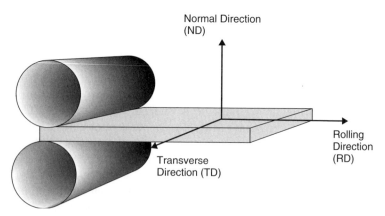

Figure 7.2   Reference coordinate system for a rolled material, consisting of one axis aligned in the rolling direction (RD), one additional axis in the rolling plane transverse to the rolling direction (TD), and a the third axis coincident with the rolled sheet's normal direction (ND).

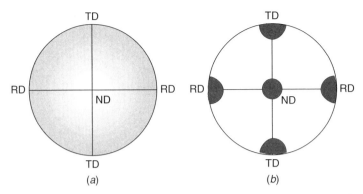

Figure 7.3   Normal direction (ND) pole figures for (a) random and (b) strongly textured sheet. Note the absence of features for the pole figure of the randomly oriented sample, while the strongly texture sample shows a concentrations of reflections along the ND as well as 90° away in the transverse (TD) and rolling (RD) as well. This alignment of the crystal orientation with the reference orientation is referred to as a *cube texture*.

information the reader is directed to a comprehensive presentation by Randle and Engler (2000).

When discussing texture, two factors should be specified: the geometric component and intensity. The intensity can be described numerically as a fraction of grains with a given orientation, or more qualitatively, as strong, moderate or weak. The geometric component is described as a set of planes that are perpendicular to the sample coordinate axis of interest, or perhaps more intuitively, the plane normal or pole that is aligned with the sample coordinate axes of interest. For an extruded component, the extrusion direction is the sample reference axis and the rolling direction is a reference axis in rolled materials. For materials that are grown (i.e, sputtered, plated, solidified), the growth direction is typically the reference direction. If the {100} planes are aligned normal to an extrusion direction, the material is said to have a {100} texture. If only one direction is oriented, a material is said to exhibit a *fiber texture*. That is, the {100} planes are normal to the extrusion direction axis, but the crystal or grain can have any rotation about this axis. As we will see, many fabrication processes will tend to form this fiber texture. In other cases, however, the process or performance of materials may dictate that orientation in another direction be closely controlled. In this case, preferred alignment of the crystal along a secondary sample reference axis is described as well. For example, a preferred orientation where all three axes of a cubic crystal are coincident with the sample reference axes would be described as {100}<100>, called a *cube texture*. Figure 7.4 compares this cube texture with a cube-on-edge texture, or {100}<110>. Textures that actually form in a given process depend on the material being fabricated, the slip systems available, and the fabrication parameters.

Texture is nearly impossible to avoid in most material processing techniques. In some cases it may be advantageous to develop a particular texture, whereas

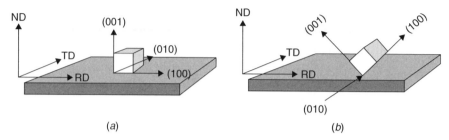

Figure 7.4 Crystal orientation relative to the rolled sheet reference axes for *(a)* {100}<100> or cube texture, and *(b)* {100}<110> cube-on-edge texture.

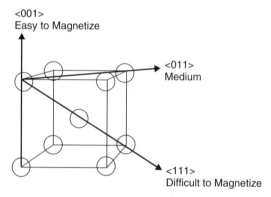

Figure 7.5 Variation of rate of approach of saturation with crystal orientation in iron (steel) lattice. <100> directions are significantly easier to magnetize.

in others the texture developed during one operation may interfere with downstream processing. One early example of the exploitation of texture to improve the performance of a material in application is the use of oriented silicon steel for electrical transformer applications. Although the saturation magnetization is the same for crystals of all orientations, the rate of approach of saturation with magnetic field varies significantly with crystal direction (Honda and Kaya, 1926; Honeycombe, 1984), as shown in Figure 7.5. For magnetic property optimization, electrical-grade steel should exhibit a strong cube texture, {100}<100>, which would place two easy-to-magnetize directions in the plane of the sheet. A cube-on-edge texture, {100}<110> or {110}<100>, would be second best. In the 1930s, the problem of forming a cube-on-edge texture in silicon steel was solved by varying metallurgical process parameters (Goss, 1935). It was not until more than 20 years later that a pure cube texture, {100}<100>, was reported in the literature by Assmus and Weiner (Assmus et al., 1957; Wiener et al., 1958). Other examples of desired textures and undesired texture formulations for various materials and applications are given in Table 7.1.

TABLE 7.1   Examples of the Influence of Crystallographic Texture on Material Performance

Material	Applications	Processing Methods	Implications for Texture Development
Silicon steels	Transformer cores, generator coils, electric motors, etc.	Rolling, magnetic fields during annealing	Increased magnetic flux density, decreased magnetization hysteresis losses, increased energy efficiency
Aluminum, copper, titanium	Sputtering targets	Rolling, forging	Atom emission trajectory, sputtered film thickness uniformity (higher uniformity leads to higher reliability in electronic devices)
Aluminum	Beverage cans	Deep drawing	Textured input material can lead to nonuniform wall thickness in the drawn can, and "ears" or variations in height that must be trimmed
Ferroelectric ceramics (such as barium titanate, lead zirconate titanate)	Sensors and actuators, electronic memory, optical applications	Tape casting, sputtering, pressing, templated grain growth	Improved dielectric and piezoelectric properties
Nickel-based superalloys	Turbine blades	Solidification (single crystal and directional)	Improved strength by aligning with maximum stress direction; run turbine hotter, more efficiently
Various ceramics	Thermal barrier coating	Deposition	Match CTE of substrate, improve performance
High-temperature superconductors	Transmission cables, transformers, current limiters, motors and generators	Rolling, direct metal oxidation, tape casting	Greater energy efficiency, higher critical current density

As with many things in the field of materials processing, development of texture is often a trade-off. Some properties are enhanced whereas others deteriorate when a strong texture is present. In the descriptions of materials processing techniques in the following sections, mechanisms for texture formation are introduced. Development of both type and degree of texture is one the factors that influence a material scientist's choice of processing route.

7.1.2. Materials Yield and Slip

Plastic deformation is defined as the permanent distortion of a body under stress. The stress at which the material deforms plastically is called the *yield stress* or *yield point*. This deformation can occur via a number of mechanisms:

- Slip from the motion of dislocations (linear defects in the crystal structure)
- Twinning, where the crystallographic orientation changes significantly in the region of plastic deformation
- Grain boundary sliding
- Grain rotation
- Creep (typically diffusional movement of material)

In nearly all metal-forming operations, slip is the dominant method of deformation, although twinning can be significant in some materials. Slip occurs when the shear stress is high enough to cause layers of atoms to move relative to one another. The critical resolved shear stress is lowered when the crystalline lattice is not perfect but contains linear defects called *dislocations*. Slip-induced plasticity was covered in Chapter 9 of the companion to this text (Lalena and Cleary, 2005) and is reviewed here only briefly. The interested reader is advised to consult Lalena and Cleary (2005), Honeycombe (1984), or Dieter (1976).

Slip occurs along specific crystal planes (slip planes) and in specific directions (slip directions) within a crystal structure. Slip planes are usually the closest-packed planes, and slip directions are the closest-packed directions. Both face-centered-cubic (FCC) and hexagonal-close packed (HCP) structure are close packed structures, and slip always occurs in a close packed direction on a closepacked plane. The body-centered-cubic (BCC) structure is not, however, close packed. In a BCC system, slip may occur on several nearly close packed planes or directions. Slip planes and directions, as well as the number of independent slip systems (the product of the numbers of independent planes and directions), for these three structures are listed in Table 7.2.

If a crystal lattice were perfect (i.e., contained no defects), slip would have to occur by the simultaneous movement of an entire plane of atoms over another plane of atoms. Frenkel (1926) developed a simple model that calculates the yield stress in a perfect crystal as $G/2\pi$, where G is the *shear modulus* of the material. Actual yield stresses are two to four orders of magnitude lower than this. The concept of a dislocation was later introduced to explain the discrepancy between the measured and theoretical shear strengths of material. The dislocation

TABLE 7.2 Active Slip Systems in FCC, BCC, and HCP Crystal Structures

Crystal Structure	Slip Plane	Slip Direction	Number of Nonparallel Planes	Slip Directions per Plane	Number of Slip Systems
Face-centered cubic	{111}	<110>	4	3	12
Body-centered cubic	{110}	<111>	6	2	12
	{112}	<111>	12	1	12
	{123}	<111>	24	1	24
Hexagonal close-packed	{0001}	<11$\bar{2}$0>	1	3	3
	10$\bar{1}$0	<11$\bar{2}$0>	3	6	18
	10$\bar{1}$1	<11$\bar{2}$0>	6	1	6

Source: T. H. Courtney, *Mechanical Behavior of Materials*, 2nd ed., McGraw-Hill, New York, 2000.

allows slip to happen by the motion of the defect in the lattice. This dislocation motion requires movement of relatively few atoms at a time and thus can occur at much lower stresses. During deformation, dislocations can interact and multiply, which can then increase the difficulty of further dislocation movement, a phenomena known as *work hardening*.

A typical stress–strain plot is shown in Figure 7.6. Initially, the material deforms elastically only and the stress is related linearly to the strain, with a proportionality constant, E, the *elastic modulus*. The point at which the stress–strain curve deviates from linearity is known as the *yield point* or *yield stress*. As stress continues to increase, the material plastically deforms uniformly. At some point, the plastic deformation localizes to one region of the stressed material, and only this region continues to deform. This is known as *necking*, since the localized area continues to decrease in area, or neck down, whereas the surrounding material does not. The stress at which necking occurs is known as the ultimate *tensile stress*. Whereas the necked area continues to work harden and carry higher stresses, the overall load on the material cannot be supported through the reduced necked area and a maximum is observed in the stress–strain curve. The material fails at the necked location soon after. The strain, or percent increase in length of a uniaxial test specimen is known as the *elongation*, which is also a measure of ductility. *Toughness*, as a measure of how much energy a material can absorb before fracture is often measured as the area under the stress–strain curve.

7.1.3. Grain-Size Effects

Movement of dislocations is a primary mechanism for plastic deformation. A dislocation's motion is impeded when they encounter obstacles, causing the stress required to continue the deformation process to increase. Grain boundaries are one of the obstacles that can impede dislocation glide, so the number of grain boundaries along a slip direction can be expected to influence the strength of a material. In the early 1950s, two researchers, Hall (1951) and Petch (1953),

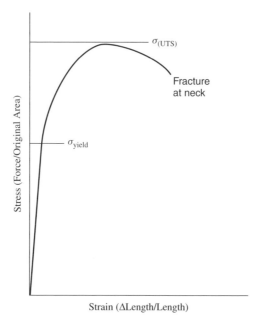

Figure 7.6 Idealized stress–strain curve. Material deforms elastically until yield stress, σ_{yield} or σ_y is reached. Load increases until the ultimate tensile strength (UTS), which marks the onset of necking. Fracture occurs at a reduced load in the necked region. Toughness is measured by the area under the stress–strain curve.

studying different mechanisms, independently developed a relationship that shows the dependence between grain size and yield strength:

$$\sigma_y = \sigma_0 + kD^{-\frac{1}{2}} \tag{7.1}$$

where σ_y the yield strength, D the grain diameter, σ_0 the stress required for dislocation movement (sometimes referred to as the lattice friction factor), and k is a fitting materials constant. Although their approaches differed, Hall and Petch arrived at essentially the same conclusion: that the grain size dependence is related to the length of a slip band and that the maximum slip band length is determined by the grain size.

The Hall–Petch relationship has been found to hold true for many FCC, BCC, and HCP metals across grain sizes ranging from millimeter scale to submicrometer (Armstrong 1971). Despite being well known, the Hall–Petch relation is not without its detractors. For example, one study has shown that the scatter in experimental data is such that a $D^{-1/3}$ or D^{-1} relationship will yield as good a fit as $D^{-1/2}$ (Baldwin, 1958). Mechanisms for supporting both of those exponents have been proposed. While the Hall–Petch relation would predict that yield strengths should continue to increase as the grain size drops into the nanometer range, a *reverse Hall–Petch effect*, or softening or weakening of material, has instead been found for many materials. Porosity and deformation via grain boundary sliding rather than dislocation movement, as well as porosity in nanoscale materials, have been proposed as mechanisms for this reverse Hall–Petch effect (Schiøtz et al., 1999). Regardless, it should be noted that grain size has a strong and well-documented effect on a material's yield strength.

Hardness and a ductile-to-brittle transition temperature (DBTT) have also been noted to follow a Hall–Petch relationship (Meyers, and Chalwa, 1984). Ductility increases as the grain size decreases. Decreasing grain size tends to improve fatigue resistance but increases creep rate. Electrical resistivity increases as grain size decreases, as the mean free path for electron motion decreases.

7.1.4. Effects of Cold Working

When materials are deformed plastically at temperatures below those at which recrystallization take place, they are said to be *cold worked*. Conversely, plastically deforming materials above their recrystallization temperature is referred to as *hot working*. Recrystallization, one of the phases of annealing, occurs not at one specific temperature but over a range of temperatures, depending on the history of a material (i.e., more heavily cold-worked materials recrystallize at lower temperatures; increased annealing times are associated with decreased recrystallization temperatures; pure metals recrystallize at lower temperatures than alloys, etc.) Thus, although there is no set rule as to what constitutes "cold" when deforming a sample, most materials are considered to be cold worked if deformed at temperatures significantly below half their melting point on an absolute scale, often referred to as a *homologous temperature* of 0.5, or simply $0.5T_m$.

Compared to a fully annealed material, a material that is cold worked is stronger and harder. The cold-worked material will also exhibit increased electrical resistivity and decreased ductility. Although some materials, such as Pb and Sn, do not cold work during commercial processes since room temperature is a significant fraction of their melting temperature (and recovery–annealing mechanisms are active at this relatively high homologous temperature), most commercially important materials will cold work at room temperature. An example of increased strength and decreased ductility in cold-worked materials is familiar to anyone who has ever tried to "reconfigure" a wire coat hanger to perform some task. Although it is relatively easy to bend the straight sections of a coat hanger into a new shape, it is extremely difficult to straighten out the tight radius (i.e., highly deformed) bends that form the coat hanger's original shape. This is work hardening in action. The steel alloy used to produce the coat hanger was chosen for its ability to work harden, so that the coat hanger would maintain its shape under load (i.e., the weight of a coat or pair of pants). If a material is annealed, a higher degree of cold work (and hence denser dislocation network) will produce a finer annealed grain structure.

It is not surprising then that for many materials the degree of work hardening must be specified when considering its properties. Conversely, the degree of work hardening imparted to a material during its processing must be taken into

TABLE 7.3   **Typical Mechanical Properties for Aluminum Alloy 3004**

Alloy Condition	Tensile Strength (MPa)	Yield Strength (MPa)	Elongation (%)
3004-O	180	69	20–25
3004-H19	295	285	2
3004-H32	215	170	10–17
3004-H34	240	200	9–12
3004-H36	260	230	5–9
3004-H38	285	250	4–6

Source: *ASM Metals Handbook*, desk ed., H. E. Boyer and T. L. Gall, Eds., American Society for Metals, Metals Park, OH, 1985 (for 3004-H19), and *ASM Metals Handbook*, 10th Ed., Vol. 2, ASM International, Metals Park, OH, 1990.

account when designing and specifying materials. The variation in yield strength and ductility for a common aluminum alloy used in the manufacture of beverage cans is shown in Table 7.3. Here "3004" refers to an alloy number and indicates that this is an Al–Mn–Mg alloy. The number–letter combination after the alloy code refers to a temper designation. The properties are seen to differ significantly depending on the degree of work hardening present in the materials.

7.2. DEFORMATION AND SOLIDIFICATION METHODS

Many of the processes used to form ductile materials have been practiced for hundreds, if not thousands, of years. As a result, considerable analytical work has been done to aid in understanding these processes. One of the more simple models is the *slab method* which assumes that the metal deforms uniformly in the deformation zone (i.e., a square grid place on a workpiece going through a reduction in area would be distorted uniformly into a rectangular grid). Increasing in complexity are the uniform deformation energy method, the slip-line field theory (where slip lines are lines of maximum shear stress), and upper and lower bound solutions, which are based on the theory of limit analyses. Slip lines are often drawn in descriptions of deformation processes. Methods beyond those of the simple slab are beyond our scope here. The interested reader is directed to excellent books by Dieter (1976) and Hosford and Caddell (1993) for more detailed analysis of these forming processes.

7.2.1. Casting

A *casting* is a shape obtained by pouring a liquid into a mold or cavity and allowing it to freeze, thus producing a solid article that has the shape of the cavity. Casting is also the name given to the process of forming a casting. Casting was introduced in Section 4.2.6. Here, we present the topic from the materials science perspective. Although not a deformation process, as are many of the other techniques introduced in this section, casting is an important materials processing method, and processing variables will have an impact on the performance of the article cast. Furthermore, cast ingots will often be the starting form for many other processing methods. In this section we examine some of the common metal casting methods, comparing and contrasting them as well as describing the types of articles for which they are best suited. A summary of this information is given in Table 7.4.

The casting process can create defects that are unique across all the materials processing methods. These defects are universal to all casting methods, so are described before presenting information about casting subtypes. Although not an exhaustive list, the major casting defects we discuss are porosity due to solidification shrinkage or trapped gas, nonmetallic inclusions, and hot tears. Solute partitioning, described in the earlier description of solidification, is not discussed here but is a concern in many applications.

Nearly all commercially important metals are denser in the solid state than they are in the liquid state. This implies that there is a volume contraction upon going from a liquid to a solid at the same temperature. This contraction, typically on the order of 3 to 8 vol%, percent, will cause a mold full of liquid metal to have insufficient material to form a fully solid casting of the same geometry. Foundrymen (a term that has resisted trends toward gender neutrality) use risers, which are reservoirs of liquid metal, attached to the desired casting shape in strategic locations in order to feed this solidification shrinkage. In a well-designed casting, the solidification shrinkage voids will be contained within the risers, which are removed from the cast shape after cool-down. Efforts to promote directional solidification within a casting, through the use of chills or mold features aimed at promoting heat withdrawal, or applying heat to selective portions of a mold, also reduce the presence of shrinkage voids. A second classification of solidification shrinkage is *microshrinkage*, also called *microporosity*. While caused by the same liquid-to-solid volume change, microporosity is porosity restricted to interdendritic spaces. To compensate for this solidification shrinkage, there must be liquid flow through these interdendritic spaces. When resistance to interdendritic liquid flow becomes too great, microporosity will form. This porosity takes the sharp-edged shape of the interdendritic spaces and can act as a stress riser. For high-integrity castings, such as jet engine components, hot isostatic pressing (HIPping) will tend to collapse these microvoids, thus "healing" the casting. Microporosity can be reduced or eliminated by reducing the physical length of the mushy zone. This can be done by designing casting alloys that have narrow mushy zone temperature ranges or by increasing the temperature gradient in the casting (which has the effect of collapsing the distance over which solid and liquid coexist). A third mechanism for generating porosity in metal casting is the evolution of gas. Gases have a higher solubility in liquids than in solids. When a liquid with a high dissolved gas content solidifies, the excess gas is rejected at the liquid–solid interface. If not left with a means of escape, this gas will

TABLE 7.4 Comparison of Several Casting Methods

Method	Mold Material	Castable Metals	Advantages	Disadvantages	Typical Parts Cast
Sand casting	Sand (expendable but recyclable), pattern is reusable	Ferrous and nonferrous	Inexpensive, even in small lots; can cast very large parts and most alloys; high-volume, automated process	Rough surface finish, may require machining; part must be designed for parting line; dimensional accuracy poorer, due to possible misalignment of cope and drag	Large components, engine blocks, valve housings, crankshafts, cookware
Lost-foam sand casting	Sand (expendable but recyclable); pattern is consumed	Ferrous and nonferrous	As above plus: no need for parting line; better surface finish than conventional sand casting; increased dimensional accuracy	Increased cost, added process steps	As above
Investment casting	Ceramic (solid or shell); wax pattern is expendable	Any, including high-melting-point materials	Intricate shapes and fine details, good dimensional control, low-or high-volume process, no need for parting line/draft in casting design; can cast thin sections	Cost, cycle time to build shell, process complexity	Intricate parts, medical/dental implants, turbine blades aircraft engine structural components, jewelry
Die casting	Reusable metal mold	Copper or low-melting-point alloys	Good dimensional tolerances, thinnest cast sections possible, quick cycle time	High die cost makes it impractical for low volume, limited alloys, need draft/parting lines, size/weight limit	Small intricate parts, housings, valves, heat sinks, toys

form voids (still filled with gas) in the solidified casting. These voids are distinguishable from solidification shrinkage by their spherical shape. Most people have seen these voids present as air bubbles in ice cubes. To minimize gas-induced porosity, most melts are degassed prior to pouring.

Nonmetallic inclusions can result from impurities with a higher solubility in the liquid metal than in solid metal, but more often result from solid particles eroded from the mold upon pouring or from the crucible during melting, or entrained slag particles from the melt surface. Embedding a ceramic filter in the castings gating (mold filling) system will help reduce inclusions. Inclusions typically have little ductility. Cracks may form in the inclusion and then propagate to the casting under load. Since the inclusion and casting have differing coefficients of thermal expansion, they also act as stress risers and potentially crack initiators if the casting is subject to thermal cycling.

Hot tears form when casting sections are constrained by the mold from contracting as they cool near the end of solidification. Hot tears are caused by a number of factors, but poor casting design is the primary cause. Castings should be designed so that solidifying sections are not subject to tensile stresses before they are fully solidified.

As described in Chapter 4, the microstructure of a solidifying material is refined when the cooling rate is increased. This is true for both primary dendrite spacings and the spacings on the secondary dendrite arms. Relationships of the form

$$d = b(GR)^{-n} \tag{7.2}$$

where d is the dendrite arm spacing, b a constant, G the thermal gradient in the mushy zone, and R the rate of advanced of the liquid–solid front (note that the product GR is the solidification rate), describe the dependence of this microstructural feature on solidification conditions. The exponent n is usually very close to 0.5 for primary dendrite arm spacings and is in the range 0.33 to 0.5 for secondary arm spacings (Flemings, 1974). These dendrite arm spacings, particularly the secondary dendrite arm spacing, can significantly influence the mechanical properties of the cast material. In general, the hardness, tensile strength, elongation, and toughness of a cast material increase as the secondary dendrite arm spacing decreases. This correlation is so strong that secondary dendrite arm spacing is an important metallurgical-quality control variable specified for many high-performance castings.

7.2.1.1. Ingot Casting   When casting an ingot, molten metal is poured into a permanent mold and allowed to solidify. Because the ingot is a casting, it is prone to the casting defects mentioned above. Ingots are the feedstock for subsequent rolling, forging, pressing, or extrusion operations. The solidified ingot structure is very inhomogeneous, as described earlier in this book and in the first chapter of our companion book (Lalena and Cleary, 2005), with variations in grain size, morphology, and orientation as well as some variations in composition profiles due to the solute partitioning described by the Scheil equation (Scheil, 1942).

Ingots tend to have an equiaxed structure at the mold surface, which gives way to a columnar grain zone that extends toward the casting's centerline. A second equiaxed region may exist along the centerline where the columnar grains converge. The equiaxed grain zones of the ingot at the mold surface and centerline tend to exhibit a random texture. The columnar grains form as a result of some of these equiaxed grains being oriented favorably for rapid growth, which allows them to crowd out other growing grains. These rapid growth directions determine the texture of the columnar region, which often makes up the bulk of the ingot. Generally, in metals with cubic structures, the rapid growth directions are <100>, and these directions are aligned with the direction of heat flow. For the nearly one-dimensional heat flow characteristic of many ingot casting operations, this translates into a <100> fiber texture in the bulk of the solidified ingot. For noncubic materials, textures will differ. For example, the body-centered tetragonal primary Sn dendrites in an Sn–Bi alloy will grow in the [110] direction (Flemings, 1974). Subsequent ingot processing will generally change the texture of the material and even out concentration gradients.

7.2.1.2. Sand Casting

7.2.1.2. Sand Casting Sand casting is used to make large parts, typically of iron but also of nonferrous metals. Molten metal is poured into a mold cavity formed out of sand, one of the earliest molten metal-processing techniques. The cavity in the sand is formed by using a pattern in the shape of the object desired (although slightly larger dimensionally, to account for contraction of the part as it cools down to room temperature). This mold, called the *flask*, is made up of two parts: the lower half is called the *drag* and the upper half, the *cope*. When assembled, the surface where the cope and drag meet is called a *parting line*. Sand is vibrated or rammed around patterns in each half of the flask to form the cavity for the part to be produced as well as its gating (to deliver molten metal) and risering (to feed solidification shrinkage) systems that are also included in the mold cavity. The cope and drag are inverted and the pattern carefully removed to avoid disrupting the sand. If it is desired to have a hole or internal passage in the casting (e.g., the cylinders of an engine block), a special sand shape called a *core* in inserted into the cavity to form those features. The cope and drag are then assembled together and the mold is ready for casting. The molten material is poured in the pouring cup, which is part of the gating system that supplies the molten material to the mold cavity. Vents in the mold provide the path for the built-up gases and the displaced air to vent to the atmosphere. After the casting solidifies, it is shaken out of the sand, which is reclaimed for subsequent use. Shakeout can occur while the part is still very hot.

Sand castings are readily recognized by the sandlike texture imparted by the mold. As the dimensional accuracy of the casting is limited by imperfections in the mold-making process, there is extra material designed in for a machining or grinding allowance. This allowance is more than is required by other, more accurate casting processes. Sand casting requires that the casting be designed to accommodate the process, as it must be possible to remove the pattern without disturbing the molding sand. A slight taper, called *draft*, must be used on surfaces perpendicular to the parting line in order to be able to remove the pattern from the mold. This requirement also applies to cores, as they must be able to be removed from their forming process.

Starting in the early 1980s, some castings (e.g., automotive engine blocks) have been made using a sand casting technique called the *lost foam process*. In this process, conceptually similar to investment casting, the pattern is made of polystyrene foam. Gating and risering are also made of foam and glued to the pattern. Sand is packed around the pattern assembly by vibration, leaving the foam in place. When the metal is poured into the mold, the heat of the metal vaporizes the foam a short distance away from the surface of the metal, leaving a cavity into which the metal flows. The lost foam process supports the sand much better than does conventional sand casting, allowing greater flexibility in the design of the cast parts, with less need for machining to finish the casting.

The use of foam patterns for metal casting was patented by Harold F. Shroyer in 1958 (Shroyer 1958). In Shroyer's patent, a pattern was machined from a block of expanded polystyrene (EPS) and supported by bonded sand during pouring. This process, known as the *full mold process*, is used to make primarily large, one-of-a kind castings. In 1964, Merton Flemings used unbonded sand, giving the current process known as *lost foam casting* (LFC), which is used by more foundries than is the full mold process. With LFC, the foam pattern is molded from polystyrene beads. In the LFC process, the foam assembly is dipped in a slurry to create a ceramic coating. The coating forms a barrier so that the molten metal does not penetrate or cause sand erosion during pouring, and helps control the transport of foam decomposition products. The castings produced using the LFC process require fewer operations to clean since there are no fins or parting lines to remove. Some of the benefits of lost foam casting are dimensional accuracy, elimination of cores and parting lines, and more flexibility in design and part consolidation, since removal of the pattern is not necessary. Lost foam casting, particularly the LFC process, has been adopted, particularly by the automotive industry, starting with Saturn in the 1980s, replacing traditional sand casting as a means of producing differential cases, crankshafts, cylinder heads, and engine blocks.

7.2.1.3. Investment Casting   Investment casting, also known as the *lost wax process*, uses a refractory mold that has been created around an expendable pattern, usually made of wax. The pattern is then melted or burned out, leaving an empty mold cavity. In contrast to sand casting, where the pattern is reusable and the mold is expendable, and ingot casting, where the mold is reusable, in investment casting both the mold and the pattern are expendable or are used one time only. If fine internal passages are desired, ceramic cores can be used (wax is typically molded around these cores). These cores are also expendable. The investment casting process is shown in Figure 7.7. Because the mold is formed around a one-piece pattern (which does not have to be pulled out from the mold as in a traditional sand casting process), very intricate parts and undercuts can be made. The mold itself can be produced via two distinct process: the solid and shell methods. In the *solid method*, the pattern assembly (the casting and

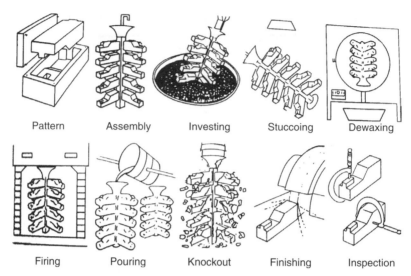

Pattern	Assembly	Investing	Stuccoing	Dewaxing

Firing	Pouring	Knockout	Finishing	Inspection

Figure 7.7 Investment casting process. Wax patterns are molded, then assembled to a gating and risering system. The wax assembly is dipped into a slurry (invested) and then stuccoed. After the shell is dewaxed, it is fired to give it strength and preheat for casting. After casting, the components are knocked out of the shell, gates and risers are removed, and the parts are inspected. (From R. A. Horton, in *ASM Handbook*, Vol. 15, Casting, D. M. Stefanescu, Ed., 1988, p. 253. Reprinted with permission of ASM International. All rights reserved. www.asminternational.org.)

associated gating and risering) is suspended in a flask, which is then filled with a refractory slurry. The slurry hardens, forming a solid mass surrounding the wax pattern. In contrast, during the *shell method* the wax pattern assembly is dipped into a coating slurry and then stuccoed with a granular refractory and allowed to dry. The process of dipping and stuccoing is repeated until a shell of the desired thickness is built up around the wax part. In both cases the wax is allowed to melt out of the mold, often using an autoclave. Although some molds can be poured at room temperature, most molds are preheated to fire or sinter the ceramic and avoid thermal shock issues when pouring molten metal into the mold. Shell molds, in particular, are designed to crumble upon cooling from casting temperature to aid in removal of the casting and avoid placing stress on the part, which may lead to hot tears. In general, jewelry making and dental implants utilize the solid mold method, while most industrial investment casting is done with the shell method.

Advantages of the investment casting process are the ability to form parts with intricate details and fine surface finishes and to hold dimensionally tight tolerances. In many cases investment casting parts replace several parts fabricated by other means. The process can permit close control of grain size and orientation and can be used to promote directional solidification or preferred grain orientation, giving a desired texture to the cast component. Using wax injection molding techniques, it lends itself well to high-volume production. When smaller parts are cast, it is typical to include multiple parts per mold. Finally, unlike may other casting techniques, there are few limitations on the metals that may be cast with this method.

7.2.1.4. Single Crystal Shaped Casting

In some extreme applications it may be desirable to have a large single crystal, to eliminate grain boundaries and take advantage of aligning the crystal axes of maximum strength with an application's major stress axes. Single-crystal casting of turbine airfoil blades is a special subset of investment casting practice performed to achieve these results. Conventionally cast turbine airfoils are polycrystalline, with an essentially random texture. Grain boundaries present problems for creep and stress corrosion, in addition to limiting strength. To optimize the efficiency of the turbine, the blades need to be able to run hotter—thus the desire to form columnar grains (imparting a specific texture to the airfoil)—and then single-crystal casting was born. In the 1960s, researchers at jet engine manufacturer Pratt & Whitney Aircraft (now Pratt & Whitney, owned by United Technologies Corp.) developed techniques to cast directionally solidified and single-crystal turbine blades and vanes. The directionally solidified columnar-grained turbine blade was invented and patented by Frank VerSnyder in 1966 (VerSnyder, 1966). This was followed by patenting of the first single-crystal turbine blade by Barry Piearcey (Piearcey, 1969; 1970).

Making a single-crystal turbine airfoil requires a carefully controlled mold temperature distribution to ensure that heat transfer occurs in one dimension only. At the base of the airfoil mold is a water-cooled chill plate. Columnar crystals form at the chill plate surface in a mold chamber called a starter. At its upper surface, the starter narrows to the opening of a vertically mounted helical channel called a *pigtail*, which ends at the blade root. The pigtail admits only a few of the fastest-growing columnar grains from the starter. As solidification proceeds up the helix, crystal elimination takes place so that only one crystal emerges from the pigtail into the blade root, to start the single-crystal structure of the airfoil itself. One-dimensional heat conduction must be maintained as the mold is withdrawn from a temperature-controlled enclosure, to ensure that no stray or secondary grains are nucleated along the airfoil. An airfoil, with the starter chamber and the pigtail, are shown in Figure 7.8.

In jet engine use, single-crystal turbine airfoils have proven to have as much as nine times more relative life in terms of creep strength and thermal fatigue resistance and over three times more relative life for corrosion resistance compared to equiaxed crystal counterparts. This technology is now being applied to land-based gas turbines, which can have airfoils that are an order of magnitude larger than a jet engine's. Single-crystal castings will help increase the operating temperature, which will increase the energy efficiency of these turbines.

7.2.1.5. Die Casting

Die casting is the process of forcing molten metal under pressure (up to 100 MPa) into the cavities of permanent steel molds, called *dies*. The pressure used induces high molten metal velocities when filling the

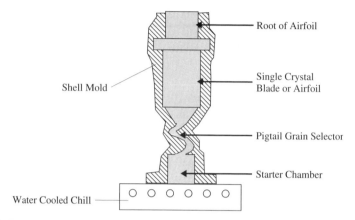

Root of Airfoil

Single Crystal
Blade or Airfoil

Shell Mold

Pigtail Grain Selector

Starter Chamber

Water Cooled Chill

Figure 7.8   Single-crystal casting of an airfoil. Casting is directionally solidified from a water-cooled chill. A pigtail grain selector ensures that only one grain, or crystal, enters the blade section of the mold cavity.

molds, which enables intricate or complex shapes to be filled before significant solidification chokes off liquid flow. Dies for producing metal parts range in complexity: from sink faucets, to toy cars, to engine blocks. Because of the thin sections and rapid cooling, die-cast parts tend to exhibit a fine-grained microstructure.

Die casting lends itself to making metal parts with tight dimensional tolerances, thin sections (sections as thin as a millimeter can be cast), high volumes, smooth surfaces, and complex shapes. Inserts of other materials may be placed in the mold cavity, where molten metal will be cast around them. Die-casting dies tend to be expensive, as they are made from hardened steel. The use of steel die limits the melting temperature of the casting alloy. Copper is generally the highest-melting metal that can be die cast, along with lower-melting aluminum, zinc, and magnesium. Due to its high melt fluidity, zinc is particularly attractive as a die-cast metal when producing castings with very thin walls. The principal limitations of die casting are cost and the size of the castings; casting weight seldom exceeds 20 kg, and castings of less than 5 kg are the norm. Dies are expensive and not economical for small production runs (Andresen, 1985). Some examples of die cast parts are shown in Figure 7.9.

Permanent mold casting is generally similar to die casting, except that gravity rather than pressure is used to fill the mold cavity. This process is employed where larger, or heavier, section castings are needed. Mold costs are still high, so this process is economical only in volume as well.

7.2.1.6. Semisolid Casting   Fundamentally, a *semisolid material* is a mixture of liquid and solid phases. In metal alloys, thermodynamics determine the temperatures at which solid, liquid, or both exist. This is usually represented by a phase diagram, which maps phases present as a function of temperature and composition. For most alloy compositions, there is a freezing range where both liquid

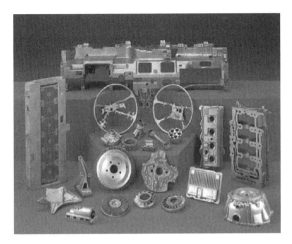

Figure 7.9 Examples of die-cast components. (Courtesy of Gibbs Die Casting, Inc., Henderson, KY. Reprinted with permission.)

and solid exist together in a semisolid state, often referred to by metallurgists as the *mushy zone*. In the context of semisolid casting, a semisolid is a mixture of rounded solid-phase particles (as opposed to the dendrites that would typically form, as discussed in Chapter 4) suspended in a liquid matrix. These mixtures are slurries, much like slush or ice cream. Semisolid casting is a special case of die casting in which a semisolid material is injected into the die, as opposed to a completely liquid material. Some of the advantages of semisolid casting include reduced porosity in the cast part (some fraction, up to ~60%, of the input material is already solid; thus, there is less solidification shrinkage), decreased cycle time (again, since the starting material is already partially solid, less heat needs to be removed to solidify the part prior to removal from the mold), and improved properties due to the unique microstructure produced by the semisolid process. An example of a semisolid cast microstructure and a conventional cast microstructure is given in Figure 7.10.

These advantages have attracted the attention of the automotive industry, where the need to produce more fuel-efficient vehicles has increased the use of lightweight materials and thinner wall components. Semisolid forming technology continues to make inroads in the manufacturing of high-performance or pressure-tight aluminum components because of its ability to provide near-net-shape components with exceptional properties. Cost pressures have prevented widescale adoption thus far; however, recent developments in the process may allow for greater usage.

The semisolid metal forming process can be applied in two different routes: thixocasting and rheocasting. These methods can be used on the same alloys and differ principally in the path employed to generate the semisolid feedstock. *Thixocasting* heats a special billet from the solid state until the appropriate volume fraction of liquid is formed. *Rheocasting* starts with a liquid melt, which

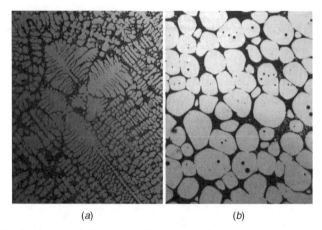

(a)   (b)

Figure 7.10 Comparison of (*a*) dendritic conventionally cast and (*b*) nondendritic semisolid formed microstructure of aluminum alloy 357 (Al–7.0 Si–0.5 Mg). The photos are at the same magnification. (From M. Kenney, in *ASM Handbook*, Vol. 15, Casting, D. M. Stefanescu, Ed., 1988, p. 327. Reprinted with permission of ASM International. All rights reserved. www.asminternational.org.)

Merton Corson Flemings (b. 1929) developed an interest in metallurgy as a sophomore at MIT. "It looked like a good hands-on field with a lot of chemistry and physics involved," he has commented in retrospect. The field of materials processing is better for this interest, not only because of his technical developments, but also due to the leadership and vision he has exhibited in strengthening this diverse field. Flemings received his S.B. degree from MIT in the Department of Metallurgy in 1951, followed by his S.M. and Sc.D. degrees in metallurgy in 1952 and 1954, respectively. Following a two-year stint as a metallurgist at Abex Corporation,

Flemings returned to MIT as an assistant professor in 1956, becoming a full professor in 1969. Flemings established an interdisciplinary Materials Processing Center at MIT in 1979 and was its director until 1982. He served as head of the Department of Materials Science and Engineering from 1982 to 1995. He is currently professor of materials processing and director of the Lemelson–MIT Program, an MIT program that aims to honor inventors and to inspire inventiveness in young people.

Professor Flemings's research and teaching concentrate on engineering fundamentals of materials processing, and on innovation of materials processing operations, with a focus on solidification and casting. He has been active nationally and internationally in strengthening the field of materials science and engineering and in delineation of new directions for the field. He is a member of the National Academy of Engineering and of the American Academy of Arts and Sciences. He is author or coauthor of over 300 papers, holds 30 U.S. patents, and has published two books in the fields of solidification science and engineering, foundry technology, and materials processing.

Among Flemings's most notable inventions are a process to use magnetic fields to improve the quality of silicon single crystals and of steel continuous castings, semisolid metal casting used to produce high-quality lightweight aluminum components for cars, and a production-friendly lost foam casting method.

Source: Department of Materials Science and Engineering, Massachusetts Institute of Technology, Cambridge, Massachusetts.

is cooled while stirring, until a slurry of the appropriate solid fraction is formed. These paths are illustrated in Figure 7.11.

The traditional method is thixocasting, where a special billet material with the desired fine-grained, equiaxed, or globular microstructure is reheated to a temperature at which some of the solid turns to liquid. The now semisolid billet, which still maintains its structural integrity and handleability, is transferred to a die-casting or injection machine and the part is formed. This process can be highly automated and does not require dealing with molten metal. On the negative side, thixocasting requires a high capital investment, and perhaps more limiting, the feedstock billet costs are high, feedstock compositions limited, and scrap produced during the casting operation cannot be recycled in-house. This makes the thixocasting process more expensive than other casting operations and has hampered its widespread adoption.

In rheocasting, the alloy is stirred during solidification to produce a semisolid slurry, which is then injected directly into the die. This approach reduces the process complexity, allows standard casting ingots to be used as raw materials, and allows for the in-house recycling of scrap, all of which lead to a more

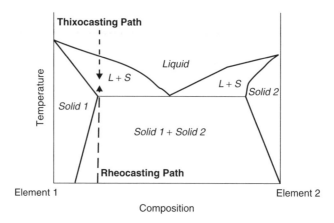

Figure 7.11   Paths to a semisolid state taken by rheocasting (heating a specially formed billet until partial melting occurs) and thixocasting (stirring a melt to break up dendrites while cooling into the liquid + solid, or mushy, region).

cost-effective process. Additionally, the fraction solid can be tailored to the application, which is more difficult to do in thixocasting, where there is a narrow temperature/fraction solid range available for working or handling the semisolid billet. Much effort has gone into developing a reliable, controlled system to produce a semisolid slurry. A process developed at MIT (Flemings et al., 2003) has recently been licensed and put into production. This process is expected to enable die casters to capture significant shares of aluminum casting growth for automotive markets because of the ability to produce thinner, higher-quality casting with shorter cycle times.

7.2.2. Rolling

The process of plastically deforming a material by passing it between rolls, known as *rolling*, has been practiced for hundreds of years and today is one of the most widely used metalworking practices, since it lends itself to high-volume production and forms the basis for many subsequent metalworking or forming operations. Rolling typically begins with ingots, which have been preheated to make them easier to shape. The ingot is first fed into a breakdown mill, where it is rolled back and forth, reversing between the rolls until the thickness has been reduced to tens of centimeters. At this point the material is referred to as *plate* and may be used in this condition (after heat treatment to develop properties, especially for precipitation-hardening material such as some aluminum alloys) or further reduced in thickness. Plate that is slated to become sheet or foil is trimmed after leaving the breakdown mill and sent through a continuous mill to reduce thickness further and is then coiled. To continue its reducing process, the coiled sheet is annealed and then cold rolled.

The rolling operation (Figure 7.12) is one of *plane-strain*: that is, that material deforms in the x and z directions, but does not deform appreciably in the y or

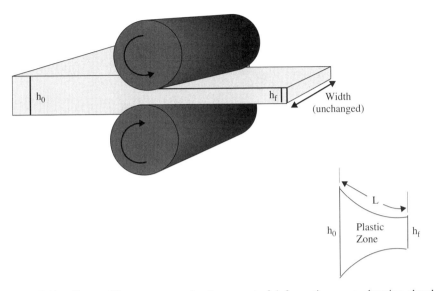

Figure 7.12   Sheet rolling process, and enlargement of deformation zone, showing chord length contact, L.

width direction. Lateral deformation is constrained by the undeformed material just entering the rolls and the deformed material exiting the rolls. As might be expected, the pressure required to induce rolling deformation can be quite large. Hosford and Caddell (1993) have presented a derivation of the average compressive stress, P_a, on the rolls: approximately

$$P_a = \frac{h}{\mu L} \exp \left(\frac{\mu L}{h} - 1 \right) (\sigma_0 - \sigma_t) \tag{7.3}$$

where h is the average thickness, μ the coefficient of friction, L the chord of the arc of contact between the rolls and the workpiece, σ_0 the yield strength of the material, and σ_t tension applied to the sheet during rolling. L may be approximated as the product of the roll diameter and the change in height of the workpiece upon passing through the rolls, Δh. Pressure required for rolling increases dramatically as h becomes small. One consequence of this high pressure is that the high forces needed to maintain these pressures cause elastic flattening of the rolls, much as a tire flattens under the weight of an automobile. The effect of this roll flattening is to increase the roll separating force such that even more compressive pressure needs to be applied to force deformation of the sheet. The increased compressive pressure induces even more roll flattening, which requires more pressure to counteract. Roll flattening can become so severe that there becomes some minimum thickness beyond which it is not possible to further reduce the thickness of a sheet or foil of material.

The minimum thickness, h_{min}, that can be rolled due to roll-flattening effects has been found by various investigators to be of the form

$$h_{min} = \frac{C\mu R}{E'}(\sigma_0 - \sigma_t) \tag{7.4}$$

where C is a constant with a value between 7 and 8, R the roll radius, and E the plane strain modulus $= [E/(1 - v^2)$, where v is Poisson's ratio]. Inspection of this equation reveals several methods that may be employed to reduce the thickness of rolled sheet. The friction coefficient may be lowered by adding lubrication, rolls may be made stiffer (higher E'), the starting stock may be annealed to lower σ_0, forward or backward tension of the rolled material may be used or increased, or the roll diameter may be decreased. Because smaller-diameter rolls are more prone to bending or flexing rather than flattening (as discussed below), backup rolls are used to counteract this tendency. This arrangement, named for its inventor, Tadeusz Sendzimir (1894–1989) (Sendzimir, 1939) and shown in Figure 7.13, is characterized by small-diameter working rolls, each backed by a pair of supporting rolls, and each pair of rolls supported by a cluster of three rolls. Sendzimir rolls are used for cold-rolling wide sheets of metal to close tolerances and low thicknesses. An additional method for reducing the amount of elastic flattening deformation when rolling thin sheet has been to sandwich the material being rolled, thus artificially increasing the thickness of the rolled product. Common aluminum foil is one example of this approach. This foil is shiny on only one side, because as it passes through the final rolling mill, two thicknesses of foil are rolled together. The sides facing each other emerge with a dull finish, while the sides in contact with the rolls are burnished to a shinier finish.

To counter the effects of roll bending or flexing, which would tend to produce a sheet of varying thickness across the rolled width, rolls are usually cambered. The degree of camber required depend upon the width of the material, the reduction per pass, and the yield strength of the material being rolled. This degree of camber in the rolls is chosen so that under load, the gap between rolls is uniform across the width, giving a flat sheet of rolled material. If the rolls are not cambered correctly and a sheet of nonuniform thickness results, those thickness nonuniformities will induce residual stresses at the edges and centerline of the rolled sheet, since conservation of material requires that thickness variations produce variations in elongations in the rolling direction. These residual stresses can cause defects ranging from edge cracking, splitting, wrinkling, or warping of the rolled sheet.

Development of texture during rolling is an area that has seen a great deal of work. In some cases the materials scientist may wish to develop a strong texture during rolling to take advantage of property anisoptropies in order to optimize the material for a given application (i.e., silicon steels for electrical applications). In other cases, such as for deep drawing (presented in the Section 7.2.3) a weak or random texture is desired. Rolling textures are defined not only by a specific

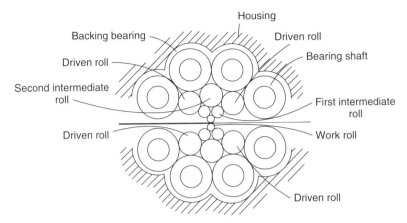

Figure 7.13 Sendzimir rolling mill. Each small-diameter work roll is backed by two sets of intermediate rolls (with two and three rolls, respectively) and four backing bearings. (From *ASM Handbook*, Vol. 14A, Bulk Metal Forming, S. L. Semiatin, Ed., 2005, p. 470. Reprinted with permission of ASM International. All rights reserved. www.asminternational.org.)

crystallographic orientation parallel to the direction of rolling, but also a preferred plane in the rolling plane. Texture tends to become more pronounced as the degree of deformation increases, and sometimes more than one texture can coexist in a rolled component.

Face-centered-cubic metals exhibit a predominant rolling texture of {110}<112>, termed a *brass texture*. In this texture, the {110} planes are in the rolling plane, and the <112> direction is parallel to the rolling direction. Some metals, such as silver, develop this simple texture easily, whereas others, such as copper, exhibit varying amounts of secondary texture. The copper texture, for instance, is irrational at low deformations, approximately {146}<211>, approaching {112}<111> with heavy deformation. Measurements indicate that silver has a substantially lower stacking fault energy than that of copper. Smallman (1955) performed a series of systematic experiments, showing that through the addition of solid solution alloying elements, which lower the stacking fault energy in copper, the {110}<112> "brass" texture will predominate over the copper texture. Elevations in temperature, on the other hand, tend to stabilize the copper rolling texture. In BCC metals the predominant texture is {100}<110> with the cube plane in the rolling plane. Other texture elements, such as {112}<110> and {111}<112>, can also form. In hexagonal close-packed structures, the rolling texture tends to take the form {0001} < 1120 >.

Cross rolling, a technique in which a material is rolled in two directions at right angles to one another in successive rolling passes, may be used for some requirements. Cross rolling tends to intensify BCC textures, whereas the opposite occurs in FCC metals (Honeycombe, 1984).

Photo Copyright © The Sendzimir Family. Reproduced with permission from Jan Sendzimir.

Tadeusz Sendzimir (1894–1989) was born in the city of Lwów in the eastern section of Poland, which is now part of Ukraine. As a child, he was fascinated by mechanical things: collecting machine parts catalogs and hanging around machine shops after school, learning to use the lathes, presses, and grinders. A couple of semesters short of a degree in mechanical engineering from the Lwów Polytechnic Institute when World War I broke out, Sendzimir left school and got a job as a manager in an automobile repair shop in Lwów. In 1915, to avoid the draft, he fled to Kiev, Ukraine, and then in 1918 to Shanghai, when the Russian Revolution threw Kiev into anarchy. There, at the age of 24, Sendzimir opened China's first mechanized nail and screw factory. Unable to obtain machines from Europe due to the war, Sendzimir innovated, improvised, and jury-rigged the necessary equipment. As his business grew, he diversified into products that required galvanizing for corrosion protection. Deciding that there must be a better way, Sendzimir set out to improve the galvanizing process then used. Arriving in New York in 1929 (three months after the stock market crash), he sought investors and spent much of the following year doing research. Inspired by Irving Langmuir's work on the interaction of gases with metal surfaces, Sendzimir realized that a hydrogen atmosphere was key to an improved galvanizing process.

Returning to Poland in 1931, Sendzimir implemented a galvanizing line that was cleaner, cheaper, and produced higher quality material than the conventional galvanizing methods. Unable to purchase the fine-gauged sheet metal in coil form required for his continuous galvanizing process, he fell back on his experience in Shanghai and invented a rolling mill. Utilizing a cascade of rolls supporting a very fine pair of work rollers, the Sendzimir mill produced uniform, thin-gauged sheet metal of unparalleled quality.

Armco Steel in Middletown, Ohio, bought the rights to Sendzimir's galvanizing process in 1935, agreeing to help him in further development of his cold-rolling mill. Armco perfected the galvanizing method, the core of which is still in use today. Sendzimir was still in the United States working on his mill in September 1939 when Germany invaded his native Poland, sparking World War II. He settled in the United States, supporting the war effort by designing a process to produce lightweight silicon steel for airborne radar transformers. He married in 1945 and became a U.S. citizen in 1946. He founded steel mills in Pennsylvania and Ohio. His rolling mills are currently in use all over the world, and more than 90% of the world's stainless steel is rolled on his mills. He received every top international award in steel, including the engineering equivalent of the Nobel Prize: the Brinell Gold Medal, from the Royal Academy of Technical Sciences in Stockholm. Overall, Sendzimir helped to completely alter the processes of fabricating steel in the twentieth century. He held over 120 patents, 73 of which were awarded to him in the United States.

Although he chose to become an American, Sendzimir maintained his identity as a Pole. His lifetime contacts and relationship with Poland earned him many honors from that country. In 1990, the largest steel mill in Krakow, known as the "Lenin" mill, was renamed "Sendzimir." Sendzimir died after a massive stroke in 1989 and was buried by his family in a galvanized coffin. After his death, following his wishes, Sendzimir's widow, Berthe, set up the Sendzimir Foundation, dedicated to helping people in Poland and other East European countries to find ways to clean the environment of industrial pollution and to find ways to live in a sustainable manner.

Source: Vanda Sendzimir, *Steel Will: The Life of Tad Sendzimir*, Hippocrene Books, New York, 1994.

7.2.3. Drawing

Drawing operations differ from most bulk-metal working operations in several respects. First, in drawing, a workpiece is under tension, whereas most other metalworking processes are predominately compressive. Second, one or both surfaces of the deforming regions may be free or unsupported by tooling. Two main drawing processes predominate: deep drawing, used to form cuplike structures, and wire drawing.

7.2.3.1. Deep Drawing   *Deep drawing* is a process by which sheet metal is formed into cup-shaped parts. A blank of diameter D is placed over a shaped die of diameter d, and pressed into the die with a punch. (Note that although diameters are being used to reflect the majority of drawn parts, other geometries, such as square cross sections, can also be drawn.) The term *drawing* or *deep drawing* implies that some drawing-in of the flange metal occurs and that the formed parts are deeper than could be obtained by stretching the material over a die. The cup

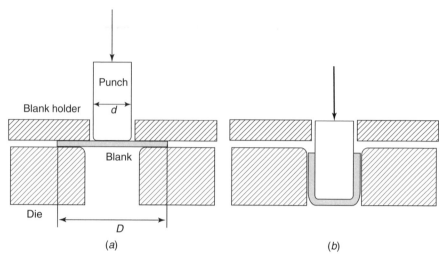

Figure 7.14 Deep-drawing process. A blank of diameter D is placed over a die of diameter d and a punch is lowered into the die, forming a cup-shaped drawn part.

shape may be fully drawn into the die cavity, creating a straight-walled part, or a flange may be retained that is either functional or trimmed in further processing. This process is used to produce pressure and vacuum vessels (advantageous due to there being no welded seam that might leak), metal flashlight and cartridge cases, soda and beer cans, and even most kitchen sinks.

The deep-drawing operation is shown schematically in Figure 7.14. Most of the deformation involved in this operation occurs in the flange material as it is both drawn in (think of an annular element of radius D transforming to an annular element of radius $D - \Delta D$ as the material is pulled in) and then flows over the die radius into the cup. The drawability of a material thus depends on two factors: the ability of the material in the flange to flow easily in the plane of the original blank, and the ability of sidewall material to support the force required to cause the flange deformation. If the original blank is too large, the force required to deform it will cause yielding and failure in the sidewall. This drawability is often expressed as a limiting draw ratio (LDR), which is defined as

$$\text{LDR} = \frac{D_{\text{max}}}{d} \tag{7.5}$$

where D_{max} is the largest diameter blank that can successfully be drawn to the diameter of the punch, d. If it is desired to draw a deeper cup than the LDR would allow for a single operation, one or more redrawing operations can be utilized. In general, LDRs are less than 2. Note that in a drawn part, the material under the punch, which forms the bottom of the cup, is not worked significantly and remains essentially unchanged in thickness. This is readily seen in beverage cans, where the drawn sides are much thinner than the can bottom.

Figure 7.15   Earing in a drawn cup. (Courtesy of VAW Aluminum AG, Bonn, Germany.)

The presence of a strong texture, or anisotropy, in the starting blank leads to a phenomenon known as *earing*, shown in Figure 7.15. In practice, enough extra metal is drawn to allow for ears to be trimmed, however, minimizing the degree of texture present is desirable in order to increase material utilization and reduce scrap.

7.2.3.2. Wire Drawing   *Wire drawing* is a process in which a wire rod is pulled or drawn through a single die, or a series of dies, thereby reducing its diameter, as shown schematically in Figure 7.16. Because the volume of the wire remains the same, the length of the wire changes with its new diameter. The degree of reduction possible for drawing through a single die is governed by the tensile strength of the material. The stress on the drawn wire,

$$\sigma_d = \frac{F}{A_1} \tag{7.6}$$

cannot exceed the tensile strength of the drawn wire, or the wire will fail by necking. When very large reductions are to be taken, it may be necessary to anneal the wire between passes to overcome the effects of cold working. Alternatively, extrusion may be employed for large reductions. Extrusion has similar geometries, but since the force is applied to the billet (the larger A_0), there is no significant load on the extruded material.

Wire-drawn FCC metals tend to exhibit a <111> or <100> fiber texture. English and Chin (1965) have shown that the ratio of the <111> and <100> fiber texture components varies depending on the stacking fault energy. Low-stacking-fault-energy metals such as silver have stronger <100> components, whereas high- and intermediate-stacking-fault-energy metals such as aluminum or copper exhibit a stronger <111> texture. Optimizing texture in extremely fine wires can be an important criterion for reliability in their use in electronics packaging.

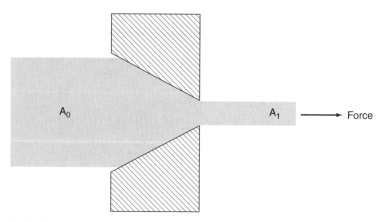

Figure 7.16   Wire drawing operation. Tension in the drawn wire causes material to flow out of the die, forming new wire.

7.2.4. Extrusion

Extrusion is a process through which a block or billet of metal is reduced in cross section by forcing it to flow through a die orifice under high pressure. Varying the die profile shape will allow production of round, rectangular, L-, T-, or I-shapes, finned shapes for heat exchangers, and many types of tubes. Hollow sections are usually extruded by placing a pin or piercing mandrel inside the die. In some cases positive pressure is applied to the internal cavities through the pin. Extrusion may be continuous (producing indefinitely long material) or semicontinuous (producing many short pieces). Extrusion may be done at room temperature (cold) or at elevated temperatures, up to a homologous temperature of 0.5 to 0.75.

Extrusion can fall into one of two categories: direct extrusion and indirect extrusion (Figure 7.17). In *direct extrusion* a metal billet is placed into a container and driven though a die by a ram, usually attached to a hydraulic press. In *indirect extrusion* the billet is again placed in a container, but the container end is closed. A hollow ram, carrying a die of the desired shape, is forced into the billet, with the extrusion exiting through the ram. Indirect extrusion may occur as a result of die movement or container movement. Because there is no movement between the container wall and billet surface, frictional forces are lower in this method, which reduces the force needed to extrude. There are, however, some limitations on the load that can be applied by the hollow ram.

The principal variables that govern the amount of force required for extrusion are the type of extrusion (direct or indirect), the ratio of input to output areas, the working temperature, the speed of deformation, and the friction between the billet and the die and container walls. At elevated temperatures the material is easier to deform, however, the surface of the billet will tend to oxidize more. This oxidized surface may be carried along shear bands to the interior of the extrusions, resulting in internal oxide stringers. Lubricant potentially becomes

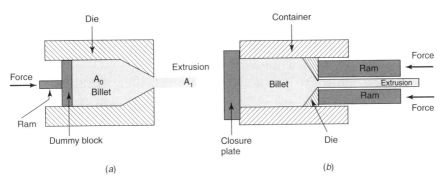

Figure 7.17 (*a*) Direct extrusion process where billet moves against a stationary die by a moving ram; (*b*) indirect extrusion, where the billet is fixed in a container and the die is at the end of a hollow, moving ram.

entrained in the same manner. Surface cracking, ranging in severity from a badly roughened surface to repetitive transverse *fir-tree cracking*, can be produced by longitudinal tensile stresses generated as the extrusion passes through the die (Dieter, 1976). Center burst or *chevron cracking* along the centerline can occur at low extrusion ratios, where the slip field lines do not extend to the center of the billet. This center area can then be under tension and rupture, a fracture that may not be visible from the surface of the extrusion.

7.2.5. Forging

Forging is the process of working a metal into a useful shape by hammering or pressing. It is perhaps the oldest of the metalworking techniques, having been practiced in blacksmith shops for over 2000 years. Today, hydraulic and mechanical presses have replaced the arm and hammer of the blacksmith. However, many of the principles of forging have remained unchanged.

The metal can be forged hot (above its recrystallization temperatures) or cold. Forgings can vary a great deal in size and can be produced from steel (automobile axles), brass (water valves), tungsten (rocket nozzles), aluminum (aircraft structural members), or any other metal. Forgability can vary with material, but generally, scales with the volume/surface area ratio. High-surface-area parts, such as strip and sheet are not good candidates for forging, due to the large pressures needed to move material.

There are two broad classifications of forging: open and closed die forging. *Open die forging* is conducted between two flat or very simply shaped surfaces. Examples of such surfaces are the platens in a press or an anvil and hammer. This method is used when the shapes to be forged are relative simple. A compression test in which a cylinder is put under uniaxial compression is one example of open die forging. Upset forging, which also increases the cross section by decreasing length, is another open die process used to put the heads on bolts, fasteneners, valves, and similar parts.

In contrast, *closed die forging* compresses the material between two die halves which contain the impression of the desired shape. Often, a series of dies that gradually transform a workpiece from a round or square billet to the desired shape is employed. As with patterns for castings, a forging die must be designed with a parting line as well as draft and generous radii to enable the forged piece to be removed. It is important to use enough metal in the forging blank so that the final die is completely filled and all the features desired can be formed. Because it is difficult to get just the right amount of metal, dies are designed for some overflow, called *flash*. When the dies come together for the finishing forging step, a small amount of flash is exuded from the die. This material is then usually removed with a trimming die.

If the deformation during forging is limited to the surface of the part, little effect will be observed on the grain structure in the bulk of the workpiece. Where complete penetration of deformation zones is present, forging refines the grain structure and improves the physical properties of the metal. With proper design, the grain flow can be oriented in the direction of principal stresses encountered in actual use. Mechanical properties such as strength, ductility, and toughness are better along the flow lines in a forging than would be present in the flow lines in identical material with a random texture. (Note, however, that there can be a property reduction in the direction normal to the flow lines.) Some common forging defects are surface cracking due to excessive working of the surface at too low a temperature; *cold shuts*, which are a discontinuity produced when deformed metal folds back onto itself without welding completely; and underfilling, resulting from an accumulation of lubricant or scale in the forging die, preventing metal from filling out a desired feature.

Forging can play an important role in developing a desired texture. In most traditional forging applications where forging is used to generate a specific component shape, the grain structure, with the grains being elongated and flowing around the contours of the part, impart additional strength to the forged component. This refined grain structure allows forgings to exhibit higher strength/weight ratios than is exhibited by a cast or machined part made from the same alloy, making them attractive for automotive or aerospace applications. This is also sometimes used as a marketing pitch for components such as hand tools, which may carry the marking indicating that the part was forged, and thus stronger than a cast part of similar dimensions.

In some other applications, such as press forging or working the blank between two platens (no die cavity), the purpose of the forging operation may be to develop a particular deformation texture. The effect of crystallographic orientation of a sputtering target on sputtering deposition rate and film uniformity has been described by C. E. Wickersham, Jr. (1987). He found that improvements in film uniformity may be achieved on a silicon wafer by controlling the deformation process for manufacturing a target. Forging has been added to the process route for a number of sputtering target fabrications in order to develop a desired texture. As one example, Zhang (2001) describes a method whereby a tantalum billet is side-forged to turn a <200> texture developed during casting into the desired

<222> texture for sputtering. Similarly, forging has been a critical element in developing texture for cast rare earth magnets (Rivoirard et al., 2005).

7.2.6. Swaging

Swaging, a special category of forging, refers to the tapering of a bar, wire, rod, or tubing by hammering or squeezing. It can be used to reduce or increase the diameter of tubes and/or rods by progressively tapering lengthwise until the entire section attains the desired dimension of the taper. Swaging can be used to compound tapers, to produce contours on the inside diameter of tubing, and to change round cross sections to square or other shapes. Swaging is done by placing the tube or rod inside a die that applies compressive force by hammering radially. This can be expanded further by placing a mandrel inside the tube and applying radial compressive forces on the outer diameter. Swaging can change the shape of an object, or allow an inner diameter to be a different shape (e.g., a hexagon) while the outer is still circular. Low strength and high ductility are required for good swagability.

7.2.7. Stamping, Blanking, and Coining

Stamping, blanking, and coining are all operations that are typically done on sheet material at room temperature. *Stamping* is a process used to form sheet metal into a desired shape or profile. Stamping is a room-temperature operation, however, because heat is generated from friction during the cutting and forming process, stamped parts often exit the dies hot. Stampings can range in size from parts that can fit in your palm, used to make microelectronics, to those that have an area of several square meters, used to make entire automobile body sides.

In most stamping operations, *bending* is the key component of deformation. After forming, some elastic springback may occur, and considerable residual stress often remains in the formed part. Designing a tool that allows for a given degree of springback in order to produce a desired shape would be a very difficult task. Tooling is therefore usually designed to place a sheet in tension while forming. The added tensile forces are enough so that the entire sheet yields in tension and springback is minimized.

Blanking is cutting, punching, or shearing a shape out of stock. In sheet metal form, blanking is typically carried out at room temperature. Usually, a blank serves as an input workpiece to some subsequent forming operation.

Coining is a closed die forging operation, usually done at room temperature, in which all surfaces of the workpiece are restrained within the die. Full contact with die surface results in a well-defined imprint of the die relief features on the workpiece. Since only a small amount of metal redistribution can take place within a single coining die, it is usually preceded by another operation, such as forging or blanking. Alternatively, a progression of coining die can be used to develop the metal profile required. During coining, the metal blank is placed within the die, which is then loaded well above the blank material's compressive yield stress. This load is often held for some dwell time in order to develop the fine features characteristic of coining. This replication of surface detail is not typically available in other metal-forming operations. For this reason, the coining process is used for coin minting, patterned flatware, medallions, metal buttons, and many small fasteners and automotive components.

7.2.8. Equal-Channel Angular Extrusion

While the primary goal of deformation processing is to change a material's dimensions in order to produce a useful shape, plastic deformation is also an effective method to refine the structure of a material and optimize texture and/or properties. There are practical limitations on the amount of deformation that can be induced in a metal in traditional rolling, forging, or drawing operations, though, since these processes inherently produce a change in cross-section thickness. To produce a shape of a specific thickness, plastic deformation is limited by either the thickness of the starting material or the capacity of the press or rolling mill to deform that material. An emerging technique in metal processing that allows for extreme plastic deformation is equal-channel angular extrusion (ECAE), which introduces uniform plastic deformation of a material through simple shear. This shearing operation does not cause significant change in geometric shape or cross section, and thus multiple extrusions can be performed on the same billet to produce severe plastic deformation.

Equal-channel angular extrusion was invented in the former Soviet Union by Vladimir Segal in 1977, for which he obtained an invention certificate of the USSR, similar to a patent. Due to restrictions on publishing, however, this technology was unknown to the West until Segal immigrated to the United States in the 1990s (Segal, 1995). In ECAE, a metal billet is extruded through a die which has two intersecting channels of equal cross-sectional area (Figure 7.18). Plastic deformation is realized uniformly by simple shear, layer after layer, in the thin zone at the crossing plane of the channels. This shear occurs across the entire billet, except for small sections at the end. The most important parameters of ECAE are contact friction and die design (Segal, 2005). Materials behavior plays a secondary role, and during severe plastic deformation the material can be considered an ideal plastic body. To attain very large strains, the extrusion should be repeated a number of times. An accumulated effective strain, ε, after N passes is given by

$$\varepsilon = 2N\frac{\cot\theta}{\sqrt{3}} \tag{7.7}$$

To understand the capability of this technique to produce deformed structures, it is useful to compare the strain induced by ECAE with equivalent strains produced during conventional metal-forming processes. A reduction in area during an ideal extrusion, R_E, produces a strain ε_{RE} that is equal to the ratio of the areas, A_0/A_f. Table 7.5 compares the strain for N ECAE passes with the equivalent reduction in area needed to produce the same strain in an ideal extrusion. After eight

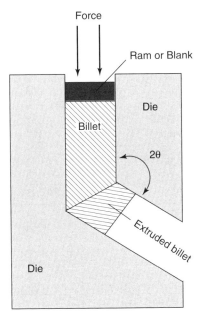

Figure 7.18 Equal-channel angular extrusion process. A metal billet is forced through an extrusion die with equal dimensioned channels separated by an angle 2θ. At the intersection of the two channels, the material undergoes intensive deformation by shear.

passes in ECAE, the billet retains its original shape, yet contains deformation equivalent to having a reduction in area of over 10,000-fold! This powerful ability to refine microstructures has garnered ECAE considerable attention in the materials community since it has been revealed to the West. Since Segal disclosed this technology, almost 2000 papers have been published on this topic by other researchers.

If billets are shaped correctly, the orientation of the billet may be changed between passes without a change in the die or tooling set. Appropriate shapes are elongated billets with square cross-section, or square, flat plates. By changing the orientation of the billet between successive extrusions, complex microstructures and textures can be developed. While numerous possibilities exist for combinations of passes, all combinations may be represented by four independent routes developed by Segal (1996; 1998). These routes have been labeled as follows:

- *Route A*. The billet orientation is unchanged in successive passes.
- *Route B*. The billet is rotated between two orthogonal directions by changing the orientation plus 90°, then minus 90°.
- *Route C*. The billet is rotated 180° between passes.
- *Route D*. The billet is turned 90° in the same direction after each pass (always + 90).

TABLE 7.5   Equivalent Strains for ECAE and Ideal Extrusion^a

Pass	ε	Equivalent Reduction in Area for Ideal Extrusion, R_E
1	1.15	3
2	2.30	10
3	3.45	32
4	4.60	101
5	5.75	321
6	6.90	1,022
7	8.05	3,329
8	9.20	10,198

Source: After Segal (2005).

$^a N$ passes, $2\theta = 90°$.

Vladimir M. Segal (b. 1936) was born in the USSR and in 1954, entered the Polytechnic University in Minsk, Belarus, graduating in 1959 with the equivalent of a master's degree in materials processing. Wanting to learn about industry from a practical perspective, Segal took a job as a production/development engineer at the Minsk Tractor Works, a sprawling facility that employed over 25,000 people.

In 1962 he went back to school, attending the Physics–Technical Research Institute of the Byelorussian Academy of Sciences in Minsk. Segal earned the Candidate of Science (equivalent to Ph.D) degree in 1965 and took a full-time research position at the same institute, where he lead a group dedicated to mathematical modeling of metal forming. He published seven books (in Russian), more than 120 papers, and received 50 invention certificates of the USSR in the area of applied plasticity and materials processing. Segal pioneered research on severe plastic deformation (SPD) and invented (1972) the most promising SPD technique, equal-channel angular extrusion (ECAE). Substantial work on development of the SPD technology was done from its inception in 1970 through the late 1980s. In 1974 he defended a science doctorate dissertation, earning his Sc.D., the highest scientific degree attainable in the Soviet Union. By 1986, ECAE was a production-ready technology, but changes and political uncertainties in the Soviet Union prevented adoption of this technology, and work on ECAE was suspended.

When Mikhail Gorbachev's policy of Perestroika allowed emigration from the Soviet Union, Vladimir Segal left the country in 1989 for the United States and sought opportunities to apply his innovations to industrial problems. With ECAE being unknown in the United States or other Western countries because of publishing restrictions in the former Soviet Union, these opportunities were not readily available. Segal started his first job in the United States as a die designer for Interstate Forging Industries, Inc. Still determined to pursue applications of ECAE, he demonstrated the new technology at Texas A&M University. This led to Segal working at Texas A&M as a contract employee, where he was able to continue his research and publish and present information on ECAE more effectively. One presentation caught the attention of Johnson Matthey Electronics (now Honeywell Electronic Materials) for ECAE's ability to produce unique deformation structures in sputtering targets. In 1996, Johnson Matthey licensed his patents and Segal went to work for the company, leading their development and scale-up of ECAE to produce ultrafine-grain sputtering targets, presently the only SPD product available on the market. The first ECAE aluminum target, used for a 300-mm wafer fab line, was sold in 2000. Going into volume production in 2002, this product line was awarded the 2004 Semiconductor International Editor's Best Choice Award.

Segal is currently chief technical officer of Engineered Performance Materials, whose mission is to lead in the development, production, and application of submicrometer and nanomaterials. In addition to being the inventor of ECAE, he has also invented a shear-extrusion process and many other innovative materials processing techniques protected by more than 50 invention certificates of the USSR and a number of U.S. patents.

Source: Personal interview, with V. Segal by N. Dean.

Routes A and C produce a deformation pattern that is two-dimensional, as the shear plane does not change. Routes B and D do, however, change the shear plane. In routes A and B the distortions increase monotonically in the flow direction, as the shear continues in the same direction. For routes C and D, the material elements are periodically sheared back to their original shape. Note that this refers to volume elements within a billet, not the billet itself. Each processing route activates a particular system of shear planes and directions, thus allowing numerous textures to be developed during ECAE with the appropriate choice of number of passes and route(s). However, there are restrictions on attainable texture strength. Flow localization within shear bands induces rotation of structural fragments, leading to a randomization of texture when more than four passes are made, regardless of the processing route taken (Segal, 2005). The force required to extrude a properly lubricated ECAE billet is not significantly greater than the material's shear yield stress and can be significantly lower than the force required for a conventional extrusion with equivalent deformation.

When performed at temperatures below the temperature for static recrystallization, multipass ECAE yields submicrometer grain sizes and sometimes nanometer grain sizes in the resulting material (Valieb et al., 2000). The extruded grain size is unaffected by the initial grain size. Under carefully controlled conditions, Segal has noted that using route D, the minimum number of passes required for grain structure refinement is four; however, due to end effects, six to eight passes may be required to produce a homogeneous billet (Segal, 2005). ECAE may also be used to disperse and break up second phases present in the microstructure. This technique has been shown to be effective in fabricating a large cross section of in situ composites by deforming equiaxed second-phase particles into filaments with the appropriate selection of extrusion route (Segal et al., 1997). The simple shear characteristic of ECAE has been shown to be the most effective deformation mode for healing of pores, voids, and cavities. For hot processing, one-pass ECAE is sufficient for elimination of large macro defects, and a few passes will eliminate micro defects (Segal, 1984). ECAE will increase hardness, ultimate tensile strength, yield strength, and toughness, with a stable increase in these properties being realized after two to four passes. Route A has been shown to be more effective than route C at strengthening. Elongation drops after ECAE, as it does with many deformation processes; however, the drop is not as severe, as other deformation methods develop and the degree of elongation stabilizes after two passes (Segal, 1995).

Although most of the work on ECAE to date has focused on research, the technology has been commercialized for the production of a series of sputtering targets, made from aluminum and copper alloys (Ferrasse et al., 2003). These sputtering targets, which are over 300 mm in diameter, present properties superior to those of standard targets: a reduced and uniform grain size 100- to 1000-fold reduction in precipitate size, a 500% increase in yield strength, and optimum texture. Benefits in use include over a 40% increase in target lifetime, a 23% increase in number of processed wafers, a low level of particles, and an absence of arcing problems during sputtering, which leads to an increased yield in the processed wafers.

With most metalworking techniques having their roots in processes developed hundreds if not thousands of years ago, it is unusual for a truly original bulk metal-forming technique to be developed. ECAE is just such a technique, and it has the potential for improving material properties across a number of applications. ECAE is a promising technique for the breakdown of cast ingot structure with a low number of passes. It was also found to be an excellent powder consolidation technique. Use of multipass ECAE to produce bulk submicrometer or even nanometer grain-structured material is an exciting concept in materials science that will continue to see research and use of this technology.

7.3. CONSOLIDATION METHODS

The dominant method of forming ceramics, as well as some refractory metals and intermetallic compounds that do not exhibit the ductility required for deformation processing, is through consolidation of fine particles or particulates, often using techniques amenable to automation in order to reduce manufacturing costs. Many of these methods may introduce a carrier fluid, or some sort of binder (usually, a polymer that generates little or no ash when burned out or removed) to ease processing of the particles. The resulting bodies of consolidated particles are then usually dried to drive off the fluid and/or burn off the binder and then sintered to join the particles together and reduce or remove porosity. In these methods, slip, or deformation, of the individual particles typically does not occur. Therefore, deformation textures do not result. When particles are asymmetrically shaped, the various methods may produce preferred orientations of the particles, which then may determine crystallographic texture.

Common consolidation methods include variations on pressing powders, plastic forming, slip casting and tape casting. The choice of which method to use for a given application is very dependent upon the shape, size and required dimensional tolerances of the final product, available capital outlay and desired productivity. Each of these methods is examined in turn below, followed by sintering, which turns the consolidated powders into a monolithic body with mechanical integrity.

7.3.1. Mechanical Pressing of Powders

Powders of most materials can be consolidated by applying mechanical pressure, usually in a die, a method called *die pressing*. While binders and/or lubricants may be used to ensure that the powders are free flowing, pressing can be done with or without a binder in either a cold or a heated die. *Uniaxial pressing* is shown schematically in Figure 7.19. A mold, which is the negative of the part to be produced, is placed in a press. The die cavity is filled with powder and pressure is applied to the powders by a ram that has tight tolerances to the die walls. The compacted part is then ejected. Cycle times can range from seconds to hours, depending on the size and nature of the part.

An important issue in die pressing is the relative motion of the ram and the die. As the ram moves, or is engaged, the resulting pressure applied to the powders forces movement of the powders as they slide locally and then deform. Since the powders nearest the moving ram must travel, or be pushed, the greatest

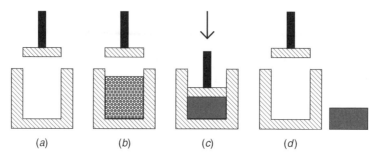

Figure 7.19 Uniaxial powder pressing process: (*a*) empty die and ram; (*b*) powders are loaded in die; (*c*) ram is lowered and pressure is applied, compacting the powders; (*d*) powder compact is ejected and die is ready for the next cycle.

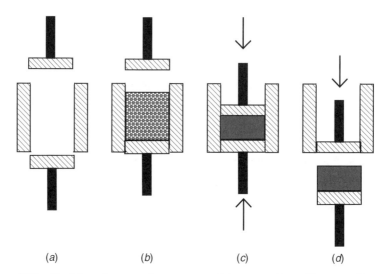

Figure 7.20 Double-action pressing process: (*a*) empty die with rams disengaged; (*b*) lower ram is raised to form the die bottom and powders are added to the die; (*c*) top ram is lowered and both top and bottom rams are engaged to apply pressure to the powders; (*d*) rams are lowered to allow powder compact to be ejected, preparing the dies for the next cycle.

distance, the pressure they experience is higher than that expercenced by powders farther away from the rams. It is not unexpected, then, that the pressure gradients that develop in the compacted powder result in particle density gradients as well. Particle density gradients can be minimized in uniaxial pressing by using opposing rams, where the bottom of the die is also a ram that moves. In this case, referred to as *double-action pressing* and shown in Figure 7.20, each ram provides equal compaction, which presses or moves particles at the top and bottom of the die, thus reducing density gradients in the pressed part.

Although it might be somewhat intuitive that the axial powder density decreases as you move farther from a moving ram, it is less clear that there also exist radial, or transverse, powder density gradients within a pressed part. The powder compacted nearest the moving ram generally reaches the highest density near the die wall and decreases to a constant density near the center axis of the part. As you look at areas away from the moving ram, variations in radial powder density lessen and then increase, often forming a mirror image of the density variations near the moving ram (i.e., higher near the center axis, lower near the periphery). Using *dual-action pressing*, from both top and bottom, will tend to reduce the density gradients, as the density profiles are mirrored from top to bottom. Examples of powder density gradients resulting from single- and dual-action pressing are shown in Figure 7.21. If not alleviated, the particle density gradients can produce porosity or uneven shrinkage in the resulting components upon sintering. Porosity is not desirable since it can reduce mechanical strength and impact heat flow by creating voids in the finished article. Uneven shrinkage upon sintering leads to shape distortions, shown in Figure 7.22.

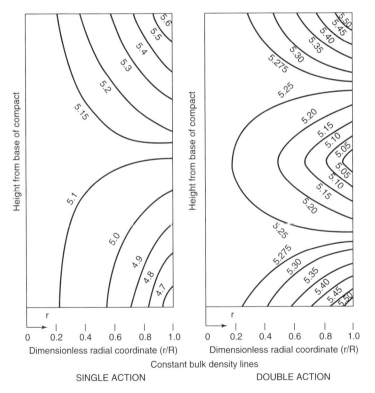

Figure 7.21 Density variations in single- and double-action pressed powder compacts. Double-action pressing gives a denser and more uniform powder compact. (From Thompson, 1981.)

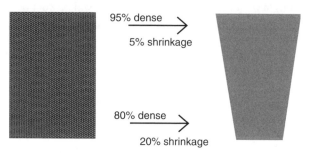

95% dense →
5% shrinkage

80% dense →
20% shrinkage

Figure 7.22 Nonuniformities in the powder compact density will lead to uneven shrinkage during sintering and distortion of pressed shape.

One might reason that if density gradients could be reduced by pressing from two directions, they might be reduced even more if you could press in a number of directions simultaneously. This can be accomplished through *isostatic pressing*, meaning, literally, equal pressure from all sides, done by encapsulating the part in a flexible container that is sealed (and often evacuated). The container is then placed in a vessel that is pressurized with a fluid (liquid or gas). The pressure is transmitted uniformly to the surface of the component. Isostatic pressing can be done at either ambient or elevated temperatures. Isostatic pressing is capable of producing parts with no gradients in particle density (Richerson, 1992). In addition to pressing powders, isostatic pressing, particularly hot isostatic pressing, referred to as HIPping, is used to densify components made by other methods, such as casting.

Variations in cross-sectional thickness of a part introduce added complexity to the die, since for maximum density uniformity each section must have its own ram (Figure 7.23). Although there are some restrictions on the complexity of parts that can be produced by die pressing, this method is suitable to automation and can be used to produce tens to possibly thousands of parts per minute (Richerson, 1992). As a set of dies usually requires a significant capital outlay, to be cost-effective, die pressing is typically limited to high-volume or high-value production parts.

The pressed part immediately after it is removed from the die is usually referred to as a *green compact* to denote that it has not yet been sintered. A green part typically does not have a lot of strength and can be damaged easily, much like an aspirin tablet or other medical pill, which are made via cold pressing of powders. Sintering of a green compact adds significant strength and structural integrity.

Powder pressing is a very flexible process in that materials that are difficult to melt or deform may be produced in a variety of shapes and sizes. Adding to the flexibility of the process may be the fact that the starting powder does not need to be uniform. Different size powders, or more commonly, different types of powders may be used to produce composite microstructures. These different powders may be distributed uniformly throughout the fabricated component, or isolated to form a functionally graded material. Powder pressing is commonly used to

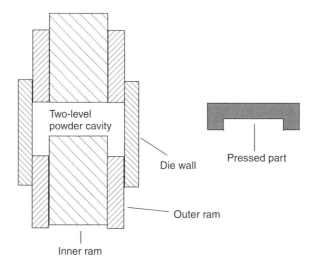

Figure 7.23 To achieve optimum density when pressing parts with variations in cross-sectional thickness requires that a die with multiple rams be utilized in order to deliver optimum pressure to each cross section.

produce ceramic and refractory metal sputtering targets, ferrite magnets, cutting tools, grinding wheels, spark plug cases (Schwartz, 1992), capacitors, dielectrics, tiles, bricks, plates, and crucibles, among other products (Richerson, 1992).

7.3.2. Slip Casting

With the exception of the potter's wheel, slip casting is probably the best known ceramics processing method. A *slip* is a suspension of fine ceramic and clay mixed with a liquid to give a creamy consistency. Slip casting has been in use for approximately 200 years (Norton, 1970) and is still practiced in a form close to the original method. The slip casting process is shown schematically in Figure 7.24. A hollow mold which has a cavity shaped like the desired part is formed from a porous material such as plaster. The slip is then poured into this mold. As the mold's pore absorbs liquid, some of the suspended particles are compacted on the mold walls by capillary forces. The thickness of the cast part increases as long as the mold continues to absorb liquid. After the desired solid layer thickness has been built up, the excess slip is poured out of the mold, leaving a hollow shell of approximately uniform thickness. The molds are then allowed to drain while the clay body dries out. The first part of drying usually takes place in the mold, where drying of the slipped part adds strength as well as causing dimensional shrinkage of the part, which promotes removal from the mold. Shrinkage that occurs during drying will break the body away from the mold so that the cast components can be removed without difficulty. The molds are then dried for reuse, while the green slip cast components are trimmed to remove any flash where the mold halves came together and are sintered to produce finished

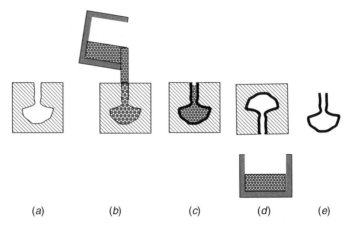

$$(a) \qquad (b) \qquad (c) \qquad (d) \qquad (e)$$

Figure 7.24 Slip-casting process: (*a*) a mold, which contains a cavity of the desired shape, is prepared; (*b*) the slip, a slurry or suspension of particles, is poured into the mold; (*c*) liquid is absorbed by the mold, compacting the slip's particles on the mold walls via capillary forces; (*d*) when a cast layer of desired thickness has been built up, the remaining slip is poured out of the mold; (*e*) after drying, the slip-cast part is removed from the mold, which is then dried for reuse.

parts. Water has historically been used as the primary liquid in a slip; however, in a manufacturing environment, stability of the slip is important and other material additions are made to the colloidal chemistry to improve slip stability. An excellent description of slip preparation has been given by Richerson (1992).

The slip casting process can be considered a type of filtration process, whereby the liquid filtrate is removed from the slip, leaving behind the filtrant, or ceramic particles, as the cast body. *D'Arcy's law* (D'Arcy, 1856), which was developed to quantify the flux of a filtrate (water) through a porous bed (sand), may be used to analyze the slip casting process:

$$Q = \frac{dV}{dt} = \frac{-\kappa A}{\mu} \frac{P_b - P_a}{X} \tag{7.8}$$

The total discharge, or fluid removed, Q (in units of volume/time), is equal to the product of the bed permeability, κ, the cross-sectional area through which flow occurs, A, and the change in pressure across the bed (the pressure drop), divided by the product of the viscosity of the slip, μ, and the length, X, over which the pressure drop is taking place. The negative sign is needed because fluids flow from high to low pressure. Note that in this analysis, we are discussing the permeability, thickness, and pressure drop across the cast slip layer, not across the mold. If extraction of liquid through the mold is the significant extraction rate limiter, mold properties must be considered instead.

Considering the cast slip layer to be the porous bed, you can then determine the growth of the layer as a function of time. Dividing both sides of Eq. 7.8 by the area gives

$$\frac{dV}{A\,dt} = \frac{\kappa\Delta P}{\mu X} \tag{7.9}$$

For each volume of slip equal to V_s entering the bed, a volume of V_c remains in the cast layer, giving a volume of filtrant, V_f, that exits the cast layer. The fraction of fluid entering the cast layer that is extracted by the mold is

$$V_f = \frac{V_s - V_c}{V_s} \tag{7.10}$$

Applying this mass balance to Eq. (7.9) and integrating gives

$$X^2 = \frac{2\kappa}{\mu}\,\Delta P\,\frac{V_s}{V_s - V_c}\,t \tag{7.11}$$

indicating that the thickness of the cast layer, X, is proportional to the square root of time. Inspection of this equation shows that slip-cast articles will tend to be thicker at the bottom of a mold, as the hydrostatic pressure of the slip in the mold increases the pressure drop across the cast layer in this area. This equation also reveals that in order to reduce the casting time to reach a given thickness, which is a concern in a production environment, one could reduce the viscosity of the slip or increase the solid content of the slip (generally opposing ideas), increase the permeability of the cast layer (by altering the size and/or distribution of the suspended particles in the slip), which may have deleterious effects on the finished component, or increase the pressure drop above that developed by capillary forces (generally on the order of 0.1 MPa). Increasing the pressure drop is usually considered the most efficient way to reduce casting time. The pressure drop across the cast layer may be increased by using a vacuum to assist in drawing the liquid to the mold and/or by applying pressure to the slip in the mold. Centrifugal casting, where the mold is spun about its central axis, has also been used to increase the pressure drop across the cast layer (Richerson, 1992). Using pressure casting, cycles times on the order of tens of seconds are achieved in many applications (Blanchard, 1988).

The biggest advantage of the slip casting process is versatility, in terms of both shape and size of parts that can be produced. While thick deposits, including solid components without cavities, can be made by topping the mold off periodically with fresh slip, slip casting is particularly well suited to manufacturing hollow, thin-walled (centimeters or less) components. Slip casting is used extensively in the sanitary ware industry as well as to make combustor baffles, scroll bodies, and connecting ducts (Schwartz, 1992). Because capital costs can be low (e.g., molds can be made from plaster of paris), slip casting is well suited for low-volume production and even prototyping of parts, although in the latter case, the differences in microstructure and defects produced by slip casting and the ultimate production method must be taken into account.

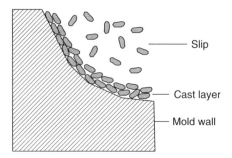

Figure 7.25 Alignment of anisometric particles in the slip-cast layer is possible due to the capillary forces acting normal to the mold wall. Particles will tend to follow the mold contours.

The slip is a suspension of particles. Control of the particle-size distribution within the slip is of utmost importance both for maintaining slip viscosity and for producing dense cast layers. As described above, the particles are compacted on the mold walls by capillary forces. As such, some common defects in slip casting are similar to those seen in particle pressing. The forces in slip casting always act on a normal to the mold wall, though, instead of in a uniaxial manner. This will tend to produce laminar defects in slip casting that follow the contour of the part being produced. Slight variations in cast layer thickness can result if a pressure gradient exists in the filled mold.

If the particles within the slip are anisometric, some preferred orientation of particles may result, with the long axis of the particles being oriented normal to the capillary forces (i.e., parallel to the cast thickness, not in a through-thickness direction), as shown in Figure 7.25. Steinlage and co-workers used centrifugal slip casting of superconducting ceramics to demonstrate a fivefold improvement in orienting the platelike particles in $(Bi, Pb)_2Sr_2Ca_2Cu_3O_x$(BSCCO) compared to identical compositions produced via powder pressing (Steinlage et al., 1994). As mentioned in Section 7.1, forces external to a mold (e.g., a magnetic or electric field) may be used to produce the desired crystallographic texture in a cast article (Suzuki and Sakka, 2002), which may then be enhanced through subsequent sintering and grain growth. In general, textures in slip-cast materials may be more pronounced than those in powder-pressed articles, as there are fewer less constraints on movements of particles in suspension as they are settling on the cast layer than on movements of particles packed within a bed.

7.3.3. Tape Casting

Since the inception of the era of modern ceramics, usually defined as starting in the late 1940s, there have been many advances in ceramics processing technology. One of the most prolific advances in terms of commercial use has been the development and implementation of tape casting, also referred to as *knife coating*

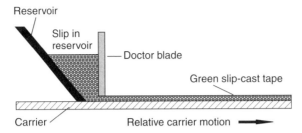

Figure 7.26 Tape-casting process. Slip is contained in a reservoir, one side of which has a stationary doctor blade. A moving carrier travels under the doctor blade, carrying with it a uniform layer of slip, which dries to a green tape.

or *doctor blading*, first disclosed by Howatt (Howatt et al., 1947; Howatt, 1952). In the tape-casting process a knife blade, or *doctor*, is used to scrape or remove excess material from a moving substrate. This process is used primarily to produce large-area thin flat sheets of ceramic and to a lesser extent, metallic parts. Such large parts are very difficult, if not impossible, to form via pressing powders and can be very difficult to extrude. Today, it is not uncommon for tapes as thin as a few micrometers to be produced (Mistler and Twiname, 2000).

This process is shown schematically in Figures 7.26 and 7.27. Slip is contained in a reservoir, one side of which has a stationary doctor blade. A moving carrier, which may be made of a polymer, metal, or glass, travels under the doctor blade, taking with it a uniform layer of slip. The wet thickness of the tape being cast is governed by the gap between the doctor blade and the carrier, the slip viscosity, the height of the slip in the reservoir, and the carrier velocity. Precision slip casting may utilize a second doctor blade downstream from the reservoir. The slip and carrier pass through a drying zone, which allows the liquids to evaporate. This typically shrinks the tape's thickness but not its width or length since these dimensions are constrained by the carrier. The dried but still green ceramic layer can be taken up on a roll for further processing, which may involve slitting, punching holes, or other such operations that are easily done before the ceramic body is sintered.

The flow of the slip under the doctor blade tends to create a preferred orientation for grains, or particulate, when those particulates are not equiaxed. Those particles tend to align with their long direction along the tape length, which in turn can lead to texture or anisotropy in properties. Application of an external force such as a magnetic or electric field may also be done to align particles in the strip as it is exiting the reservoir, in order to generate a preferred texture.

Although tape casting is in many ways similar to slip casting, there are a few important differences. Since the goal is to produce thin—in some cases, extremely thin—sheets, the particulate or powder used to make up the slurry is typically finer than you would see in most slip-casting operations. Second, tape-casting processes typically utilize a nonporous substrate. This necessitates

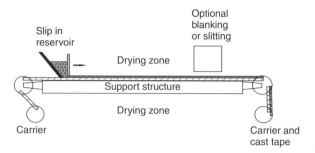

Figure 7.27   Slip-casting process showing the addition of a drying zone, to remove the slip liquid, cutting or slitting and tape-takeup reels.

that the carrier fluid be removed by surface evaporation rather than absorption into the carrier. Also, whereas in slip casting capillary forces help compact the particles in the built-up slip layer, in tape casting there are no forces other than gravity and shrinkage of the organic system due to liquid evaporation that aid the creation of a dense cast tape. Particle size control then becomes of utmost importance in producing high-density tapes. Generally a distribution of sizes is utilized to maximize the density of the tape since small particulates can fit into the interstices (spaces) between larger particles, thus partially filling the void space.

During drying of the tape, two processes occur: solvent evaporation from the surface of the cast tape, and solvent diffusion through the thickness of the tape to the drying layer. Of these two processes, diffusion tends to be the rate-limiting step (Mistler and Twiname, 2000). As the drying process progresses, the tape shrinks, creating the dense, packed bed of particles that is the goal of tape casting, although this bed limits the escape paths for any solvents trapped below it (i.e., solvents near the carrier–tape interface). As the particle size decreases or the particle size distribution is optimized to improve the packing density, this problem is exacerbated in that the voids through which solvent can diffuse are minimized.

Solvent profiles for best- and worst-case conditions are illustrated in Figure 7.28. Ideally, the solvent profile would drop uniformly throughout the tape as a function of drying time. This would be the case if evaporation were difficult compared to diffusion. At the other extreme, if evaporation is much easier than diffusion (which is generally the real-world case), the top surface of the tape will dry, forming a skin, before any solute has been removed from the interior of the tape. Ulti-mately, this is likely to leave trapped solute in the dried tape, which will lead to defects such as pores and/or surface depressions upon firing. For effective drying and highest-quality tapes the solvent must be removed quickly, but not so quickly that a solid skin forms on the free surface. Means to improve the diffusion of sol-vent away from the bottom of the tape, such as heating the carrier but not the air above the tape (which would accelerate evaporation at the tape's free surface), are typically employed to produce high-density tapes.

In addition to being able to form thin parts, one of the primary advantages of tape casting is the ability to pass the tape through many forming operations while

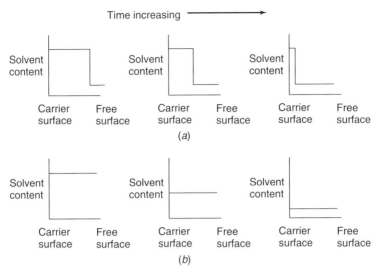

Figure 7.28 Solvent profiles in a tape-cast layer. (*a*) Worst case, with high free surface evaporation rates. Solvent may be trapped in layers near the carrier surface. (*b*) Ideal case, where solvent decreases uniformly through the tape-cast layer. Heating the carrier side of the tape will promote ideal behavior.

it is still in the green state. Many of these operations are directly analogous to those used to form individual parts from metal sheet or strip. Blanking is used to form the outside shape of the desired part. For complex shapes this may be done with a punch, which cuts cast tape, similar to a cookie cutter forming a shaped cookie, or with something as simple as a slitting roller to cut the tape in the direction of the cast, synchronized with a knife that cuts the tape widthwise. Holes may be punched into the tape during blanking or in another step.

For many years, nearly all the substrate materials used in the electronics industry were ceramics produced by tape casting (Mistler, 2000). These substrates consist of multiple layers of tape-cast material acting as the insulative carrier, with layers of metal deposited by sputtering, CVD, or evaporation processes. Holes punched into the tapes fill with metal and act as *vias* or pathways between layers when the tapes are stacked. Stacked layers of tape-cast blanks are laminated together and then sintered to produce a monolithic substrate. This process is detailed in many handbooks on electronics packaging (e.g., Harper, 2000). One of the prime benefits is that the layers of the final part can be different: different hole patterns, different metallization patterns, even different dimensions on each layer.

The first reported use of a multilayered ceramic package was by RCA (Liederbach et al., 1961). Since that time, memory and logic devices, diode and transistor packages, as well as small circuit boards have been produced by this technology. Multilayered ceramic capacitors are another big use of tape casting, taking advantage of multiple layers (and their corresponding additive capacitance) to form a highly capacitive device that occupies little space. Performance and reliability of these stacks of tape-cast layers depends on the tape-cast ceramic performing as an insulator. Excessive grain size (greater than 20% of the tape thickness) and/or voids in the tape left by trapped solvent or improper particle-size distribution could lead to dielectric breakdown between layers. Many ceramics are used in single-layer tape-cast ceramic devices. These sheets of piezoelectric materials are used in sensors, buzzers, and actuators. Multilayered chip inductors are a special use of ferrites. These are surface-mounted devices that are miniature in size and light in weight, manufactured by tape casting and lamination and used in high-frequency applications such as cell phones. They are considered to be very reliable since they have monolithic construction (i.e., it is all one piece).

Thin layers of materials are needed for most fuel cells, and tape casting has been used extensively there, too (Mistler and Twiname, 2000). Because the layers formed are thin, tape casting can also be used to form functionally graded materials: that is, materials that have a controlled gradient in composition to address some functional need (e.g., wear resistance on one surface and high toughness on another; gradients in coefficient of thermal expansion to avoid warping of a structure when one side is heated). This is accomplished by laminating and sintering together layers of different compositions.

7.3.4. Consolidation via Plastic Forming

Many of the methods of plastic or deformation processing discussed earlier can be adapted to consolidation through the addition of a binder material that allows the resulting mixture to be deformed plastically. Examples of this technology date as far back as to 7000 B.C., with water being added to traditional ceramic materials to form clays that can be formed by hand into bricks, pipes, or pottery vessels (Norton, 1970, 1974). Later, the potter's wheel was invented to allow more flexibility in pottery forming. Today, variations on the potter's wheel are still in use for forming cookware and china. Two other important methods for plastically forming ceramics are extrusion and injection molding. Extrusion of ceramics is used in forming filters, ceramic tubes, rods, and other components with constant cross sections. Injection molding is typically used to form small, intricately shaped parts.

The clay–water system serves as a basis for many traditional ceramics. When suspended in water or another polar fluid, clay particles exhibit a platelike morphology consisting of negatively charged faces and positively charged edges. This leads to a tendency to form a "house of cards" type of network, where the face of one particle is attracted to the edge of another, as shown in Figure 7.29. The clay particle network allows the material to exhibit a yield stress, which results in a component shape being maintained after a deforming pressure or force is removed. This natural plasticity exhibited by clay-based ceramics provides for excellent shape-forming capabilities.

Unlike clay-based systems, modern ceramics require additives, termed *binders*, to provide the plasticity required for ductile-forming methods to be used. These organic additives serve to modify the rheological behavior of the ceramic suspensions and impart handling strength to the green, as-formed ceramic bodies. Their

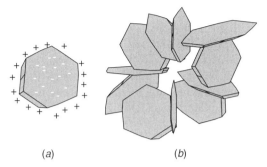

(a)   (b)

Figure 7.29 (*a*) Clay particles suspended in a polar fluid exhibit a platelike morphology with negatively charged faces and positively charged edges; (*b*) "house of cards" structure for an agglomeration of clay particles, where one platelet is attracted to the edge of another platelet. This structure allows for plastic deformation and shape retention.

presence, however, poses significant challenges, especially when high binder loadings are used. Undissolved organic species may generate defects during binder removal (Lange et al., 1986).

Addition of binder to these systems adds another potential influence on the microstructure and hence the properties of the resulting component. While the binder acts somewhat as a lubricant for the ceramic particles to move to accommodate deformation, the plastic flow of the binder can have an important impact on the resulting structure. For example, in extrusion, where material is pushed through a die to yield a long structure of constant cross section, the binder itself is also extruded and elongated axially. This will leave elongated areas of porosity after binder removal. While sintering may reduce the degree of interconnectivity of this porosity, some studies have shown that the porosity in extruded ceramics is consistent with a tubular, or elongated, structure (Rice, 1996, 1998). Another microstructural consideration is the orientation of the ceramic particles themselves. When particles are spherical in nature, there will be no preferred orientation, since there will be no net force on the particles causing them to orient in one particular crystallographic direction. If the particles are anisotropically shaped, or anisometric, however, they will tend to orient with their longest dimension in the direction of material flow. In extrusion this is along the axis of the extrusion. Significant orientation of particles developed during extrusion and magnified during sintering has been observed in the BeO system, where experiments were conducted with both equiaxed and elongated particles (Fryxell and Chandler, 1964). Although both types of particles gave isotropic bodies of sintered BeO when consolidated by pressing, the extruded BeO was isotropic only when equiaxed particles were used. The elongated particles were oriented substantially along the extrusion axis, resulting in anisotropy in physical properties of the extruded component. More recently, several others have noted that extrusion will orient anisometric particles, or even whiskers, in the extrusion direction (Chen and Crawley, 1992; Muscat et al., 1992; Rice, 2000).

In injection molding, material flow can depend strongly on the die design. Materials tend to exhibit a preferred orientation while in the inlet nozzle but a more turbulent path after the entering the die cavity. Careful attention must be paid to the design of the material delivery system, as well as the particle size distribution, which will greatly affect the viscosity of the material injected. Some attractions of injection molding are the fast cycle time, the ability to form intricate parts, and the ability to maintain tight tolerances.

7.3.5. Sintering and Grain Growth

The green bodies formed via consolidation processing do not exhibit sufficient strength or toughness to be used in most applications. The individual particles are held together loosely, either by an interlocking of particles or more commonly with the aid of an organic binder. Total porosity is often on the order of 25 to 60 vol% (Kingery et al., 1976). To develop the properties desired in the article being fabricated, the individual particles must be bonded together and the interparticle porosity removed to yield a dense polycrystalline solid. This is done through a process called *sintering*, one of the oldest material processing techniques known to man, having been performed on ceramic materials for thousands of years (Kingery, 1992). Some of the first sintered products were bricks heated in open-pit fires to strengthen them. As you might expect from such an ancient technique, operationally sintering can be quite simple to perform, even though the mechanisms and microstructural changes that impart the added strength can be quite complex. Work over the last 60 to 70 years has led to an understanding of these fundamental mechanisms that control densification and turned sintering from an art into a science that can be controlled to optimize the properties of the finished component. Today, sintering is employed in a diverse range of products, including dental implants, rocket nozzles, aircraft wing weights, ultrasonic transducers, golf clubs, and semiconductor substrates, in addition to the traditional applications of whitewares, refractories, bricks, and porcelains (German, 2001).

For sintering to proceed, two conditions must be met. First, there must be a means or mechanism for materials transport that will allow material to flow to fill pores and create bonds between particles. Second, there must exist a source of energy to drive or activate that materials transport. There are several different forms of sintering, each with its own combination of conditions, as shown in Table 7.6. We examine solid-state sintering in depth as one of the most general methods, but similar principles can be seen to apply to all sintering methods.

Sintering is a thermodynamically irreversible process whose driving force is a reduction in the total free energy of the system. The free-energy change that arises during densification results from the decrease in surface area and lowering of the surface free energy by the elimination of solid–vapor interfaces. In solid-state sintering, solid-state diffusion is the dominant materials transport mechanism. Diffusion is a thermally activated process whose rate is usually significant only when the homologous temperature (fraction of a material's melting point) is in excess of 0.5. Considering the two adjacent particles shown in Figure 7.30, in

TABLE 7.6   Driving Force and Material Transport Mechanism for Various Types of Sintering Processes

Sintering Method	Driving Force	Material Transport Mechanism
Solid state	Reduction in surface free energy	Solid-state diffusion
Vapor phase	Differences in vapor pressure (high on surfaces, low in pores)	Evaporation condensation cycle
Liquid or viscous	Capillary pressure	Viscous flow, liquid-state diffusion
Reactive	Capillary pressure, surface tension	Viscous flow, precipitation of reaction products

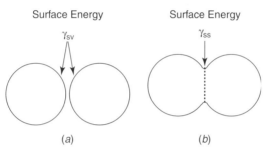

Figure 7.30   Two adjacent particles. (*a*) Before sintering, the particles have two solid–vapor surfaces, each with a surface energy γ_{sv}. (*b*) After the onset of sintering, a necked region forms between particles, replacing two solid–vapor surfaces with a lower-energy solid–solid surface. $\gamma_{ss} < \gamma_{sv}$, providing the thermodynamic driving force for sintering.

the green compact state there exist two solid–vapor surfaces with associated energy $2\gamma_{sv}$. If the particles are joined, these two surfaces would be replaced by one solid–solid surface with lower energy γ_{ss}. It is this reduction in energy that provides the driving force for the diffusion that joins the particles. Since ceramics are composed of at least two elements, both ionic species must diffuse together to maintain the electrical neutrality of the system. Therefore, it is the diffusion coefficient of the slowest-moving ion that controls mass transfer, and thus densification, during sintering.

Sintering is often thought to occur in three stages, corresponding to the physical changes that occur in the material. During the initial stage, particles may rearrange and "necks" of joining material form at particle-to-particle contact points. During the second stage, also referred to as *intermediate sintering*, these necks grow in dimension, forming grain boundaries between individual particles, or *grains* (note the change in terminology as the particles become grains). During this intermediate stage, the grain boundaries lengthen as the grains grow. A large

TABLE 7.7 Effect of Porosity on Properties of Consolidated Ceramic Materials

Compact Property	Symbol	Equation Incorporating Effect of Porosity on Compact Property
Electrical conductivity	Ω	$\Omega = \Omega_0 A f^2$
Magnetic saturation	B	$B = B_0(A_1 + A_2 f)$
Strength	σ	$\sigma = \sigma_0 A_1[1 - A_2(1 - f)^{2/3}]$
Elastic modulus	E	$E = E_0 f^{3.4}$
Shear modulus	G	$G = G_0 f^3$
Poisson ratio	v	$v = 0.068 \exp(1.37f)$

Source: German (1991).

[a] Property calculated as a function of the bulk material property (indicated by the subscript zero), f is the fraction density, and A_1 and A_2 are adjustable constants.

degree of bulk shrinkage of the component occurs as porosity is reduced, due to transport of material along the pore boundaries. The pores remain interconnected or continuous during this stage. During the third or final stage of sintering, significant grain growth occurs. The pore phases become discontinuous and tend to a spherical shape to minimize pore surface area. These pores may be isolated at grain boundaries, or if grain growth is rapid, the grain boundaries can break away from the pore, leaving them isolated in the grain interior. Pores in the grain interior will not typically densify or fill. Grain growth continues with large grains growing at the expense of smaller grains as the system seeks to reduce the grain boundary area and hence, grain boundary energy. This coarsening of the microstructure (i.e., increased grain size) reduces the grain boundary area and hence energy, which lowers the driving force for densification, thus slowing the sintering rate. Larger grain or coarser bodies are more difficult to sinter than those that are finer grained. Further details on sintering can be found in the text by Kingery et al. (1976).

Green compacts are sintered to improve a material's properties. German (1991) has tabulated the quantitative effects of the degree of sintering on many important engineering properties (Table 7.7). In this table, each material property is given as a function of the bulk material property (subscript 0) and the fractional density, f (which is $1 - $ porosity). Sintering or degree of densification is shown to have a significant effect on these properties.

7.3.5.1. Templated Grain Growth In many instances it is desirable to produce a specific texture in an article produced by consolidation processing, to take advantage of crystalline-based property anisotropies. In addition to the mechanical forces acting to orient particles during consolidation, electrical, magnetic, and thermal gradients can be applied to induce rotation of anisotropic particles, introducing texture to a fabricated article. These processes for obtaining texture tend to work best for larger, anisometric particles. A component composed entirely of these types of particles can be difficult to densify, as initial porosity can be large and the driving force for sintering not as high as for smaller particles with larger surface area per volume (Messing, 2001). Therefore, there is often a trade-off between the degree of texture desired in a component and the optimization of other properties (including cost).

Messing and co-workers found that texture can be enhanced, sometimes significantly, through a process they called *templated grain growth* (TGG) (Seabaugh et al., 1997). TGG takes advantage of the fact that during a grain growth process, large grains grow at the expense of smaller ones to reduce the free energy of the system. In the TGG process a minor portion of large anisometric grains, or seed particles, are dispersed in a matrix of finer equiaxed particles. Initially, the seed particles are oriented randomly in the unconsolidated powder, but they develop a preferred orientation, or texture, during processing of the green ceramic component. Upon sintering and grain growth, these larger grains grow preferentially, with a concomitant increase in the fraction of material that exhibits the same crystallographic orientation as that of the seed grains. In this manner, the seed grains serve as templates for crystallographic texture development. This process is shown, very schematically, in Figure 7.31.

When the template and matrix have the same composition, the kinetics of abnormal grain growth determine the TGG process. TGG can also be used, however, where the matrix is a precursor to the final ceramic phase and the template particles act as nucleation sites to seed the transformation of the matrix phase. This process is referred to as *reactive templated grain growth* (RTGG) to reflect the dual nature of the templates (Hong and Messing, 1999; Takeuchi et al., 1999). It is worth noting that although anisometry is a useful property for particle orientation during consolidation processing, it is not necessary for reactive TGG if some other method that does not rely on shape, such as electric or magnetic fields, can be used to align the template particles.

Template particles must possess a variety of properties for successful TGG (Kwon et al., 2003). If mechanical forces are acting to align the seed particles,

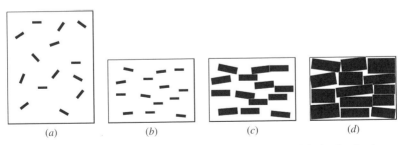

(*a*)	(*b*)	(*c*)	(*d*)

Figure 7.31 Templated grain growth process: (*a*) templates (black platelets) are randomly oriented in powder; (*b*) during consolidation, templates achieve a preferred orientation; (*c*) larger templates grow at the expense of smaller particles during sintering; (*d*) after sintering, the majority of the final part can be composed of grains with the preferred orientation or texture.

templates must have a high aspect ratio so that they can be aligned during consolidation processes. Morphologically, most of these templates are platelets or whiskers, and the desired crystallographic orientation must be aligned along the template axis. The template grains must be large enough so that there is a thermodynamic driving force for their growth at the expense of the matrix grains. The template seeds must also be distributed uniformly throughout the matrix to be most effective.

Templated grain growth has been demonstrated to be effective for a number of applications. The amount of growth, hence the degree of texture, depends on the number, size, and distribution of the template particles. The "quality" of the texture produced depends on how well the seed texture particles are oriented during processing. Grains impinging on each other limits the attainment of a 100% textured material; however, x-ray diffraction studies have shown that obtaining the desired texture in more than 90% of the material is possible (Messing, 2001).

7.3.5.2. Selective Laser Sintering   One fabrication method that combines consolidation and a form of sintering into one step is a process first developed by Carl Deckard at the University of Texas called selective laser sintering (Deckard, 1989), or SLS (SLS is a registered trademark of 3D Systems, Inc., Rock Hill, South Carolina). SLS is a rapid prototyping and manufacturing technique that uses a high-powered laser to fuse small particles or powders of metal, ceramic or polymers into a three-dimensional object. Using a three-dimensional digital representation (i.e., from a CAD file), a part is divided electronically into a large number of cross sections, whose thickness corresponds to one layer of particles or powder (from one to several particulates thick). A layer of powder is laid down on a bed within the machine and a laser fuses the powdered material selectively by following a scanning pattern that traces the cross sections generated from a CAD file on the surface of a powder bed, defining the regions that are sintered. After each cross section is scanned, the powder bed is lowered by one layer thickness, a new layer of material is applied on top, and the process is repeated until the part is completed. After cooling, the unbonded powder is removed, leaving a complete three-dimensional representation of the electronic CAD file. This process is shown schematically in Figure 7.32.

Selective laser sintering can produce parts from a relatively wide range of commercially available powder materials, including polymers (nylon, also glass-filled or with other fillers, and polystyrene), metals (steel, titanium, alloy mixtures and composites) and green sand (3DSystems, 2007). True solid-state sintering does not occur in the SLS process, so it is not applicable to most ceramic materials. The laser will soften or locally melt the particles, which then flow or bond together by viscous flow, analogous to viscous sintering. Further processing, such as hot isostatic processing, may be required to fully densify the SLS-produced parts (Knight, 1994). Although developed originally as a rapid prototyping method for use early in the design stage, in many cases large numbers of parts can be packed within the powder bed, allowing increasingly for use of SLS as a limited-volume manufacturing method.

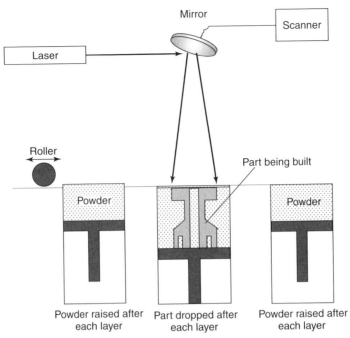

Figure 7.32 Selective laser sintering process. A laser is bounced off a mirror, which is controlled by a scanner to selectively trace the outline of a cross section of a part on a powder bed. When one layer has been traced, the part is lowered and fresh powder is rolled into place. After finishing, the part is removed from the loose, unfused powder.

7.4. SUMMARY

Our aim in this chapter was to develop some appreciation of the characteristics of several key fabrication techniques. Processing techniques are linked inexorably to the materials properties and structure desired for performance in the ultimate application. It should be readily apparent that the performance of a given product or component cannot be optimized without concurrent optimization of the fabrication parameters of that component. The development of new materials processing techniques, such as equal-channel angular extrusion, semisolid casting, templated grain growth, and selective laser sintering, have helped drive improvements in the properties of fabricated components used in everyday life.

REFERENCES

3DSystems. http://www.3dsystems.com/products/sls/sinterstation_pro/index.asp, **2007**.

Andresen, W. T. In *Metals Handbook*, desk ed. H. Boyer and T. Gall, Eds., American Society for Metals, Metals Park, OH, **1985**, pp. 23.32–23.41.

Armstrong, R. W. In *Advances in Materials Research*, Vol. 5, R. F. Bunshah, Ed., Wiley-Interscience, New York, **1971**, p. 101.

ASM International. *Machining*, ASM International, Materials Park, OH, **1989**.

Assmus, F.; Dietert, K.; et al. *Z. Metallkd.* **1957**, *48*, 344.

Baldwin, J. W. M. *Acta Metall.* **1958**, *6*, 14.

Blanchard, E. G. *Am. Ceram. Soc. Bull.* **1988**, *68*(10), 1680.

Boothroyd, G.; Knight, W. W. *Fundamentals of Machining and Machine Tools*, Marcel Dekker, New York, **1989**.

Chen, A. Y.; Crawley, J. D. *J. Am. Ceram. Soc.* **1992**, *75*(3), 575–579.

D'Arcy, H. (1856). *Les Fontaines Publiques de la Ville de Dijon*, Dalmont, Paris.

Deckard, C. U.S. patent 4,863,538, Sept. 5, **1989**.

Dieter, G. E. *Mechanical Metallurgy*, McGraw-Hill, New York, **1976**.

English, A. T.; Chin, G. Y. *Acta Metall.* **1965**, *13*(9), 1013–1016.

Ferrasse, S.; Alford, F.; et al. *Semicond. Fabr.* **2003**, *4*, 76.

Flemings, M. C. *Solidification Processing*, McGraw-Hill, New York, **1974**.

Flemings, M. C.; Martinez-Ayers, R. A.; et al. U.S. patent 6,645,323, **2003**.

Frenkel, J. *Z. Phys.* **1926**, *37*, 572.

Fryxell, R. E.; Chandler, B. A. *J. Am. Ceram. Soc.* **1964**, *47*(6), 283–291.

German, R. M. In *Engineered Materials: Ceramics and Glasses*, Vol. 4, S. J. Schneider, Ed., ASM International, Materials Park, OH, **1991**, pp. 260–269.

German, R. M. In *Encyclopedia of Materials: Science and Technology*, Vol. 9, K. H. J. Buschow, R. Cahn, M. Flemings, et al., Eds., Elsevier, New York, **2001**, pp. 8641–8647.

Goss, N. P. *Trans. Am. Soc. Met.* **1935**, *23*, 515.

Hall, E. O. *Proc. Phys. Soc. B.* **1951**, *64*, 747–753.

Harper, C. A., Ed. *Electronic Packaging and Interconnection Handbook*, McGraw-Hill, New York, **2000**.

Honda, H.; Kaya, S. *Sci. Rep. Tohoku Univ.* **1926**, *15*, 721.

Honeycombe, R. W. K. *The Plastic Deformation of Metals*, Edward Arnold and American Society for Metals, Metals Park, OH, **1984**.

Hong, S. H.; Messing, G. L. *J. Am. Ceram. Soc.* **1999**, *78*, 867–872.

Hosford, W. F.; Caddell, R. M. *Metal Forming: Mechanics and Metallurgy*, Prentice Hall, Englewood Cliffs, NJ, **1993**.

Howatt, G. N. U.S. patent 2,582,993, Jan. 22, **1952**.

Howatt, G. N.; Breckenridge, R. G.; et al. *J. Am. Ceram. Soc.* **1947**, *30*(8), 237–242.

Kingery, W. D. In *Sintering 1991*. A. C. D. Chaklader and J. A. Lund, Eds., TransTech, Enfield, NH, **1992**, pp. 1–10.

Kingery, W. D.; Bowen, H. K.; et al. *Introduction to Ceramics*. New York, **1976**.

Knight, R. E. A. In *Proc. Solid Freeform Fabrication Symposium*, University of Texas, Austin, TX, **1994**.

Kwon, S.; Sabolsky, E. M.; et al. In *Handbook of Advanced Ceramics*, S. Somiya, Ed., Vol. I., Materials Science, Elsevier Academic Press, Oxford, **2003**, pp. 459–469.

Lalena, J. N.; Cleary, D. A. *Principles of Inorganic Materials Design*, Wiley-Interscience, New York, **2005**.

Lange, F. F.; Davis, B. I.; et al. *J. Am. Ceram. Soc.* **1961**, *69*(1), 66–69.

Liederbach, W. H.; Stetson, H. *Am. Ceram. Soc. Bull.* **1961**, *40*(9), 584.

Messing, G. L. In *Encyclopedia of Materials: Science and Technology*. Vol. *10*, K. H. J. Buschow, R. Cahn, M. Flemings, et al., Eds., Elsevier, New York, **2001**, p. 9129.

Meyers, M. A.; Chawla, K. *Mechanical Metallurgy: Principles and Applications*. Prentice-Hall, Englewood Cliffs, NJ, **1984**.

Mistler, R. E.; Twiname, E. R. *Tape Casting: Theory and Practice*, American Ceramics Society, Westerville, OH, **2000**.

Muscat, D.; Pugh, M. D.; et al. *J. Am. Ceram. Soc.* **1992**, *75*(10), 953–957.

Norton, F. H. *Fine Ceramics: Technology and Applications*, McGraw-Hill, New York. **1970**.

Norton, F. H. *Elements of Ceramics*, Addison-Wesley, Reading, MA, **1974**.

Petch, N. J. *J. Iron Steel Inst.* **1953**, *174*, 25–28.

Piearcey, B. J. U.S. patent 3,485,291, **1969**.

Piearcey, B. J. U.S. patent 3,536,121, **1970**.

Randle, V. In *Encyclopedia of Materials: Science and Technology*, Vol. *10*, K. H. J. Buschow, R. Cahn, M. Flemings, et al., Eds., Elsevier, New York, **2001**, pp. 9119–9129.

Randle, V.; Engler, O. *Introduction to Texture Analysis: Macrotexture, Microtexture and Orientation Mapping*, Gordon and Breach, Amsterdam, **2000**.

Rice, R. W. *J. Mater. Sci.* **1996**, *31*, 1509–1528.

Rice, R. W. *Porosity of Ceramics*, Marcel Dekker, New York, **1998**.

Rice, R. W. *Mechanical Properties of Ceramics and Composites: Grain and Particle Effects*, Marcel Dekker, New York, **2000**.

Richerson, D. W. *Modern Ceramic Engineering*, Marcel Dekker, New York, **1992**.

Rivoirard, S.; Popa, I.; et al. *Solid. State Phenom.* **2005**, *105*, 291–296.

Scheil, E. *Z. Metallkd.* **1942**, *34*, 70.

Schiøtz, J.; Vegge, T.; et al. *Phys. Rev. B* **1999**, *60*(17), 11971–11983.

Schwartz, M. M. *Handbook of Structural Ceramics*, McGraw-Hill, New York, **1992**.

Seabaugh, M. M.; Kerscht, I. H.; et al. *J. Am. Ceram. Soc.* **1997**, *80*, 1181–1188.

Segal, V. M. *J. Appl. Mech. Tech. Phys.* *1*, 127.

Segal, V. M. *Mater. Sci Eng. A* **1995**, *197*(2), 157.

Segal, V. M. U.S. patent 5,513,512, **1996**.

Segal, V. M. U.S. patent 5,850,755, **1998**.

Segal, V. M. *ASM Handbook*, Vol. *14A*, Metalworking: Bulk Forming, S. L. Semiatin, Ed., ASM International, Materials Park, OH, **2005**.

Segal, V. M.; Hardwig, K. T.; et al. *Mater. Sci. Eng. A* **1997**, *224*, 107.

Sendzimir, T. U.S. patent 2,169,711, **1939**.

Shroyer, H. F. U.S. patent 2,830,343, **1958**.

Smallman, R. E. *J. Inst. Met.* **1955**, *83*, 10.

Steinlage, G.; Roeder, R.; et al. *J. Mater. Res.* **1994**, *9*(4), 833–835.

Suzuki, T. S.; Sakka, Y. *Chemi. Lett.* **2002**, *12*, 1204–1205.

Takeuchi, T.; Tani, T.; et al. *Jpn. J. Appl. Phys.* **1999**, *38*, 5553–5556.

Thompson, R. A. *Am. Ceram. Soc. Bull.* **1981**, *60*(2), 237–243.

Valieb, R. Z.; Islamgaliev, R. K.; et al. *Prog. Mater. Sci.* **2000**, *45*, 103.
VerSnyder, F. L. U.S. patent, 3, 260, 505, **1966**.
Wickersham, C.E., Jr. *J. Vac. Sci. Technol. A* **1987**, *5*(4).
Wiener, G.; Albert, P. A.; et al. *J. Appl. Phys.* **1958**, *29*, 366.
Zhang, H. U.S. patent 6,193,821, **2001**.

APPENDIX A1
General Mechanical Engineering Terms

Compliance	The reciprocal of stiffness; low resistance to elastic deformation.
Compressive strength	A measure of the ability to withstand compressive loads without crushing or deforming.
Creep	Time-dependent plastic deformation at constant stress and temperature.
Ductility	The ability to stretch under tensile loads without rupture. Measured as percent elongation.
Elastic moduli	Elastic constants for elastically isotropic polycrystalline materials. A measure of the manner in which a polycrystalline body responds to small external forces in the elastic regime.
Elasticity	The ability of a material to deform under loads and to return to its original dimensions upon removal of the load.
Fatigue	Damage or failure due to cyclic loads.
Fracture	Crack propagation involving permanent rupture of the chemical bonds binding together the atoms of a substance. The mechanisms are different in ductile and nonductile materials. Ductile fracture occurs in the direction of the primary slip system.
Hardness	The ability of a material to withstand permanent (plastic) deformation on its surface (e.g., indentation, abrasion).
Malleability	The ability to deform under compressive loads without rupture.
Plasticity	The ability of a material to deform permanently without rupture under application of a load.
Rigidity modulus	Shear stress divided by shear strain.

Inorganic Materials Synthesis and Fabrication, By John N. Lalena, David. A. Cleary, Everett E. Carpenter, and Nancy F. Dean
Copyright © 2008 John Wiley & Sons, Inc.

Shear strength	A measure of the ability to withstand transverse loads without rupture.
Slip	The gliding motion of closed-packed atomic planes. The primary mechanism of plastic deformation in ductile materials. Slip involves chemical bond breaking *and* reformation.
Stiffness	A measure of resistance to elastic deformation.
Strain	Deformation due to stress. Equal to the change in length divided by the initial length. Strain is a second-rank tensor.
Strength	A measure of resistance to plastic deformation.
Stress	A state of a material in which mechanical forces are acting on it. Defined as force per unit area. Stress is a second-rank tensor.
Tensile strength	A measure of the ability to withstand tensile loads without rupture.
Toughness	The ability to withstand shatter. A measure of how much energy can be absorbed before rupture. Easily shattered materials (small strain to fracture) are termed *brittle*.
Ultimate tensile strength	The point at which fracture occurs.
Yield strength	The point at which the stress–strain curve ceases to be linear and at which plastic deformation begins to occur. Also known as the *elastic limit*.
Young's modulus	Normal stress divided by normal strain.

APPENDIX A2
Green Materials Synthesis and Processing

Throughout this book we have discussed materials synthesis and processing from the standpoint of reaction energetics and kinetics, with little mention of a fairly recent initiative that has taken hold in many areas of science and engineering known as *green* or *eco-friendly processing*. Green processing is an integral part of society's broader movement toward *sustainability*. By definition, sustainability implies the ability to meet the needs of the current generation without compromising the needs of future generations—but this is no small feat. The movement has several dimensions, including political, economic, and cultural, as well as environmental aspects. It is the latter area where physical scientists and engineers are especially suited for making an impact.

The environmental aspect actually has somewhat early roots. The metallurgical industry, for one, has for many years been cognizant of the fact that our planet does not have an inexhaustible supply of natural resources. Accordingly, it has recycled and extracted metals from scrap materials for decades. Scientists of various other backgrounds working in the field of inorganic materials are also presently contributing to the development of more energy-efficient materials and materials for use with alternative energy sources. For example, researchers are working to fully develop high-temperature superconductor technology so that electrical power can be transmitted with no power loss. It is possible that even materials with a relatively low superconducting transition temperature (e.g., the boiling point of liquid nitrogen) can feasibly be used to reduce current resistive power losses by half, the remaining power difference being needed to supply the necessary refrigeration costs. Solid-oxide fuel cells (SOFCs), which use ceramic electrolytes, are another important area of materials science research. SOFCs can use many types of hydrocarbon fuels while producing very little pollution, and they operate at high temperatures (up to $1000°C$), which presents an opportunity to recycle the waste thermal energy by generating steam that can be used for heating or in a steam turbine.

Inorganic Materials Synthesis and Fabrication, By John N. Lalena, David. A. Cleary, Everett E. Carpenter, and Nancy F. Dean
Copyright © 2008 John Wiley & Sons, Inc.

However, green materials and processing are concerned not only with the prudent use of energy and natural resources, but also with avoidance of processes that use or generate harmful substances (including end products and by-products) that are ultimately released into the environment and/or pose immediate exposure hazards to laboratory personnel or process operators. Human-induced climate change and environmental damage from chemical pollution have been highlighted in the news for years. In the inorganic sector, lead, asbestos, and radioactive substances immediately come to mind as examples of materials now highly regulated due to environmental concerns. There is still room for much work, however. The highly active area of nanomaterials research has not yet produced sufficiently detailed studies outlining the potential risk of these materials, which possess unique chemical and physical properties, on human health and the environment. In select cases, of course, nanostructured biodegradable substances (e.g., calcium phosphate) have been targeted for use in the prevention and treatment of certain diseases. But the broader, more general problem is only now being addressed by some groups, including governmental agencies.

Often, in practice, green processing translates to [volatile] solvent-free processing or, at least, a very judicious choice of solvent. Hence, some of the synthetic methods that we discussed in this book—microwave, mechanochemical, solvothermal processes using eco-friendly supercritical fluids (e.g., water, carbon dioxide), and ionic liquid synthesis—can be considered green processing. In nanomaterials production, purification is a major concern. It is often necessary to wash products with relatively large volumes of organic solvent. In the future there will undoubtedly be a push toward alternative greener routes, such as, perhaps, filtration through nanoporous membranes.

Unfortunately, chemical processes can be tailored only to a limited extent, and furthermore, the use of some hazardous substances by society is unavoidable. There is much work to be done. It is certain, however, that an integral component of the global sustainability movement is the need for greater respect for our environment. Sustainability will require a concerted effort by all science and engineering professionals and political bodies, as well as by an informed and responsible public.

INDEX

Inorganic Materials Synthesis and Fabrication, By John N. Lalena, David. A. Cleary,
Everett E. Carpenter, and Nancy F. Dean
Copyright © 2008 John Wiley & Sons, Inc.